# Study Guide and Solutions Manual

Marvin L. Hackert • Roger K. Sandwick
Michael W. Pelter • Libbie S. Pelter

# Chemistry and Life
### Sixth Edition
#### An Introduction to General, Organic, and Biological Chemistry

John W. Hill
Stuart J. Baum
Rhonda J. Scott-Ennis

Upper Saddle River, NJ 07458

Executive Editor: John Challice
Special Projects Manager: Barbara A. Murray
Production Editor: Dawn Murrin
Supplement Cover Manager: Paul Gourhan
Supplement Cover Designer: PM Workshop Inc.
Manufacturing Buyer: Michael Bell
*Cover Photo Credit: Steve Gettle/Ellis Nature Photography*

© 2000 by Prentice Hall
Upper Saddle River, NJ 07458

All rights reserved. No part of this book may be
reproduced, in any form or by any means,
without permission in writing from the publisher.

Printed in the United States of America

10 9 8 7

ISBN 0-13-085385-2

Prentice-Hall International (UK) Limited, London
Prentice-Hall of Australia Pty. Limited, Sydney
Prentice-Hall Canada, Inc., Toronto
Prentice-Hall Hispanoamericana, S.A., Mexico
Prentice-Hall of India Private Limited, New Delhi
Pearson Education Asia Pte. Ltd., Singapore
Prentice-Hall of Japan, Inc., Tokyo
Editora Prentice-Hall do Brazil, Ltda., Rio de Janeiro

# Foreword

This Study Guide is a learning tool written to supplement the text, *Chemistry and Life*, by John Hill, Stuart Baum, and Rhonda Scott-Enis. However, this Study Guide should prove useful to any student in a course covering General, Organic, and Biological Chemistry with an allied health science perspective.

Part I of this Study Guide is designed to supplement the textbook. It can serve as a set of "lecture notes," chapter summaries, a tutorial to help with difficult subject areas, and as a source of practice problems with answers to help check your mastery of the material. Each unit begins with a list of KEY WORDS and a SUMMARY. This course covers topics normally presented to chemistry or biochemistry majors over a full three-year sequence with separate courses in General Chemistry, Organic Chemistry, and Biochemistry. Even though your course will not go into as much depth, you will be exposed to much of the same terminology and language used in those courses. Mastering the language of chemistry and biochemistry is an essential first step toward understanding the concepts of chemistry. (*I often have students who feel that they should receive credit for a foreign language for taking this course.*) Each of the KEY WORDS listed at the beginning of a unit is highlighted in bold italics when it is first used in the SUMMARY. This should help you identify and understand the meaning of these terms. The SUMMARY is a section-by-section outline of the textbook. It can serve as the basis for your lecture notes and as a quick review of the material presented in the chapter. Also included are most of the important structures that you might be asked to know or be familiar with during this course. In the DISCUSSION section that follows, difficult concepts are explained in more detail and study hints are presented. Each unit ends with a SELF-TEST. Only after you think that you have mastered the material, should you take the SELF-TEST. It will provide a final check on your understanding of the chapter's material. If you are still having difficulty, seek assistance from your instructor or teaching assistant. Part II of this Study Guide presents the answers to selected problems from the text.

We can provide you with a good textbook and this Study Guide to help you master the material in this course. However, your success in this course will depend mostly on *you* and what *you* do with these materials. Here are a few good study suggestions. Review your syllabus to obtain an overview of the material and what topic will be covered before each lecture. Read the SUMMARY in this Study Guide, and also read your text material *before* the lecture so you will be prepared to receive information and insight from your instructor. Remember, learning is an incremental process; *"We can only learn that which we almost already know."* Real learning also requires repetitive exposure to the material. Before each class, review your lecture notes taken since the start of that unit. This will help you gradually build your accumulated knowledge of the subject. Keep up, and try not to cram just before exams. Look over the KEY WORDS at the beginning of each chapter and then review them at the end of a unit to make sure that you understand their meaning. Finally, this Guide and the text provide a wealth of practice problems. Work through these exercises, do not simply look at them. Only by doing the problems, will you be able to measure your own level of understanding of the material.

We hope that you will find the new text and this Study Guide to be helpful. We invite your suggestions and comments so that we might improve this Guide for the aid of future students. Please send your comments on the Guide by e-mail to M.HACKERT@mail.utexas.edu.

Marvin L. Hackert  
Professor of Chemistry and Biochemistry  
The University of Texas at Austin

Michael and Libbie Pelter  
Purdue University-Calumet

## * * *Acknowledgments* * *

I would like to thank John Hill, Stuart Baum, and Rhonda Scott-Enis for the privilege of putting together this study guide for their text, ***Chemistry and Life***. I would also like to thank my wife, Bretna, for her proofreading and patience while this work was completed. Finally, I am grateful to the staff of Prentice Hall for their assistance and to all the students who have used this Study Guide and supplied helpful comments on its revisions.

# Contents

## PART I - Chapter Summaries and Self Tests

| Chapter | Title | Page |
|---|---|---|
| 1 | MATTER and MEASUREMENT | 1 |
| 2 | ATOMS | 10 |
| 3 | CHEMICAL BONDS | 20 |
| 4 | CHEMICAL REACTIONS | 32 |
| 5 | OXIDATION and REDUCTION | 41 |
| 6 | GASES | 47 |
| 7 | LIQUIDS and SOLIDS | 54 |
| 8 | SOLUTIONS | 61 |
| 9 | ACIDS and BASES I | 70 |
| 10 | ACIDS and BASES II | 77 |
| 11 | ELECTROLYTES | 91 |
| Selected Topic A – Inorganic Chemistry | | 100 |
| 12 | THE ATOMIC NUCLEUS | 107 |
| 13 | HYDROCARBONS | 114 |
| 14 | ALCOHOLS, PHENOLS, and ETHERS | 128 |
| 15 | ALDEHYDES and KETONES | 135 |
| 16 | CARBOXYLIC ACIDS and DERIVATIVES | 143 |
| Selected Topic B – Drugs: Some Carboxylic Acids, Esters, and Amides | | 154 |
| 17 | AMINES and DERIVATIVES | 159 |
| Selected Topic C – Brain Amines and Related Drugs | | 166 |
| 18 | STEREOISOMERISM | 173 |

| | | |
|---|---|---|
| *Selected Topic D – Chemistry of the Senses* | | *177* |
| 19 | CARBOHYDRATES | *178* |
| 20 | LIPIDS | *187* |
| *Selected Topic E – Hormones* | | *196* |
| 21 | PROTEINS | *199* |
| 22 | ENZYMES | *210* |
| *Selected Topic F – Vitamins* | | *218* |
| 23 | NUCLEIC ACIDS and PROTEIN SYNTHESIS | *224* |
| *Selected Topic G – Viruses* | | *234* |
| 24 | METABOLISM and ENERGY | *238* |
| 25 | CARBOHYDRATE METABOLISM | *248* |
| 26 | LIPID METABOLISM | *256* |
| 27 | PROTEIN METABOLISM | *263* |
| 28 | BODY FLUIDS | *272* |

## *PART II - Answers to Questions* — *283*

# PART I
## Chapter Summaries and Self Tests

# 1 Matter and Measurement

## KEY WORDS

| | | | | | |
|---|---|---|---|---|---|
| milli | structure | potential energy | liter | chemistry | element |
| kilo | matter | kinetic energy | heat | calorie | phyical change |
| centi | mass | work | temperature | joule | physical property |
| deci | weight | meter | Fahrenheit | density | chemical change |
| micro | energy | kilogram | Celsius | hypothesis | chemical property |
| mega | liquid | mixture | Kelvin | gravity | specific gravity |
| nano | solid | molecule | composition | formula | homogeneous |
| pico | gas | chemical symbol | solution | SI units | heterogeneous |
| giga | variable | scientific law | specific heat | compound | conversion factor |
| atom | second | substance | precision | accuracy | exponential notation |
| force | errors | significant figures | theory | factor-label | dimensional analysis |
| | | unit-conversion | one equation | one unknown | |

## SUMMARY

1.1 Science and the Human Condition
  A. Modern chemistry has roots in the alchemy of the Middle Ages, but was born in the 17$^{th}$ century with an emphasis on experimentation (Francis Bacon: 1561–1626).
  B. Modern drugs for medicine, fertilizers for crops, nutritional supplements, plastics, and synthetic fibers are all products of modern chemistry.

1.2 Problems in Paradise
  A. Such modern problems as toxic wastes, pollution, and carcinogens are also products of chemistry.
  B. The answers to these problems will also have to come from chemistry and will require an educated, informed society to ensure that chemistry is used for the human good.

1.3 The Way Science Works – Sometimes
  A. Science is cumulative. The "body of knowledge" is constantly growing and changing.
  B. It is not the "body of facts" that characterizes science but the organization of those facts according to concepts that can be tested by experimentation.
  C. Data is sometimes summarized as a *scientific law*
  D. The scientific method involves making observations, gathering data, organizing that data, forming a *hypothesis*, and subjecting that hypothesis to experimental verification.
  E. Careful measurements require recognizing *variables* and noting the effects when they change.
  F. A hypothesis can sometimes be expanded into a *theory* as a more comprehenive explanation.

1.4 Some Fundamental Concepts
  A. *Chemistry* – a study of the composition, structure, and properties of matter and the changes that occur in matter. Chemistry is often called the central science because a knowledge of chemistry is essential to understanding the other sciences such as the life sciences, physics, medicine, etc.
  B. *Matter* occupies space and has mass.
    1. *Mass* measures a quantity of matter that is independent of its relative position.
    2. *Weight* measures a force such as the gravitational force of attraction between an object and Earth. Weight varies with gravity; mass does not.
    3. *Atoms* are the smallest units we associate with the chemical behavior of matter.

1

## Chapter 1 - Matter and Measurement

      4. *Molecules* are composed of bonded groups of atoms.
      5. *Composition* refers to the types of atoms present and their relative proportions.
      6. *Structure* refers to the arrangement of atoms in space.
   C. Properties
      1. *Physical properties* (color, odor, hardness) can be observed without reference to other substances. A *physical change* occurs without a change in composition of the substance (such as melting, grinding).
      2. *Chemical properties* describe how one substance reacts with another substance. *Chemical changes* (such as combustion) are associated with changes in compostion or structure.
   D. States of matter: *solid*, *liquid*, or *gaseous* states
      1. *Solids* – maintain their shape and volume.
      2. *Liquids* – occupy a definite volume but assume the shape of their container.
      3. *Gases* – maintain neither their shape or volume but expand to fill their container.

1.5 Elements, Compounds, and Mixtures
   A. A *substance* (element or compound) has a defined or fixed composition.
      1. *Elements* are pure fundamental substances represented by a *chemical symbol*, e.g. H = hydrogen, O = oxygen, Al = aluminum, Fe = iron, etc.
      2. *Compounds* are pure substances made up of two or more elements that are chemically combined in a fixed ratio and represented by a *chemical formula*, e.g. glucose = $C_6H_{12}O_6$.
   B. The composition of a *mixture* is variable.
      1. A *homogeneous mixture* has the same composition and properties throughout the *solution*.
      2. A *heterogeneous mixture* varies in composition and or properties (sand/water mixture).

1.6 Energy and Energy Conversion
   A. *Energy* – the capacity for doing work.
      1. *Potential energy* of an object is that which is stored by virtue of its position or composition.
      2. *Kinetic energy* is the energy of motion ($KE = 1/2\ mv^2$).
      3. Energy is often classified in other ways depending on some characteristic of the energy considered, such as radiant, solar, thermal (*heat*), electrical, chemical, or nuclear energy.
      4. Chemical potential energy changes are associated with changes in the electronic arrangements of atoms.
   B. *Work* – force acting through a distance. Work is the expenditure of energy to make things happen.

1.7 Electric Forces
   A. *Force* – the push or pull that sets objects in motion.
   B. *Gravity* is the force of attraction between two masses that determines the weight of the object.
   C. Electrical forces involving charged particles are important in chemistry. Like electrical charges repel one another. Unlike charges attract one another.

1.8 The Modern Metric System
   A. Basic *SI units* (*Systeme International*)

| | | | |
|---|---|---|---|
| 1. Length – *meter* (m) | 1 m | = 1.09 yd | = 39.37 in. |
| 2. Mass – *kilogram* (kg) | 1 kg | = 1000 g | = 2.20 lb |
| 3. Volume – *liter* (L) | 1 L | = 1.06 qt | |
| 4. Time – *second* (s) | 1 s | = 1 s | |

# Chapter 1 - Matter and Measurement

B. Prefixes – The decimal nature of the metric system.
  1. *giga* – (G)     × 1,000,000,000.
  2. *mega* – (M)    × 1,000,000.
  3. *kilo* – (k)     × 1,000.
  4. *deci* – (d)     × 0.1
  5. *centi* – (c)    × 0.01
  6. *milli* – (m)    × 0.001
  7. *micro* – (μ)   × 0.000001
  8. *nano* – (n)    × 0.000000001
  9. *pico* – (p)    × 0.000000000001

C. Combining prefixes and units
   0.001 meters = 1 milli (m) = 1 mm;   1000 meters = 1000 m = 1 km;   0.000001 seconds = 1 μs

D. Sizes of objects

   | Object:   | atom    | sugar | protein  | baseball                       | Earth   |
   |-----------|---------|-------|----------|--------------------------------|---------|
   | Diameter: | 0.1 nm  | 1 nm  | 5–10 nm  | $8 \times 10^{-2}$ m = 8 cm    | 6000 km |

## 1.9 Precision and Accuracy in Measurements

A. Counting can give exact numbers, but experimental measurements are subject to error.

B. *Precision vs. Accuracy*
  1. *Precision* refers to how closely individual measurements agree with one another.
  2. *Accuracy* refers to how close the average measurement is to the "correct" value. Measurements can be precise but still not accurate.

C. *Errors* in measurements can arise from instrument errors (miscalibration), human error (misreading display), sampling errors (not using reproducible samples), etc.

D. *Significant Figures* – all digits with known certainty plus the first uncertain one in a *measurement*.
   e.g. number (# of significant figures);   74(2);  1102 (4);  100,000 (1);  100,010 (5);  0.000012 (2)
   1. Multiplication and Division: The answer has no more significant figures than the factor with the fewest significant figures.
   2. Addition and Subtraction:  1) Express in common units.
                                 2) The result should not increase the number of significant figures.

E. *Exponential* (scientific) *Notation* – base 10
  1. Chemistry deals with the very large (602,000,000,000,000,000,000,000) and very small (0.000000018) numbers. It is convenient to express these in exponential notation.
     $602,000,000,000,000,000,000,000 = 6.02 \times 10^{23}$;   $0.000000018 = 1.8 \times 10^{-8}$
  2. When numbers are expressed in scientific notation, the number of digits in the coefficient is the number of significant digits.
     $702 = 7.02 \times 10^2$    $41,000 = 4.1 \times 10^4$    $0.000011 = 1.1 \times 10^{-5}$

## 1.10 Unit Conversions

A. Quantities can be expressed in a variety of different units (6 ft, 2 yd, 72 in). It is often necessary to convert a measurement from one kind of unit into another type of unit. *Conversion factors* are "unit value ratios" that permit the conversion from one type of unit to another.

   | 1 ton = 2000 lb | 1 mi = 1760 yd | 1 gal = 4 qt  | 1 Cup = 8 fl oz |
   |-----------------|----------------|---------------|-----------------|
   | 1 lb  = 16 oz   | 1 yd = 3 ft    | 1 qt  = 2 pt  | ½ fl oz = 1 tbs |
   |                 | 1 ft = 12 in.  | 1 pt  = 2 cups| 1 tbs = 3 tsp   |

B. In the metric system, conversion factors for a given type of unit are all multiples of ten.

   | 1 kg = 1000 g   | 1 km = 1000 m  | 1 L  = 1000 mL  |
   |-----------------|----------------|-----------------|
   | 1 g  = 1000 mg  | 1 m  = 100 cm  | 1 mL = 1000 mL  |

## Chapter 1 - Matter and Measurement

C. It is important to keep track of the units to ensure that the conversion factors are being applied correctly.

1. 5 g  = ? kg        5 g $\times \dfrac{1 \text{ kg}}{1000 \text{ g}}$ = 0.005 kg

2. 3.5 kg = ? g        3.5 kg $\times \dfrac{1000 \text{ g}}{1 \text{ kg}}$ = 3500 g

D. Conversion to a different system of units may seem like working in a foreign language until you become familiar with the new units. Remember, all conversion factors must express an identity relationship; mass to mass, length to length, volume to volume, etc.

| 1 kg = 2.2 lb | 1 mi = 1.61 km | 1 L = 1.06 qt |
|---|---|---|
| 1 lb = 454 g | 1 m = 39.37 in | 1 qt = 0.946 L |
| 1 oz = 28.4 g | 1 in. = 2.54 cm | |

E. Some problems require several conversion factors. These can be carried out one step at a time or "chained" together using the *unit-conversion method* (also called the *factor-label method*, or *dimensional analysis*) as shown in this example.

A man runs the 100-m dash in 10.0 s. What is his speed in miles per hour?
(Note: 1 m = 39.37 in., 12 in. = 1 ft, 5280 ft = 1 mi; 60 s = 1 min, 60 min = 1 hr )

$$\dfrac{100 \text{ m}}{10 \text{ s}} \times \dfrac{39.37 \text{ in.}}{1 \text{ m}} \times \dfrac{1 \text{ ft}}{12 \text{ in.}} \times \dfrac{1 \text{ mi}}{5280 \text{ ft}} \times \dfrac{60 \text{ s}}{1 \text{ min}} \times \dfrac{60 \text{ min}}{1 \text{ hr}} = 22 \dfrac{\text{mile}}{\text{hr}}$$

( Given )   ×   (_____meters to miles_____)   ×   ( __s to hr__ )   =   New value

### 1.11 Density (density = mass/volume)

A. *Density* is mass per unit of volume (g/mL or g/cm$^3$).
   Densities of some common substances in (g/mL) (Note - density varies with temperature.)

| air | ethanol | ice (0 °C) | **water** | sugar | aluminum | iron | mercury | gold |
|---|---|---|---|---|---|---|---|---|
| 0.0012 | 0.79 | 0.92 | **1.0** | 1.6 | 2.7 | 7.2 | 13.6 | 19.3 |

B. *Specific gravity* is a ratio of density of the substance to the density of water. Specific gravity has no units.
   1. Specific gravity can be measured with a hydrometer.
   2. The specific gravity of mercury is 13.6.

C. Density problems
   1. Many terms used in chemistry involve ratios of several quantities; thus the units of such quantities also involve ratios of other units. *Density* = g/mL
   2. *One equation – one unknown*: When working with equations of several parameters, you can solve for the value of any one of the variables, provided you know the values of all the others simply by rearranging the equation to solve for the missing parameter.
      Density = mass/volume      Mass = density x volume      Volume = mass/density

   a) Calculate density, given a mass of 225 g and a volume of 30 mL.
      Density = $\dfrac{225 \text{ g}}{30 \text{ mL}}$ = 7.5 g/mL

   b) Calculate mass, given a density of 7.5 g/mL and volume of 30 mL.
      Mass = (7.5 g/1 mL) × 30 mL = 225 g

# Chapter 1 - Matter and Measurement

1.12 Energy: Temperature and Heat
  A. *Temperature* is an intensive physical property that tells us the direction of heat (energy) flow when two bodies are brought into contact. There are three commonly used temperature scales.
   1. *Celsius* (°C)       °C = (°F − 32) × (5/9)
   2. *Kelvin* (K)         K  = °C + 273
   3. *Fahrenheit* (°F)    °F = (9/5) × (°C) + 32
  B. Reference temperatures

| | °C | K | °F |
|---|---|---|---|
| 1. Freezing point of water | 0 | 273 | 32 |
| 2. Boiling point of water | 100 | 373 | 212 |
| 3. Absolute zero | −273 | 0 | −459 |

  C. Heat Energy – *joule* (J) or *calorie* (cal). A calorie is the amount of heat energy required to increase the temperature of 1 gram of water from 14.5 °C to 15.5 °C.
   1. 1 cal = 4.184 J    1 Calorie (food measure) = 1000 cal or 1 kcal
   2. A typical cookie has about 50–100 Cal (50–100 kcal) of heat energy. Fast walking for one hour burns about 350 Cal (350 kcal), i.e. one hour of exercise burns off about 3 cookies.
  D. The *specific heat* of a substance is the amount of heat required to increase the temperature of 1 g of that substance by 1.0 °C.
   1. The specific heat of water is very high, about 1.0 cal/(g·°C), while that of iron is only 0.108 cal/(g·°C).
   2. The general formula for specific heat problems is:
      **Heat absorbed or released = (mass) × (specific heat) x $\Delta T$**

# DISCUSSION

Much of the first chapter in the text is intended to place chemistry in both a historical and a contemporary perspective–to give you a feeling for chemistry as it affects society. If, after reading the chapter, you recognize chemistry as something more than just a course required for your particular academic program, then you have indeed understood what we were trying to say. In addition to this overview, Chapter 1 also introduces several concepts important to our further study of chemistry. These include the international system of measurement; the meaning of terms such as matter, force, and energy; different temperature scales; density and specific gravity.

You will find that the effort you invest in becoming familiar with dimensional analysis will help you a great deal later in the course. The problems that follow are relatively simple but illustrate the basic principle of dimensional analysis using factors with which you are more familiar. Keep in mind that the basic idea is to set up an equation such that the value given in one set of units multiplied by appropriate conversion factors will yield the desired value in the new set of units.

   **Original value  x  Conversion factors  =  New value sought**
   (old units)                                 (desired units)

In the case of the sprinter described above, his speed was given in m/s, but we wanted to know his speed in mi/h. This required two sets of conversion factors, one to convert meters to miles and another to convert seconds to hours. Note that the latter conversion appears awkward (reciprocal) because the second units appeared in the denominator of the given. However, if you set up the conversion factors so that the appropriate units cancel, you will find that this will take care of itself.

           (m/s)  ×  (meters to miles)  ×  (1/s to 1/h)  =  (mi/h)

   or      (m/s)  ×  (mi/m)             ×  (s/h)         =  (mi/h)

# Chapter 1 - Matter and Measurement

The problems at the end of the chapter are meant to check your understanding of this material. The following questions offer another opportunity for you to test yourself on Chapter 1. This set of problems provides practice in converting among British, SI, metric, and apothecary units. You may need to refer to Chapter 1 for needed conversion factors. Note: Many of these problems are more than tests of your memory. A number of them require preliminary calculations before an answer can be selected. You are expected to know the metric prefixes and units of measure, but you may also need to refer to the metric units presented in Tables 1.3 and 1.4.

## SELF-TEST

1. Which of the following represents a hypothesis?
    a. Mary read her textbook assignment, and Sally did not.
    b. Mary did her homework, and Sally did not.
    c. Mary attended class and took good notes, Sally did not.
    d. Mary got an A in the course, and Sally did not.
    e. Students who read their assignments, do their homework, and attend lectures earn higher grades than those students who do not do those things.
2. Which of the following represents a physical change?
    a. burning this study guide     b. tearing a page from this book
3. Identify the element(s) among the following.
    a. aluminum   b. milk   c. NaCl   d. Pb   e. mercury
4. Identify the compound(s) among the following.
    a. aluminum   b. milk   c. NaCl   d. Pb   e. mercury
5. Identify the mixture(s) among the following.
    a. aluminum   b. milk   c. NaCl   d. Pb   e. mercury
6. Which of the following abbreviations stands for a unit of length?
    a. mL   b. mg   c. mm   d. cc
7. The prefix nano is equivalent to:
    a. $10^{-2}$   b. $10^{3}$   c. $10^{-3}$   d. $10^{6}$   e. $10^{-6}$   f. $10^{-9}$
8. Which of the following units of measure is equivalent to a cc?
    a. mL   b. cm   c. mm   d. mg   e. g
9. How long is 0.2 cm?
    a. 0.02 mm   b. 2 mm   c. 20 mm   d. 200 mm   e. 0.002 mm
10. How many millimeters are there in 10 cm?
    a. 1   b. 10   c. 100   d. 1000   e. 10,000
11. How many cubic centimeters are there in a deciliter?
    a. 0.01   b. 0.1   c. 1   d. 10   e. 100
12. An object that weighs 10 µg also weighs:
    a. 0.001 mg   b. 0.01 mg   c. 0.1 mg   d. 1 mg   e. 100 mg   f. $10^{-6}$ g
13. If a container holds 5 mL, it will hold:
    a. 5000 L   b. 0.05 L   c. 50 cm$^3$   d. 0.5 cc   e. 0.005 L
14. Approximately how wide in cm is a coin that is 2.0 in. in diameter?
    a. 2   b. 4   c. 5   d. 6   e. 7   f. 8
15. Lorraine is 150 cm tall and weighs 82 kg. She is:
    a. skinny   b. just about perfect   c. a bit chubby   d. obese
16. If 10 mL of A has a mass of 8 g, the density of A is:
    a. 0.5 g/mL   b. 1 g/mL   c. 1.2 g/mL   d. 4 g/mL   e. 0.8 g/mL
17. If a container that can hold 5 g of water is filled with 10 g of another liquid, what is the density of the other liquid?
    a. 0.5 g/mL   b. 2 g/mL   c. 10 g/mL   d. 0.5   e. 2   f. 10

# Chapter 1 - Matter and Measurement

18. What is the specific gravity of a compound if 5 mL of it weighs 15 g?
    a. 0.3      b. 0.3 g/mL      c. 3      d. 3 g/mL      e. 6      f. 12
19. If a bottle will hold 10 g of water or 30 g of bromine, the specific gravity of bromine is:
    a. 0.33      b. 3 g/mL      c. 3      d. 30 g/mL      e. 30
20. If a liter of a substance weighs 1250 g, its specific gravity is:
    a. 1250      b. 1.2 g/cc      c. 0.80      d. 2 g/mL      e. 1.25      f. 0.08
21. If the density of a substance is 8 g/mL, what volume would 40 g of the substance occupy?
    a. 0.32 mL      b. 0.5 mL      c. 2 mL      d. 5 mL      e. 20 mL
22. Will a bar of soap (volume of 250 cc; mass of 300 g) float on water?
    a. yes      b. no
23. The boiling point of water is:
    a. 100 °C      b. 212 °F      c. 373 K      d. all of these      e. none of these
24. On the absolute temperature scale, temperatures are reported in:
    a. °Celsius      b. °Centigrade      c. °Fahrenheit      d. Kelvin
25. A temperature of 98.6 °F is the same as:
    a. 0 °C      b. 32 °C      c. 37 °C      d. 100 °C      e. 212 °C
26. A temperature of -40 °C is equivalent to:
    a. −7 °F      b. −40 °F      c. −57 °F      d. −81 °F      e. −113 °F
27. A temperature of 273 °C is equal to:
    a. zero K      b. 100 K      c. 273 K      d. 546 K      e. −273 K
28. A temperature of 77 °F is equal to how many degrees Celsius?
    a. 25      b. 61      c. 81      d. 171      e. 196
29. One food Calorie equals:
    a. 1000 cal      b. 1000 kcal      c. 1 cal      d. 0.001 kcal
30. A cola that contains 150 Calories also contains:
    a. 0.150 cal      b. 0.150 kcal      c. 1500 cal      d. 1.5 kcal      e. 150 kcal
31. We wish to heat 40 g of water from 15 °C to 25 °C. The amount of heat energy required is:
    a. 4 cal      b. 40 cal      c. 100 cal      d. 400 cal      e. 1000 cal
32. If 40 calories of energy are added to 10 g of water originally at 50 °C, the final temperature of the water will be:
    a. 4 °C      b. 10 °C      c. 46 °C      d. 54 °C      e. 60 °C      f. 90 °C
33. How many cal are released as 50g of water cools from 100 °C to room temperature (25 °C)?
    a. 50      b. 750      c. 1250      d. 3750      e. 7500
34. Which has the greatest potential energy?
    a. a small rock moving at high speed at sea level
    b. a large rock moving at slow speed at sea level
    c. a large rock balanced at the edge of a mountain top
    d. a small rock balanced on a ledge halfway down the same mountain
    e. none of the choices is clearly the best
35. This form of matter is characterized by its tendency to maintain its volume but not its shape.
    a. gas      b. liquid      c. solid
36. High compressibility is a property associated with:
    a. gases      b. liquids      c. solids
37. If two charged particles attract one another,
    a. one particle must be negatively charged.
    b. both particles must be negatively charged.
    c. both particles must be positively charged.
38. How many significant figures are there in a measurement of a liquid of 50.3 mL?
    a. 1      b. 2      c. 3      d. 4      e. 5

## Chapter 1 - Matter and Measurement

39. How many significant figures are there in a measurement of $5.062 \times 10^{-3}$?
    a. 1  b. 2  c. 3  d. 4  e. 5
40. Express the following to the proper number of significant figures: $15 + 10.2 + 0.2$ ?
    a. 25.4  b. 25  c. 26  d. 25.2  e. 35
41. Express the following to the proper number of significant figures: $4.7 \times 834.78$?
    a. 3900  b. 3923  c. 3920  d. 4000  e. 3800
42. Brian is 6ft 2in. tall. How tall is he in meters? (1 in.= 2.54 cm)
    a. 74  b. 6.20  c. 1.45  d. 1.88  e. 2.10
43. Bretna is 1.80 m tall. How tall is she in feet and inches?
    a. 5ft 3in.  b. 6ft 2in.  c. 5ft 9in.  d. 5ft 5in.  e. 5ft 11in.
44. How wide in inches is 35-mm film?
    a. 0.35 in.  b. 0.71 in.  c. 1.4 in.  d. 7.1 in.  e. 14 in.
45. We may define a premature baby as one who weighs less than 5 lb at birth. What is the corresponding birthweight in the metric system?
    a. 11 kg  b. 2.3 kg  c. 5.0 kg  d. 0.2 kg  e. 50 kg
46. Chris is 1.94 m tall and weighs 70 kg. Chris would be described as:
    a. slim  b. average  c. heavy  d. grossly overweight
47. If a prescription calls for 400 mg of Edrisal and each Edrisal tablet contains 0.40 g, how many tablets should be taken?
    a. 1  b. 2  c. 3  d. 4  e. 5
48. You are to administer 0.20 g of a medication by injection. The container label reads 50 mg/mL. How many cubic centimeters should you administer?
    a. 40 cm$^3$  b. 50 cm$^3$  c. 2 cm$^3$  d. 4.0 cm$^3$  e. 20 cm$^3$
49. Seven grams of aspirin is enough to fatally poison a small child. If aspirin tablets contain 5 grains per tablet, how many tablets would there be in a fatal dose? (There are 15 grains per gram.)
    a. 35  b. 21  c. 7  d. 75  e. 1125
50. How fast in km/h is the 65 mph speed limit?
    a. 34  b. 550  c. 27.5  d. 68  e. 105
51. How many liters will a 15.0 gal gas tank in a Ford pickup hold?
    a. 15.0 L  b. 56.6 L  c. 60.0 L  d. 45.3 L  e. 63.6 L
52. What is the volume of 1 pd (28.4 g/oz) of gold (density = 19.3 g/mL)?
    a. 18.6 mL  b. 1.3 mL  c. 454 mL  d. 3.5 mL  e. 23.5 mL
53. Gas in Europe was for sale at one place for $3.89/gal and across the street for $1.00/L. Which is the better deal?
    a. $3.98/gal  b. $1.00/L  c. both are equal in price
54. A square plot of land one mile on each edge is 640 acres. Given that there are 5280 ft/mi, approximately how many feet on each edge is needed for a square, one acre homesite?
    a. 50 ft  b. 100 ft  c. 150 ft  d. 200 ft  e. 400 ft
55. If on average a child is born in the United States every 10 s, how many births would occur in the United States each year? (365 days/year)
    a. 365,000  b. 3.2 million  c. 5.23 billion  d. 211 million  e. 43 million
56. T / F  The boiling point of water is 100 °F at 1 atm pressure.
57. T / F  Chemistry is the study of matter and the changes it undergoes.
58. T / F  Chemistry is not concerned with changes in energy.
59. T / F  The United States is a leader among nations using the metric system.
60. T / F  Manufactured chemical products have greatly affected our lifestyle.
61. T / F  The ultimate source of nearly all energy on Earth is the sun.
62. T / F  If measured on the Moon, your mass would be different from your mass measured on Earth, although your weight would be the same.
63. How tall are you in meters?
64. How much do you weigh in kg?

Chapter 1 - Matter and Measurement

**ANSWERS**

| | | | | | | | |
|---|---|---|---|---|---|---|---|
| 1. e | 9. b | 17. b | 25. c | 33. d | 41. a | 49. b | 57. T |
| 2. b | 10. c | 18. c | 26. b | 34. c | 42. d | 50. e | 58. F |
| 3. a, d, e | 11. e | 19. c | 27. d | 35. b | 43. e | 51. b | 59. F |
| 4. c | 12. b | 20. e | 28. a | 36. a | 44. c | 52. e | 60. T |
| 5. b | 13. e | 21. d | 29. a | 37. a | 45. b | 53. b | 61. T |
| 6. c | 14. c | 22. b | 30. e | 38. c | 46. a | 54. d | 62. F |
| 7. f | 15. d | 23. d | 31. d | 39. d | 47. a | 55. b | 63. __ |
| 8. a | 16. e | 24. d | 32. d | 40. b | 48. d | 56. F | 64. __ |

# 2 Atoms

**KEY WORDS**

| | | | | |
|---|---|---|---|---|
| atom | atomic theory | | electron | orbital cons. of mass |
| mass number | quantized | protons | ground state | definite proportions |
| element | chemical property | neutron | excited state | multiple proportions |
| compound | isotope | nucleus | atomic number | atomic mass |
| molecule | radioactivity | periodic table | period | energy level |
| law | cathode ray | VSEPR | group | electron configuration |
| amu | chemical reaction | aufbau principle | alkali metal | alkaline earth |
| atomic radius | electron affinity | ionization energy | halogen | main-group element |
| metal | metalloid | nonmetal | noble gas | transition element |
| ion | valence shell | | | |

**SUMMARY**

The prevailing view of matter held by early Greek philosophers in the 5$^{th}$ century B.C. was that of endless divisibility. This was the view supported by Aristotle. Democritus, a student of Leucippus, believed there must be a limit to the divisibility of matter and called his ultimate particles "atomos" meaning "indivisible." We now define an *atom* as the smallest characteristic particle of an *element*. The Greeks believed that there were only four elements: earth, air, fire, and water.

The scientific and political revolution in France led to the acceptance of experimental data as the basis for accepting or rejecting scientific theories. Boyle (1661) proposed that pure substances capable of being broken down into simpler substances were compounds, not elements. Lavoisier (1782) helped establish chemistry as a quantitative science and used systematic names for elements. He proposed that matter is neither created nor destroyed during a chemical change – *law of conservation of mass*. Proust (1799) concluded from compositional analyses that elements combine in definite proportions to form compounds – *law of definite proportions*.

2.1 Dalton's Atomic Theory (1803)
  A. Dalton extended the ideas of Lavoisier and Proust with the *law of multiple proportions*.
  B. Dalton proposed his atomic theory to explain the "laws" of chemistry. A chemical *law* is a statement that summarizes data obtained from experiments. A *theory* is a model that consistently explains observations.
    1. All matter is made up of small indestructible and indivisible atoms.
    2. All *atoms* of a given *element* are identical, but different elements have different atoms.
    3. *Compounds* are formed by combining atoms.
    4. A *chemical reaction* involves a change in the way atoms combine to form *molecules* but does not involve a change in the atoms themselves.
  C. Dalton setup a table of relative masses for the elements based on hydrogen. These relative masses are expressed in *atomic mass* units or *amu*.
  D. Dalton was later shown to be wrong on the first two assumptions (atoms are divisible, and some elements have *isotopes*) but Dalton's atomic theory served to explain a large body of experimental data. An important consequence of his theory was the emphasis on relative atomic masses.

# Chpater 2 - Atoms

## 2.2 The Nuclear Atom
A. The electrical nature of the atom was discovered in the 19th century.
1. Crookes (1875) – invented the vacuum *cathode ray* tube (Crookes tube).
2. Goldstein (1886) – used an apparatus similar to the Crookes tube to study the positive atomic particles. The positively charged particles were found to be more massive (1837 ×) than electrons and also varied with the type of gas used in the experiment. The lightest positive particle obtained was derived from hydrogen and was called a *proton*.
3. Thomson (1897) – cathode rays were deflected in an electric field and thus must be charged particles (*ions*). These negatively charged particles were called *electrons* and were found to be the same for all gases used to produce them.
4. The French physicist Becquerel discovered that some atoms fall apart. Polish-born chemist, Marie Curie, named this *radioactivity.*
5. Millikan (1909) – measured the charge on an electron.
   **Electrons,** (−) negative charge,  $9.1 \times 10^{-28}$ g;
   **Protons,** (+) positive charge,  $1.7 \times 10^{-24}$ g
B. Rutherford directed a beam of alpha particles at a thin sheet of gold foil and, to his amazement, observed that some of the alpha particles were deflected sharply and some even bounced back toward the source. In 1911 Rutherford formulated his nuclear theory of the atom, which postulated that all of the positive charge and virtually all of the mass of an atom were concentrated in a tiny nucleus surrounded by the negatively charged electrons. Rutherford (1914) proposed that protons constitute the positively charged matter of all atoms, not just those of hydrogen.
C. Chadwick (1932) discovered an uncharged nuclear particle called a *neutron*, which has about the same mass as a proton. In chemistry all atoms can be thought of as having a small positively charged *nucleus* containing protons and neutrons, which is surrounded by negatively charged electrons. An oversimplified but useful chemical view of the atomic nucleus is that it consists of protons, neutrons, and the force that holds them together.
1. proton      (p)      1.007276 amu      charge 1+
2. neutron     (n)      1.008665 amu      charge 0
3. electron    (e−)     0.000549 amu      charge 1−
Note: The mass of the neutron is slightly greater than the mass of a proton plus an electron, corresponding to the binding energy required ($\Delta E = \Delta mc^2$).
D. The *atomic number* of an element is determined by its number of protons. An element is a substance in which all atoms have the same atomic number.

## 2.3 Isotopes, Atomic Masses, and Nuclear Symbols
A. Atoms with the same number of protons (and atomic number) but a different number of neutrons are called *isotopes*. All isotopes of a given element have very similar chemical properties.
   **Deuterium** (1p + 1n) and **Tritium** (1p + 2n) are isotopes of **Hydrogen** (1p).
B. All isotopes can be represented by a nuclear symbol.

$^{A}_{Z}X$ where ($^{12}_{6}C$)    $A$ = *mass number* = protons (p) + neutrons (n);  6p + 6n  or $A = 12$
                                    $X$ = chemical symbol ; C for $Z = 6$
                                    $Z$ = atomic number = number of protons (p); 6p or $Z = 6$

C. Atomic masses (or atomic "weights") listed in the periodic table are the isotopically weighted average value for that element. An amu is defined as 1/12 the mass of a C-12 atom.
   $^{12}C = 12.000$; $^{13}C = 13.003$; but C = 12.01 since C-12 is much more abundant than C-13

# Chapter 2 - Atoms

2.4 The Bohr Model of the Atom
   A. The light emitted by atoms excited by a flame is not a continuous spectrum but contains instead only a few discrete lines. Light of different energies has different colors ($E_{red} < E_{blue}$).
   B. Niels Bohr (1913) postulated that the discrete spectra arose because the energy of an electron in an atom was *quantized* and could take on only certain discrete values of energy.
   C. The Bohr model of the atom was based on the known laws of planetary motion.
      1. Electrons orbit the nucleus in a manner analogous to planets orbiting the sun.
      2. Different *energy levels* correspond to different orbits with the lowest orbit (*ground state*) being the closest, and the higher energy orbits (*excited states*) being more distant from the nucleus.
      3. The maximum number of electrons in any given energy level (shell) was represented by the formula $2n^2$ where $n$ = shell number.

   D. Bohr diagrams (Mg  Z = 12)

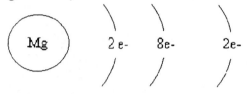

2.5 Electronic Configurations
   A. de Broglie (1925) postulated that electrons were particles with wave-like properties.
   B. Quantum mechanics is a highly mathematical discipline that treats electrons as waves and gives their locations in terms of probabilities.
      1. Schrödinger (1929) used wave equations to describe the modern picture of the atom.
      2. Electrons were no longer described as being in definite planetary orbits but instead, according to a high degree of probability, were confined to three-dimensional charge clouds referred to as *orbitals*.
      3. The letters "*s, p, d* and *f*" are used to refer to different types of orbitals:

| Orbital | Number of Orbitals | Number of Electrons |
|---|---|---|
| s  *(spherical)* | 1 | 2 |
| p  *(perpendicular)* | 3 | 6 |
| d  *(diffuse)* | 5 | 10 |
| f  *(fundamental)* | 7 | 14 |

   C. Energy levels and orbitals
      1. Each orbital can be occupied by 1 "lone" electron or by 2 "paired" electrons.
      2. Each energy level "$n$" contains $n^2$ orbitals that can contain up to $2n^2$ electrons.
      3.

| Energy Level | Orbital | (Number) | Electrons | (Subtotal) |
|---|---|---|---|---|
| 1 | 1s | 1 | 2 | (2) |
| 2 | 2s | 1 | 2 | |
|   | 2p | 3 | 6 | (8) |
| 3 | 3s | 1 | 2 | |
|   | 3p | 3 | 6 | |
|   | 3d | 5 | 10 | (18) |
| 4 | 4s | 1 | 2 | |
|   | 4p | 3 | 6 | |
|   | 4d | 5 | 10 | |
|   | 4f | 7 | 14 | (32) |

D. The *electron configuration* of an atom is determined by "filling" the lower energy levels first (*aufbau principle*). The electrons in the outermost shell are called the *valence electrons*.

The general order for filling orbitals is:

| $1s^2$ | $2s^2 2p^6$ | $3s^2 3p^6$ | $4s^2 3d^{10} 4p^6$ | $5s^2 4d^{10} 5p^6$ |
|---|---|---|---|---|
| 1 | 2 | 3 | 4 | 5 |

(-------------------------------period-------------------------------)

Examples

| | | | | | |
|---|---|---|---|---|---|
| N  ($Z = 7$) | $1s^2$ | $2s^2 2p^3$ | | | |
| Cl  ($Z = 17$) | $1s^2$ | $2s^2 2p^6$ | $3s^2 3p^5$ | | |
| Ba  ($Z = 56$) | $1s^2$ | $2s^2 2p^6$ | $3s^2 3p^6$ | $4s^2 3d^{10} 4p^6$ | $5s^2 4d^{10} 5p^6 6s^2$ |

2.6 Mendeleev's Periodic Table

A. By the mid-1800s there were several known elements, but no successful way had been determined to classify the elements.

B. Mendeleev's *periodic table* (1869) grouped elements together with properties and is the basis for the modern periodic table.

2.7 Electronic Structure and the Modern Periodic Table

A. Groups and Periods

1. *Period* – *horizontal row* in the table.
2. *Group* – *vertical column* in the table; members of the same group have similar chemical properties because of similar outer electron configurations. Some of these *main-group elements* are given family names.

| | | | |
|---|---|---|---|
| Group IA | - | *alkali metals* | - | $ns^1$ |
| Group IIA | - | *alkaline earths* | - | $ns^2$ |
| Group VIIA | - | *halogens* | - | $ns^2 np^5$ |
| Group VIIIA | - | *noble gases* | - | $ns^2 np^6$ |
| Groups B | - | *transition elements* | - | filling $nd^2$ |

B. The *chemical properties* of an element are defined by its number of electrons (or electron structure) and by its number of protons, not by its atomic mass. All atoms of the same element have the same number of protons, which defines its atomic number. Elements in the modern periodic table are arranged in order of increasing atomic number (not atomic mass) and are grouped according to their electronic structure.

2.8. Periodic Atomic Properties

A. *Periodic properties* and trends in the periodic table. Most elements are metals (lower left). *Non-metals* are in the upper right of the periodic table, *metals* (most abundant) towards lower left, and *metalloids* (B, Si, As, Te, At) along the diagonal between the nonmetals and metals. Metallic character, *atomic radius*, and *ionization energy* vary as shown. *Electron affinity* refers to the addition of an electron to a neutral atom, it is not simply the reverse of the ionization energy.

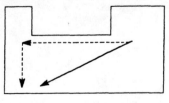
Metallic character

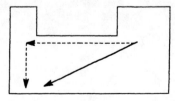
Size

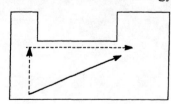
Ionization Energy

Chapter 2 - Atoms

    B. Which model to use? Use the simplest model that fits the task at hand.
        1. Dalton's billiard ball model of the atom is useful in describing the behavior of gases.
        2. Bohr's planetary model helps one understand the sizes of atoms and energy levels. The Valence Shell Electron Pair Repulsion (VSEPR) model is an extension of the Bohr theory, which we will use to explain the shape of molecules (Chapter 3).
        3. Schrödinger's quantum mechanical model can be used to better understand the principles of chemical bonds, magnetic properties, and spectral properties.

## DISCUSSION

    A number of terms are introduced in Chapter 2. These terms are basic to the language of chemistry. As in any language, it is important to learn the vocabulary or it will be difficult to understand the material. Before you proceed, review the KEY WORDS. A number of individuals also were introduced in this chapter. It will be our habit to associate developments in chemistry with the people who were responsible for them. The emphasis that individual instructors place on such biographical information will vary greatly. However, the contributions of a number of individuals mentioned in this chapter are so fundamental to the development of chemistry as a science that many of their names will become familiar to students of chemistry. At the end of the SELF-TEST, you will have an opportunity to match the most prominent individuals mentioned with the fundamental concepts they helped to develop.

    It is important that you be able to write down the electron configurations of the Group A type elements. You should be able to do this by looking at the position of the element in the periodic table and remembering a few simple steps.

    1. The atomic number gives you the number of electrons in the neutral atom. Add electrons for any negative charge on anions, and subtract electrons for the positive charge on cations.

    2. Use the order-of-filling given below to write the electronic configuration.

$$1s^2 \quad 2s^2 2p^6 \quad 3s^2 3p^6 \quad 4s^2 3d^{10} 4p^6 \quad 5s^2 4d^{10} 5p^6 \quad 6s^2 4f^{14} 5d^{10} 6p^6$$

Note: You do not have to memorize this general order of filling since this information correlates directly with the buildup of the periodic table. The figure that follows relates the subshell configuration to the positions of various groups in the periodic chart. Once you master this relationship, it will be much easier to write down electronic configurations. Note the position of zinc (Zn) in the figure. Zinc would have a ground state electronic configuration of $1s^2 \; 2s^2 2p^6 \; 3s^2 3p^6 \; 4s^2 3d^{10}$. The sum of the exponents in the electronic configuration describing the neutral atom must equal the atomic number (30 for Zn). Finally, it is often useful to represent all filled inner shell electrons by the symbol of the corresponding noble gas. Thus a short-hand way of expressing the electronic configuration for Zn would be: Zn = [Ar] $4s^2 3d^{10}$.

Chpater 2 - Atoms

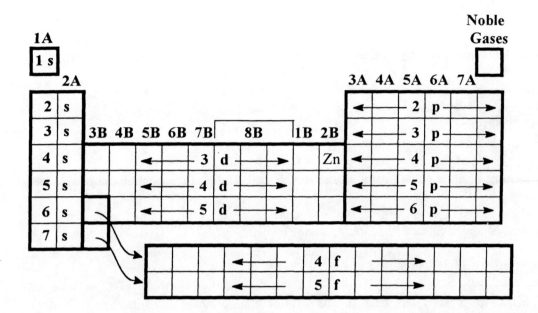

**Electronic configurations and the periodic table.**

Being able to write an electron configuration may seem like an interesting exercise but may be of little practical importance to you at this time. However, you will find that this information is of great value to you later on in predicting preferred ion forms of the elements and in writing molecular formulas.

Example 1. What ion form would you predict for the element fluorine?
$$F : 1s^2\ 2s^2 2p^5 \qquad\qquad F^- : 1s^2\ 2s^2 2p^6$$
Fluorine will want to gain 1 electron to form a stable 1– anion, which yields the noble gas electron configuration of Ne.

Example 2. What ion form would you predict for the element magnesium?
$$Mg : 1s^2\ 2s^2 2p^6\ 3s^2 \qquad\qquad Mg^{2+} : 1s^2\ 2s^2 2p^6$$
Magnesium will lose 2 electrons to form a stable 2+ cation with the same electron configuration as the noble gas Ne.

These examples show how knowledge of the periodic table and electron configurations can help you predict the common ion form of an element. This information also helps you to predict that the correct molecular formula for magnesium fluoride would be $MgF_2$ (one magnesium ion for every two fluoride ions).

Numerical problems are few in this chapter. Only the law of definite proportions lends itself to arithmetic manipulation. Remember that a molecule is a fixed combination of at least two atoms. For example, all molecules of water ($H_2O$) incorporate exactly two hydrogen atoms and one oxygen atom, and all molecules of carbon dioxide ($CO_2$) have two atoms of oxygen for each carbon atom. After reviewing the examples in the chapter, try the following problems.

## Chapter 2 - Atoms

### Problems

The analysis of the compound magnesium chloride shows it to consist of 25% magnesium and 75% chlorine by weight.
1. How many grams of magnesium are present in 10 g of magnesium chloride?
2. How many pounds of magnesium are present in a 100-lb sample of the compound?
3. A sample of the compound weighed 20 g. How many grams of magnesium and how many grams of chlorine are present in this sample?
4. How large a sample of the compound is required to have 5 g of magnesium?
5. How many grams of chlorine will combine completely with 100 g of magnesium to produce the compound?
6. An analysis shows that a sample of magnesium chloride contains 1.5 g of magnesium and 6 g of chlorine. Is this consistent with the law of definite proportions?

### SELF-TEST (Refer to a periodic table.)

1. Which term, as used by scientists, best fits this definition: a statement that summarizes the data obtained from observations?
    a. law          b. model          c. theory
2. Which term best describes Einstein's thoughts on relativity?
    a. law          b. model          c. theory
3. Which of the following facts does not fit Dalton's atomic theory?
    a. All atoms of oxygen are different from all atoms of nitrogen.
    b. Atoms are not destroyed in chemical reactions.
    c. Combinations of atoms rearrange during a chemical reaction.
    d. All oxygen atoms have the same number of protons but may have different atomic masses.
4. Which of the following postulates of Dalton's atomic theory conflicts with the existence of radioactivity?
    a. All matter is made up of small indestructible and indivisible atoms.
    b. All atoms of a given element are identical, but different elements have different atoms.
    c. Compounds are formed by combining atoms.
    d. A chemical reaction involves a change in the way atoms combine to form molecules but does not involve a change in the atoms themselves.
5. According to the law of definite proportions, if a sample of compound A contains 10 g of sulfur and 5 g of oxygen, another sample of A that contains 50 g of sulfur must contain:
    a. 100 g of oxygen     b. 50 g of oxygen     c. 25 g of oxygen
6. If 100 g of sulfur dioxide contains 50% sulfur by weight, 0.5 g of sulfur dioxide will contain what percentage of sulfur by weight?
    a. 0.5%          b. 25%          c. 50%          d. 75%
7. You have 10 g of element A and 10 g of element B. Compound X is known to consist of 40% A and 60% B. According to the law of definite proportions, if you mix all of your A and B together to form compound X:
    a. You will get 20 g of X.
    b. After all possible X has formed, some A will be left unreacted.
    c. After all possible X has formed, some B will be left unreacted.
    d. The compound X that forms will consist of 50% A and 50% B.
8. What is the maximum amount (g) of X that could be produced in question #7?
    a. 10.0          b. 20.0          c. 15.0          d. 16.7          e. 12.5

Chpater 2 - Atoms

9. Which of the following illustrates the law of multiple proportions?
    a. A sample of a compound contains 3 g of carbon and 4 g of oxygen; a second sample of the same compound contains 6 g of carbon and 8 g of oxygen.
    b. A sample of one compound contains 6 g of carbon and 8 g of oxygen; a sample of another compound contains 6 g of carbon and 2 g of hydrogen.
    c. A sample of one compound contains 6 g of carbon and 8 g of oxygen; a sample of another compound contains 6 g of carbon and 16 g of oxygen.
10. The anode and a cation are:
    a. positively charged    b. negatively charged    c. neutral
11. When an evacuated Crookes tube is discharged,
    a. negative electrons jump from cathode to anode.
    b. negative electrons jump from anode to cathode.
    c. positive electrons jump from anode to cathode.
12. Rutherford's gold foil experiment offered evidence in support of the theory that:
    a. an atom has a very compact nucleus.
    b. the atoms of an element can have different masses.
    c. gold is a radioactive element.
13. According to the Bohr model, how many electrons are there in the highest energy level of a ground state chlorine atom?
    a. 1    b. 2    c. 3    d. 6    e. 7    f. 8    g. 13    h. 18
14. In the Bohr model of the sodium atom the lowest energy level contains how many electrons?
    a. 1    b. 2    c. 3    d. 4    e. 5    f. 6    g. 7    h. 8
15. Which element should be chemically similar to sulfur, S ?
    a. Na    b. O    c. C    d. Ar
16. Which element would you expect to be chemically similar to Sr ?
    a. Co    b. Ba    c. K
17. Which is the proper electron configuration for Si ?
    a. $1s^2\ 2s^22p^6\ 3s^23p^4$    b. $1s^2\ 2s^22p^6\ 3s^2\ 4s^2$
    c. $1s^2\ 2s^22p^6\ 3s^2$    d. $1s^2\ 2s^22p^6\ 3s^23p^2$
18. Predict the major ion form for aluminum.
    a. $Al^{2+}$    b. $Al^{2-}$    c. $Al^{3+}$    d. $Al^{3-}$    e. $Al^{4+}$
19. Which is the proper electron configuration for $Cl^-$ ?
    a. $1s^2\ 2s^22p^6\ 3s^23p^5$    b. $1s^2\ 2s^22p^6\ 3s^2\ 4s^24p^6$
    c. $1s^2\ 2s^22p^6\ 3s^2$    d. $1s^2\ 2s^22p^6\ 3s^23p^6$
20. How many electrons and neutrons are in a neutral atom of $^{123}Sb$?
    a. 38, 85    b. 50, 23    c. 51, 72    d. 51, 21    e. 50, 73
21. A particle contains 16 protons, 18 electrons, and 17 neutrons.
    A. The atomic mass number is:       16    17    18    33    34    51
    B. The symbol for the element is:    Si    P    S    Cu    Zn    Ga
    C. The particle has a charge of:     0    1+    2+    1–    2–    3–
    D. If an electron were removed, which of the above answers would have to be changed?
        A    B    C
    E. If a neutron were added to the original particle, which of the answers would have to be changed?    A    B    C
    F. If a proton were added to the original particle, which of the answers would have to be changed?    A    B    C

Chapter 2 - Atoms

22. Given:          Atom A  Atom B  Atom C  Atom D  Atom E  Atom F
    No. of electrons   13      6       8       16      8       7
    No. of neutrons    14      7       7       16      6       8
    No. of protons     14      5       7       16      8       7
     a. Which atoms have the same atomic mass?    A  B  C  D  E  F
     b. For which atom(s) is the atomic number 5?  A  B  C  D  E  F
     c. Which atom(s) is(are) neutral?             A  B  C  D  E  F
     d. Which is(are) nitrogen?                    A  B  C  D  E  F

23. Refer to the diagram to answer the following questions:

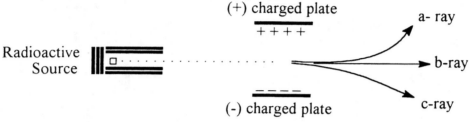

A. The a-ray
   a. could be a stream of electrons.
   b. could be a stream of protons.
   c. could be a stream of neutrons.
B. The b-ray
   a. could be a stream of γ-rays.
   b. could be a stream of neutrons.
   c. could be a stream of protons.
C. The c-ray
   a. could be a stream of protons.
   b. could be a stream of electrons.
   c. could be a stream of neutrons.

**Matching** (A name may be associated with more than one concept.)

**Name of Person**               **Concept**
a. Becquerel      __ 24. proponent of the atomic structure of matter among the ancient Greeks
b. Bohr           __ 25. proposed the modern atomic theory
c. Chadwick       __ 26. law of conservation of mass
d. Curie          __ 27. discovered radioactivity
e. Dalton         __ 28. law of definite proportions
f. Democritus     __ 29. law of multiple proportions
g. Lavoisier      __ 30. made quantitative measurements the standard of chemical
                        experimentation
h. Mendeleev      __ 31. periodic table of the elements
i. Proust         __ 32. the nuclear atom
j. Thomson        __ 33. the atom as a miniature solar system
k. Rutherford     __ 34. measured the charge on an electron
l. Millikan       __ 35. discovered the electron
                  __ 36. discovered the neutron
                  __ 37. explained the line spectra of the elements
                  __ 38. discovered new radioactive elements

# ANSWERS

## Problems

1. 2.5 g
2. 25 lb
3. 5 g of magnesium and 15 g of chlorine
4. 20 g of magnesium chloride
5. 300 g of chlorine
6. No

## Self-Test

1. a
2. c
3. d
4. a
5. c
6. c
7. b
8. d
9. c
10. a
11. a
12. a
13. e
14. b
15. b
16. b
17. d
18. c
19. d
20. c

21. A) 33
    B) S
    C) −2
    D) C
    E) A
    F) A, B, C
22. a) C, E
    b) B
    c) D, E
    d) C, F

23. A) a
    B) a, b
    C) a
24. f
25. e
26. g
27. a
28. i
29. e

30. g
31. h
32. k
33. b
34. l
35. j
36. c
37. b
38. d

# 3 Chemical Bonds

**KEY WORDS**

| | | | | |
|---|---|---|---|---|
| cations | electron configuration | ionic bond | linear | bonding pair |
| anions | electron dot symbols | covalent bond | bent | nonbonding pair |
| metals | molecular formula | polar covalent | salt | polar |
| nonmetals | structural formula | electronegativity | pyramidal | nonpolar |
| valence shell | polyatomic ion | octet rule | tetrahedral | stoichiometry |
| single bond | double bond | triple bond | Lewis structure | VSEPR |
| lone pair | electronegative | trigonal | chemical bond | ionic compound |
| crystalline | free radical | dipole | molecule | ion |

**SUMMARY**

There are over 100 elements but *"billions and billions"* of possible compounds. Here we introduce the concept of chemical bonding, the forces that enable atoms to join together to form compounds.

3.1 Stable Electronic Configurations
  A. Noble gasses are typically stable, undergoing few reactions.
  B. Atoms can gain or lose electrons to form *ions*.
    1. *Cations* – positively charged ions.
    2. *Anions* – negatively charged ions.
  C. The more stable ions have an *electron configuration* with a filled outer electron level (like the noble gases). Na looses an electron to become Na+ with 10 electrons arranged similarly to Ne.

3.2 Lewis Structures
  A. *Electron dot symbols* (*Lewis structures*) are a useful way to represent atoms or ions. The nucleus and the inner levels (core) are represented by the chemical symbol; the outer or *valence* electrons are represented by dots.

   Atoms: sodium Na•,  chlorine •C̈l:  ;   Ions:  sodium ion Na$^+$ ; chloride ion  [ :C̈l: ]$^-$

  B. Electron dot formulas are more convenient to use than energy level diagrams. For the Group A elements, the number of valence electrons is equal to the group number. The valence electrons are represented by dots.
  C. For the first 18 elements, it is useful to represent the first four electrons as lone dots (unpaired electrons) on each side of the symbol before pairing any of the dots.

| 1A | 2A | 3A | 4A | 5A | 6A | 7A | Noble Gases |
|---|---|---|---|---|---|---|---|
| H• | | | | | | | He: |
| Li• | •Be• | •B• | •C• | :N• | :O• | :F• | :Ne: |
| Na• | •Mg• | •Al• | •Si• | :P• | :S• | :Cl• | :Ar: |

20

# Chapter 3 - Chemical Bonds

3.3 The Sodium–Chlorine Reaction
   A. Sodium is a very soft, reactive metal normally stored under oil.
   B. Chlorine is a greenish-yellow reactive gas that is used as a disinfectant for swimming pools.
   C. When sodium metal is added to chlorine gas, a violent reaction takes place and produces a white stable, water-soluble solid *salt* known as sodium chloride (table salt).
   D. Sodium reacts with chlorine by giving up an electron to chlorine.

$$Na\cdot \; + \; :\!\ddot{Cl}\!\cdot \; \longrightarrow \; Na^+ \; + \; [:\!\ddot{Cl}\!:]^-$$

   E. The ion products of this reaction are stable because both ions have stable, noble gas type electron configurations. The ions have opposite charges, so sodium chloride forms a *crystalline* solid stabilized by ionic bonds, $Na^+ Cl^-$.

3.4 Ionic Bonds: Some General Considerations
   A. *Metals* (Na, K, etc.) are located on the left side of the periodic table, while *nonmetals* (N, O, Cl, etc.) are located on the right side of the periodic table. Metals tend to form cations by giving up electrons to nonmetals which then form anions. The resulting *salt* that is produced is held together by *ionic bonds*.
   B. The number of electrons given up or taken on can usually be predicted from the group number of the element, being equal to the group number for metals and equal to eight minus the group number for nonmetals.

| Na | Mg | Al | Si | P | S | Cl |
|----|----|----|----|----|----|----|
| +1 | +2 | +3 | (+4,-4) | (+5,-3) | -2 | -1 |

3.5 Names of Simple Ions and Ionic Compounds
   A. Cations – add "*ion*" to the name of the parent element: $Na^+$, sodium ion.
   B. Anions – change the ending to "-*ide*" and add "*ion*": $Cl^-$, chloride ion.
   C. Some elements have more than one stable ion form. In those cases, Roman numerals are used to indicate the charge of the ion: $Fe^{2+}$, iron(II) ion; $Fe^{3+}$, iron(III) ion
   D. The least common multiple is used to determine the correct *stoichiometry*, or numbers of cations and anions needed for the correct formula for an *ionic compound*.
   E. Examples

| Name | Formula |
|------|---------|
| sodium chloride | NaCl |
| magnesium chloride | $MgCl_2$ |
| iron(II) bromide | $FeBr_2$ |
| iron(III) fluoride | $FeF_3$ |
| aluminum oxide | $Al_2O_3$ |

3.6 Covalent Bonds: Shared Electron Pairs
   A. Many compounds are composed of elements that are unable to transfer electrons completely between them to form stable ionic bonds. These *molecules* achieve stable electronic configurations by sharing pairs of electrons.
   B. A *chemical bond* formed by sharing a pair of electrons is called a *covalent bond*. Many covalently bonded atoms, except hydrogen, seek an arrangement that surrounds them with eight electrons (*octet rule*) with filled outer shell *s* and *p* orbitals. It is important to distinguish shared or bonded pairs of electrons from unshared or lone pairs of electrons do not contribute to bonding. For convenience, a shared pair of electrons is often represented as a dash, and lone pairs are omitted entirely. $Cl_2$ has 1 shared or bonding pair (*single bond*), the rest are *nonbonding pairs* of electrons.

## Chapter 3 - Chemical Bonds

:Cl· + ·Cl: ⟶ :Cl::Cl: or :Cl—Cl: or $Cl_2$

·N· + ·N· ⟶ :N:::N: or :N≡N: or $N_2$

### 3.7 Multiple Covalent Bonds
A. Some atoms are able to share two pairs (**double bond**) or three pairs (**triple bond**) of electrons to achieve the stable octet. $N_2$ has 3 bonding pairs of electrons, a triple bond.
B. Examples: Double bond - sharing 2 pairs of electrons. Triple bond - sharing 3 pairs of electrons.

:Ö: :C: :Ö: or O=C=O          :N:::N: or :N≡N: or $N_2$

### 3.8 Names for Covalent Compounds
A. Prefixes: mono (1), di (2), tri (3), tetra (4), penta (5), hexa (6), hepta (7), octa (8), etc.
$NO_2$ = nitrogen **di**oxide;   $N_2O_4$ = **di**nitrogen **tetra**oxide;   $CCl_4$ = carbon **tetra**chloride
CO = carbon **mon**oxide;   $CO_2$ = carbon **di**oxide;   $SF_6$ = sulfur **hexa**fluoride
B. Many covalent compounds have common names.
$H_2O$ water   /   $CH_4$ methane   /   $C_6H_{12}O_6$ glucose

### 3.9 Unequal Sharing: Polar Covalent Bonds
A. In many compounds the electron pair is unequally shared between the two bonded atoms. This type of bonding is referred to as a **polar covalent** bond.
B. The element in a polar covalent bond that more often has the electrons is said to carry a "partial" negative charge and to be more "*electronegative*" than the other element. For example, fluorine is more electronegative than hydrogen. Thus, HF has a polar bond ($^{\delta+}$H—F$^{\delta-}$).

### 3.10 Electronegativity
A. **Electronegativity** refers to the tendency of an atom to attract electrons to itself.
B. Nonmetals are more electronegative than metals. Fluorine (upper right in the periodic table) is the most electronegative element, and metals like cesium (lower left) are the least electronegative.

| Electronegativities (*EN*) | Li | Be | B | C | N | O | F |
|---|---|---|---|---|---|---|---|
| | 1.0 | 1.5 | 2.0 | 2.6 | 3.1 | 3.5 | 4.0 |

C. The difference in electronegativities is a measure of bond polarity. $EN_{Na}$ = 0.9, $EN_{Cl}$ = 3.2)
  1. NaCl, an ionic compound        $\Delta EN$ = 2.3
  2. $N_2$, a covalent, nonpolar compound    $\Delta EN$ = 0.0
D. Bonds with DEN > 2 are considered ionic; Bonds with $\Delta EN$ < 0.4 are nonpolar, covalent.

### 3.11 Rules for Writing Lewis Structures
A. Different atoms form different numbers of bonds. The number of covalent bonds an atom can form is called its *valence*.
B. Hydrogen  H    monovalent     Oxygen  O   divalent
   Nitrogen  N    trivalent      Carbon  C   tetravalent
C. Steps in writing an electron dot formula.
  1. Calculate the total number of valence electrons.
  2. Write the skeletal structure.
  3. Apply the "octet-rule" - place 8 electrons around all outer atoms and 2 electrons around H atoms.

4. Any valence electrons that remain are assigned to the central atom.
5. If the central atom has fewer than 8 electrons, a multiple bond is likely. Move non-bonding electron pairs to form double and/or triple bonds as needed. (see Discussion.)

3.12 The VSEPR Theory
   A. Polyatomic molecules have three-dimensional shapes which can determine polarity and biological activity. The shape of many polyatomic molecules can be predicted by a procedure called the valance-shell electron-pair repulsion (**VSEPR**) theory.
      1. Draw the Lewis structure (identify bonding pairs [BPs] and lone pairs [LPs] of electrons).
      2. Determine the steric number = number of atoms bonded to central atom + number of Lps.
      3. Sketch the molecule by placing all electron pairs (BPs and LPs) as far apart as possible
         (2 – linear; 3 – trigonal planar; 4 – tetrahedral)
      4. All BPs and LPs determine the electonic arrangement, but the shape of the molecule is determined by the arrangement of bonded atoms.
   B. The structural formula indicates the bonding arrangement of its atoms.

   *Molecular Formula*        *Structural Formula*

   $H_2O$    |    $C_2H_6$    ||    H—O\\H    |    H—C—C—H (with H's above and below each C)

   C. The valence shell electron pair repulsion (*VSEPR*) theory is often used to predict the arrangement of atoms about a central atom. The theory postulates that groups of electrons surrounding a central atom will repel one another and arrange to move as far apart as possible.

   (2) *Linear*      (3) *Trigonal*      (4)*Tetrahedral*      Water      Ammonia
                                                               (*bent*)   (*pyramidal*)

   Water is surrounded by four pairs of electrons (2 *bonded pairs* + 2 *lone pairs*). These four pairs of electrons will point toward the corners of a tetrahedron, but the water molecule will have a *bent* shape since only two hydrogens occupy two of the four positions. Similarly ammonia ($NH_3$) has four pairs of electrons (three bonded pairs and one lone pair) tetrahedrally surrounding the central N, but its shape is referred to as *pyramidal*. Shape describes the placement of atoms, not electron pairs.
   D. If a molecule consists of a central atom surrounded by 2, 3, or 4 other atoms or combination of atoms and lone pairs, it is possible to predict its shape from the VSEPR rules (above).
   E. Some examples

| Number of Bonded Atoms | Number of Lone Pairs | Example | Shape (see below) |
|---|---|---|---|
| 2 | 0 | $CO_2$ | linear |
| 3 | 0 | $BF_3$ | trigonal |
| 4 | 0 | $CH_4$ | tetrahedral |
| 3 | 1 | $NH_3$ | pyramidal |

## Chapter 3 - Chemical Bonds

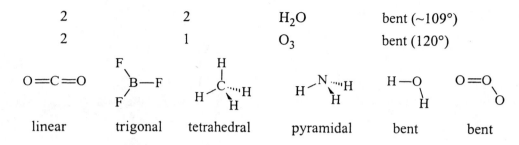

| | | | |
|---|---|---|---|
| 2 | 2 | $H_2O$ | bent (~109°) |
| 2 | 1 | $O_3$ | bent (120°) |

linear   trigonal   tetrahedral   pyramidal   bent   bent

C. Water: a bent (polar) molecule
  1. The *molecular formula* for water is $H_2O$.
  2. The *electron dot formula* for water is:

  3. Water is shown as a *bent* molecule, rather than a linear molecule, because we know that water has a permanent *dipole*.

  net dipole , not   dipoles cancel

D. Ammonia: a *pyramidal* molecule
  1. The molecular formula for ammonia is $NH_3$.
  2. The electron dot formula for ammonia is:

  or   pyramidal shape

E. Methane: a *tetrahedral* molecule
  1. The molecular formula for methane is $CH_4$.
  2. The electron dot formula for methane is:

  or   tetrahedral shape

3.13 Polar and Nonpolar Molecules
  A. Diatomic (linear) molecules are polar (having a *dipole*) if their bonds are polar ($\Delta EN > 0.4$).
  B. For polyatomic molecules, both shape and $\Delta EN > 0.4$ are important since opposing dipoles can cancel eachother.

  net dipole   vs.   dipoles cancel

  Water: a bent (polar) molecule     Carbon dioxide: a linear (nonpolar) molecule

Chapter 3 - Chemical Bonds

3.14 Polyatomic Ions
  A. A number of *polyatomic ions* are so common that they have been given names and are usually treated as one entity.
  B.

| Name | Formula | Name | Formula |
|---|---|---|---|
| Ammonium ion | $NH_4^+$ | Nitrate ion | $NO_3^-$ |
| Carbonate ion | $CO_3^{2-}$ | Phosphate ion | $PO_4^{3-}$ |
| Bicarbonate ion | $HCO_3^-$ | Phosphite ion | $PO_3^{3-}$ |
| Hydroxide ion | $OH^-$ | Sulfate ion | $SO_4^{2-}$ |
| Cyanide | $CN^-$ | Nitrite | $NO_2^-$ |

3.15 Exceptions to the Octet Rule
  A. Most compounds of C, N, O, and F always obey the octet rule.
  B. There are many exceptions to the octet rule.
   1. *Free radicals* – atoms and molecules with an unpaired electron: $NO$, $NO_2$.
   2. Boron and beryllium compounds: $BF_3$, $BeBr_2$.
   3. Compounds involving elements in the 3$^{rd}$ period and beyond: $PCl_5$, $SF_6$.

# DISCUSSION

Chapter 3 marks a turning point in our study of chemistry. If, instead of chemistry, English literature were our area of study, we would just be at the point of having learned to read. You can't appreciate fine literature unless you understand that those little squiggles on paper are letters of the alphabet, that letters of the alphabet are symbols representing sounds we make to communicate with one another, and that the right combinations of letters make words that have meaning. In the first three chapters we learned to use chemical symbols to represent bits of matter. In Chapter 3 we've finally begun to read and write the "words" of chemistry–the formulas for compounds. In Chapter 4 we'll be composing "sentences," i.e., equations, and eventually we'll be able to read and understand some of the most complex "literature" of chemistry, including the chemical version of the story of life.

Right now it's necessary that you become comfortable in dealing with the structures of compounds. The fact that sodium chloride is an ionic compound is of great importance to its role in living systems. The fact that water is a polar covalent molecule is just as important. You must understand what ionic, covalent, and polar mean in order to understand what makes these structural features important.

First, how do you know whether a compound is ionic or covalent? A fairly reliable rule states that ionic compounds are formed when a Group 1A or 2A element combines with a Group 6A or 7A element. (When using this rule, consider hydrogen as a Group 7A element.) Group 3A and 5A elements sometimes form ionic compounds and sometimes do not. Group 4A elements tend to form covalent bonds. Since we're interested in general trends, we tend to use clear-cut cases in our examples. Ionic compounds are formed when the positive and negative ions form compounds. We emphasize recognizing ionic compounds because they are less common, and if the compound isn't ionic, it's covalent or polar covalent.

Let's assume that you are able to recognize which combinations of elements will not form ionic compounds. You are therefore dealing with covalent bonding. Putting together covalent molecules is much like working jigsaw puzzles. You move pieces around until everything fits. Once again, you may start

## Chapter 3 - Chemical Bonds

with electron dot structures for the elements involved or with a valence bond parts list. If you use electron dot symbols, here are some rules of thumb to follow:

*a. Any single (unpaired) electron should be paired with a single electron on another atom.*
*b. The objective is to give each atom an octet of electrons (except hydrogen, which is satisfied with a duet).*
*c. This is a hint more than a rule: Elements that can form only one bond (hydrogen and the Group 7A elements) should be fitted into the structure last.*

Here are three examples to supplement those given in the chapter.

### Example 1. Write the electron dot and valence bond structures for $H_2S$.

Step 1: Write down the Lewis structures of the parts  ·S̈:   H·   H·

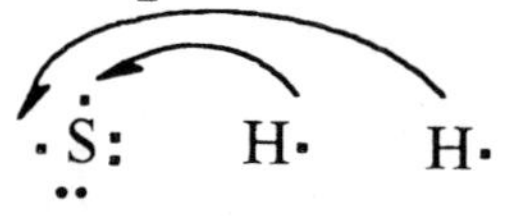

Step 2: Note that a single electron on S and a single electron on H can join to form a single, covalent bond.

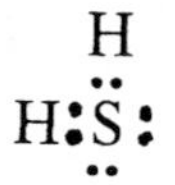

H:S̈:
 H

Note, in the valence bond structure of this molecule, the shared pair of electrons are replaced with a line (dash) between the ato

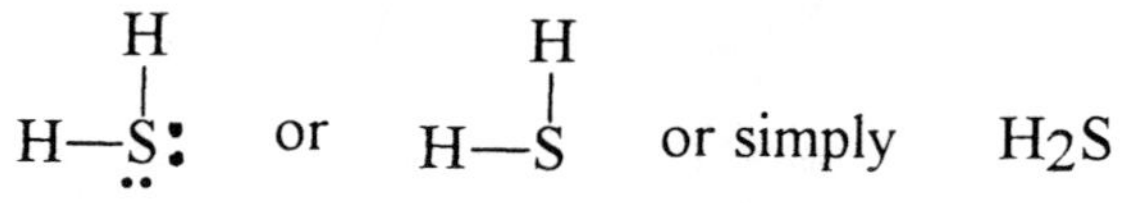

$$H-\ddot{S}: \quad \text{or} \quad H-\overset{H}{\underset{|}{S}} \quad \text{or simply} \quad H_2S$$

*Voila !!!* (unshared electrons not shown)

### Example 2. Construct the $CH_4S$ molecule.

Step 1:   ·C̈·   H·   H·   H·   H·   ·S̈·

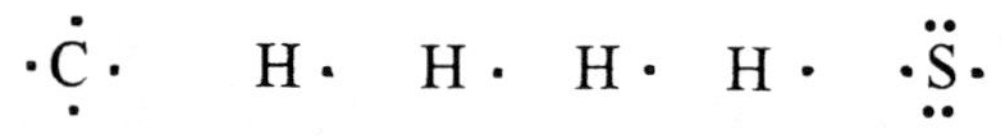

Step 2: Combine C and S.

·C̈:S̈·
4 bonds still needed

·C̈:S̈·
4 hydrogens also need bonds

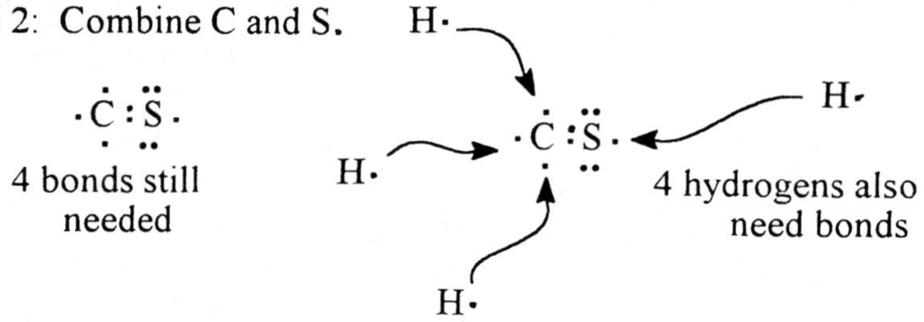

Step 3: Add hydrogens.

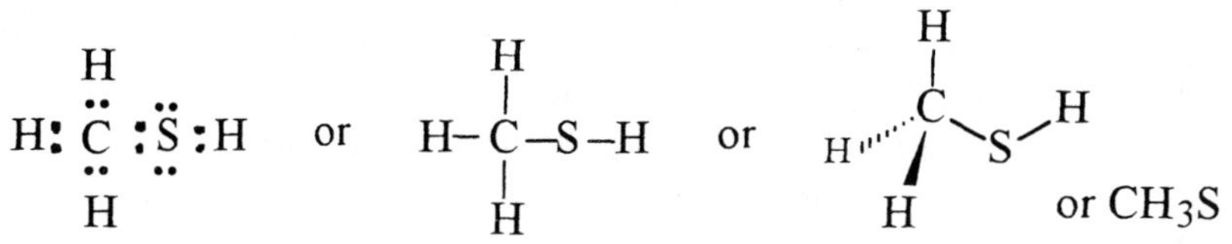

$$H:\overset{H}{\underset{H}{\ddot{C}}}:\ddot{S}:H \quad \text{or} \quad H-\overset{H}{\underset{H}{\overset{|}{C}}}-S-H \quad \text{or} \quad H^{\backslash}\overset{H}{\underset{H}{C}}\diagup S\diagdown H \quad \text{or} \quad CH_3S$$

26

Chapter 3 - Chemical Bonds

**Example 3.** Construct the $N_2H_2$ molecule.

Step 1:   ·N̈·   ·N̈·   H·   H·

Step 2:   Pair the 2 N.

·N̈:N̈·            H·   H·

4 bonds still            only 2 hydrogens
needed                  that need bonds

Step 3: Give one H to each N.

H:N̈:N̈:H

Step 4: Each N must form one more bond (octet rule),
so they share another pair of electrons.

H:N̈::N̈:H     or     H—N=N—H

Polarity is the last of the general concepts dealing with chemical bonding that was introduced in this chapter. One speaks of polar bonds or polarity only for covalently bonded molecules, and even then, not for all covalently bonded molecules. A covalent bond is polar only if it forms between elements of unequal electronegativity. A molecule is polar only if it contains polar dipoles that do not cancel one another out.

The bond in Br-Br is not polar because the bonding atoms are of identical electronegativity; the bond in Br-F is polar because the two elements are of different electronegativities. Water is a polar molecule, but carbon dioxide is not, despite the fact that both molecules have bonds joining atoms of very different electronegativity. Water is "bent," but carbon dioxide is "linear." The two sets of unequally shared electrons (represented by the arrows) in water yield a polar molecule, in carbon dioxide they cancel each other out so even though carbon dioxide has polar bonds it is a nonpolar molecule.

$\delta^+$ ⟶ $\delta^-$
H—O
       \
        H $\delta^+$

net dipole

vs.

$\delta^-$ ⟵ $\delta^+$ ⟶ $\delta^-$
O=C=O

dipoles cancel

We end our discussion with the importance of molecular shapes, because the shape of a molecule helps determine whether the molecule is polar or nonpolar, and that influences its solubility and interaction properties. You will learn how to predict shapes of molecules using the VSEPR theory.

Many of the problems at the end of the chapter are like grade-school spelling drills or vocabulary tests. They simply ask you to practice over and over drawing ions or putting together molecules or naming compounds. This practice should establish firmly the rules governing chemical structure in your mind. In case it hasn't, here is some additional help.

## Chapter 3 - Chemical Bonds

**Problems**

1. You should be able to draw electron dot symbols for any element in the A groups (IA, IIA, etc.). To do this, write the symbol for the element and surround it with dots representing the valence (outermost) electrons. The number of valence electrons is given by the group number. For practice, draw electron dot structures for atoms of these elements:
    - a. nitrogen
    - b. carbon
    - c. argon
    - d. calcium
    - e. silicon
    - f. hydrogen
    - g. potassium
    - h. fluorine
    - i. sulfur
    - j. aluminum

2. Electron dot structures for ions are drawn either by adding electron dots to complete the octet or by removing electron dots to empty the outermost level. Electrons are added to elements with 5 or more valence electrons; they are subtracted from elements with 3 or fewer valence electrons. For the following elements, how many electron dots would you add to or remove from the electron dot symbols of the atoms to form the ions?
    - a. magnesium
    - b. chlorine
    - c. oxygen
    - d. calcium
    - e. potassium
    - f. fluorine
    - g. sulfur
    - h. aluminum

3. The charge is written to the upper right of the electron dot symbol for an ion. The charge is equal to the number of electrons added to or removed from the neutral atom to form the ion. The charge is positive if electrons are removed and negative if electrons are added. Write the proper ionic form for the stable ions that are formed from the eight elements listed in problem 2.

4. Compounds are formed from the following sets of elements. Indicate whether the compound would be ionic or covalent. (Note that you're not being asked to draw the compounds, just to evaluate their tendency to form ionic or covalent bonds.)
    - a. Ba and O
    - b. C and S
    - c. N and O
    - d. Br and Cl
    - e. Na and H
    - f. C and H
    - g. Ca and Cl
    - h. F and Ca
    - i. Li and O
    - j. Rb and F

5. Write structural formulas (electron dot and valence bond) for the following compounds:
    - a. $PCl_3$
    - b. $SiH_4$
    - c. $CO_2$
    - d. $H_2CO$
    - e. HNCS

**SELF-TEST** (Refer to the periodic table.)

1. The charge on an ion formed from sulfur would be:
    - a. $S^+$
    - b. $S^{2+}$
    - c. $S^{3+}$
    - d. $S^-$
    - e. $S^{2-}$
    - f. $S^{3-}$
2. The aluminum ion is:
    - a. $Al^+$
    - b. $Al^{2+}$
    - c. $Al^{3+}$
    - d. $Al^{4+}$
    - e. $Al^{2-}$
    - f. $Al^{4-}$
3. The ion formed from iodine is:
    - a. $I^+$
    - b. $I^{2+}$
    - c. $I^{6+}$
    - d. $I^{7+}$
    - e. $I^-$
    - f. $I^{7-}$
4. The name of the ion $CN^-$ is:
    - a. acetate
    - b. ammonium
    - c. carbonate
    - d. cyanide
    - e. nitrate
5. How many protons are there in the $Mg^{2+}$ ion?
    - a. 2
    - b. 6
    - c. 10
    - d. 12
    - e. 20
    - f. 40
6. The electron dot symbol for lithium is:
    - a. Li·
    - b. ·Li·
    - c. ·Li
    - d. ·L̇i
    - e. :Li·
    - f. :Li:
    - g. :L̇i:
    - h. :L̈i:
7. The electron dot symbol for a carbon atom is:
    - a. C·
    - b. ·C·
    - c. ·Ċ·
    - d. ·C̣·
    - e. :Ċ·
    - f. :C̣:
    - g. :C̈:
    - h. :C̈:

## Chapter 3 - Chemical Bonds

8. Which is the best formula for magnesium oxide?
   a. MgO    b. $MgO_2$    c. $Mg_2O$    d. $Mg_2O_2$
9. Which is the correct formula for a compound of boron and sulfur?
   a. BS    b. $BS_2$    c. $B_2S$    d. $BS_3$    e. $B_3S$    f. $B_2S_3$
10. Which is the correct formula for a compound of aluminum and oxygen?
    a. AlO    b. $Al_3O$    c. $AlO_3$    d. $Al_2O_3$
11. Ferrous chloride is:
    a. FeCl    b. $FeCl_2$    c. $FeCl_3$    d. $Fe_2Cl$    e. $Fe_3Cl$    f. $Fe_2Cl_3$
12. Copper(I) sulfide is:
    a. CuS    b. $Cu_2S$    c. $CuS_2$    d. $Cu_2S_2$    e. $Cu(SO_4)_2$
13. Calcium sulfate is the name of:
    a. $CaSO_4$    b. $Ca_2SO_4$    c. $Ca(SN)_2$    d. $Ca_2NO_3$    e. $Ca(SO_3)_2$
14. Which is the correct formula for aluminum sulfate?
    a. $AlSO_4$    b. $Al_2SO_4$    c. $Al_2(SO_4)_3$    d. $Al_2SO_3$    e. $Al(SO_3)_2$
15. Name this compound: $Ca(NO_2)_2$
    a. calcium nitrite    b. copper nitrate    c. calcium nitrate    d. nitrous calcite
16. What is the correct formula for dinitrogen tetraoxide?
    a. $NO_2$    b. $N_2O$    c. NO    d. $N_2O_3$    e. $N_2O_4$
17. Which is the most electronegative element among the following?
    a. O    b. S    c. Se    d. Te
18. Which element is most electronegative?
    a. C    b. N    c. O    d. F
19. Which is the most likely structure for a compound incorporating Be and F?
    a. $Be^+F^-$    b. $Be^{2+}F^{2-}$    c. $Be^{2+}2F^-$    d. $2Be^+F^{2-}$    e. $Be^{2+}F^-$
20. Which would be expected to have ionic bonding?
    a. $CO_2$    b. $NCl_3$    c. $SiCl_4$    d. NaBr
21. The compound formed from these elements would not be ionic:
    a. calcium and fluorine    b. sodium and sulfur
    c. nitrogen and oxygen    d. lithium and bromine
22. Which is the best description of the bonding in ClBr?
    a. Cl—Br    b. Cl—Br $\delta^+ \delta^-$    c. Cl—Br $\delta^- \delta^+$    d. $Cl^+Br^-$    e. $Cl^-Br^+$
23. Which compound contains a polar covalent bond?
    a. NaF    b. HF    c. $F_2$
24. Which of the following bonds would be most polar?
    a. C–C    b. C–N    c. C–O    d. C–F    e. C–H
25. Which of the following would have the most polar bond?
    a. F–Cl    b. Cl–Cl    c. I–Br    d. F–I
26. The valence bond formula of hydrogen sulfide is:
    a. $H_2S$    b. H–S–H    c. $2H^+S^{2-}$
27. Based on bonding rules, which is not a reasonable formula?
    a. $N_2$    b. $O_2$    c. $F_2$    d. $Ne_2$
28. Which molecule is polar?
    a. Cl – Be – Cl    b. $BF_3$ (trigonal)    c. $OF_2$ (bent)    d. F – F
29. The $CF_4$ molecule is:
    a. linear    b. bent    c. tetrahedral    d. pyramidal

## Chapter 3 - Chemical Bonds

30. The $NH_3$ molecule is:
    a. linear    b. bent    c. tetrahedral    d. pyramidal

31. A triple bond involves the sharing of a total of how many electrons?
    a. 1    b. 2    c. 3    d. 4    e. 5    f. 6    g. 7    h. 8

32. The correct electron dot formula for carbon dioxide is:
    a. :C:O:C:    b. :O:O:C:    c. :O::C::O:    d. :O:C:O:

33. Which is the most reasonable valence bond formula for $C_2F_2$?
    a. F–C–C–F    b. C–F–F–C    c. F=C–C=F    d. F=C=C=F    e. F–C≡C–F

34. The correct electron dot formula for HNO is:
    a. H:N::O:    b. :H:N::O:    c. H:O:N:    d. :H:O:N:

35. A reasonable structure for $C_2H_3OCl$ is:
    a. H-C=O-C-Cl (with H's)    b. H-C-C-Cl (with H, H, O)    c. H-C-Cl-C-H (with H, O)    d. H-C≡C-O-H (with Cl, H)

36. The molecule $H_2$ is _____ in shape.
    a. linear    b. bent    c. tetrahedral    d. trigonal    e. pyramidal

37. The molecule $H_2S$ is _____ in shape.
    a. linear    b. bent    c. tetrahedral    d. trigonal    e. pyramidal

38. The molecule $CO_2$ (two double bonds) is _____ in shape.
    a. linear    b. bent    c. tetrahedral    d. trigonal    e. pyramidal

39. The molecule $H_2O$ is _____ in shape.
    a. linear    b. bent    c. tetrahedral    d. trigonal    e. pyramidal

40. The molecule $SO_2$ is _____ in shape.
    a. linear    b. bent    c. tetrahedral    d. trigonal    e. pyramidal

41. The nitrate ion ($NO_3^-$) is _____ in shape.
    a. linear    b. bent    c. tetrahedral    d. trigonal    e. pyramidal

42. The $CF_4$ molecule is _____ in shape:
    a. linear    b. bent    c. tetrahedral    d. trigonal    e. pyramidal

43. $H_2S$ is a _____ molecule.
    a. polar    b. nonpolar

44. $CO_2$ (two double bonds) is a _____ molecule.
    a. polar    b. nonpolar

45. $NH_3$ is a _____ molecule.
    a. polar    b. nonpolar

46. $CF_4$ is a _____ molecule:
    a. polar    b. nonpolar

47. Fluorine ($F_2$) is a _____ molecule.
    a. polar    b. nonpolar

# Chapter 3 - Chemical Bonds

## ANSWERS

**Problems**

1. a. :N̈·   b. ·C̈·   c. :Är:   d. Ca:   e. ·S̈i·   f. H·   g. K·   h. :F̈:   i. ·S̈:   j. ·Äl·

2. 
   a. remove two   c. add two   e. remove one   g. add two
   b. add one   d. remove two   f. add one   h. remove three

3. 
   a. $Mg^{2+}$   c. $O^{2-}$   e. $K^+$   g. $S^{2-}$
   b. $Cl^-$   d. $Ca^{2+}$   f. $F^-$   h. $Al^{3+}$

4. Ionic: a, e, g, h, i, j   Covalent: b, c, d, f

5. a. :C̈l:P̈:C̈l:   b. H:S̈i:H   c. :Ö::C::Ö:   d. H:C̈::Ö:   e. H:N̈::C::S̈:
      :C̈l:           H                                H

   Cl–P–Cl        H–Si–H        O=C=O        H–C=O        H–N=C=S
    |              |                           |
    Cl             H                           H

## Self-Test

| | | | | | |
|---|---|---|---|---|---|
| 1. e | 9. f | 17. a | 25. d | 33. e | 41. d |
| 2. c | 10. d | 18. d | 26. b | 34. a | 42. c |
| 3. e | 11. b | 19. c | 27. d | 35. b | 43. a |
| 4. d | 12. b | 20. d | 28. c | 36. a | 44. b |
| 5. d | 13. a | 21. c | 29. c | 37. b | 45. a |
| 6. a | 14. c | 22. c | 30. d | 38. a | 46. b |
| 7. d | 15. a | 23. b | 31. f | 39. b | 47. b |
| 8. a | 16. e | 24. d | 32. c | 40. b | |

# 4 Chemical Reactions

**KEY WORDS**

| | | | | |
|---|---|---|---|---|
| *reactions* | *balanced equation* | *activated complex* | *exothermic* | *collisions* |
| *equation* | *coefficients* | *energy of activation* | *endothermic* | *energy* |
| *reactant* | *Avogadro's number* | *reversible reaction* | *reaction rate* | *orientations* |
| *product* | *formula mass* | *reaction mechanism* | *LeChatelier's* | *enzyme* |
| *catalyst* | *molecular mass* | *dynamic equilibrium* | *principle* | *rate law* |
| *molar mass* | *stoichiometric* | *Avogradro's hypothesis* | *atomic mass* | *element* |
| *mole* | *photosynthesis* | *law of combining* | *heat of reaction* | |
| | *enthalpy change* | *volumes* | | |

**SUMMARY**

4.1 Balancing Chemical Equations
   A. A chemical *equation* is shorthand notation for describing chemical *reactions* as the *reactants*, A, (shown on the left) are converted to *products*, P, (shown on the right).   A → P.
   B. Chemical equations involve electronic rearrangements as chemical bonds are broken and formed. The nuclei of all atoms remain unchanged during chemical reactions.
   C. Because atoms are conserved in chemical reactions, the same number of each kind of atom must appear on each side of a *balanced* chemical *equation*. Hint: If an element occurs in just one compound on each side of the equation, try balancing that element first (i.e., H). Balance all elements last (i.e., $O_2$).

$$2\,H_2 + O_2 \rightarrow 2\,H_2O$$

4.2 Volume Relationships in Chemical Equations
   A. In 1809 Gay-Lussac determined the "*law of combining volumes,*" that the volumes of gaseous reactants and products were always in a small whole number ratio.
   
           2 vol. hydrogen + 1 vol. oxygen    →    2 vol. water
           3 vol. hydrogen + 1 vol. nitrogen   →    2 vol. ammonia
   
   B. *Avogadro* explained the law of combining volumes on the basis of a *hypothesis* that equal volumes of all gases (at the same temperature and pressure) contain the same number of molecules. The reactions above can be reexpressed as a chemical equation in which chemical formulas represent the substances involved and *stoichiometric coefficients* reflect the combining volumes.

$$2\,H_2 + (1)\,O_2 \rightarrow 2\,H_2O$$
$$3\,H_2 + (1)\,N_2 \rightarrow 2\,NH_3$$

4.3 Avogadro's Number: **$6.02 \times 10^{23}$**
   A. Chemists cannot weigh individual atoms. Dalton proposed a table of relative atomic masss based on hydrogen being assigned a relative mass of 1. Thus, carbon is 12, oxygen is 16, and so forth.
   B. The *atomic mass* of any element, expressed in grams, contains $6.02 \times 10^{23}$ atoms. The number $6.02 \times 10^{23}$ is called *Avogadro's number*. (*Avogadro's number is so big that a computer that can count to 10 million in 1 second would still need 2 billion years to count up to $6.02 \times 10^{23}$!!!*)

4.4 Molecular Masses and Formula Masses
   A. Each *element* has a characteristic *atomic mass*.
   B. The *molecular mass* is the average mass of a molecule (sum of the masses of all the atoms represented in a molecular formula) relative to that of a carbon-12 atom.

# Chapter 4 - Chemical Reactions

C. The *formula mass* of a substance is the average mass of a formula unit (sum of the atomic masses of all the atoms in the formula unit) relative to that of a carbon-12 atom..

4.5 Chemical Arithmetic and the Mole
   A. Chemists count atoms and molecules by the *mole*, not by the dozen. A *mole* of a substance (atoms, molecules, ions) contains $6.02 \times 10^{23}$ units of that substance, the same as the number of atoms of C-12 in 12 g of C-12.
   B. A *molar mass* of a substance equals the mass of one mole of that substance. The molar mass is numerically equal to the formula mass, but expressed as g/mol.
   C. The mole/mass relationships can be used as conversion factors between gram amounts and mole amounts (1 mol C = 12 g C or 12g C/ mol C).

4.6 Mole and Mass Relationships in Chemical Equations
   A. Atoms react in small whole number ratios. In the laboratory we cannot count numbers of atoms or molecules; instead, we measure out an amount (grams) of a substance.
   B. The *stoichiometric coefficients* of a chemical equation directly express the atom (or mole) ratios of reactants and products.
   C. Combining formula mass information with these stoichiometric coefficients means that a chemical equation also indirectly expresses the mass ratios of reactants and products.
   D. The mole method illustrates the steps in going from mole ratios to mass ratios.
      1. Write the balanced chemical equation.
      2. Determine the molar masses (molecular masss) of interest.
      3. Convert from grams to number of moles of reactant using the molar mass.
      4. Use the stoichiometric coefficients of the balanced chemical equation to convert from the mole amount of the given substance (reactant) to the mole amount of the desired substance (product).
      5. Finally, use the molar mass of the desired substance to convert from moles to grams of the desired substance.

$$\text{consider} \quad a\,A \rightarrow b\,B$$

$$g\,A \; \times \; \left[\frac{1 \text{ mole A}}{g\,A}\right] \; \times \; \left\{\frac{b \text{ moles B}}{a \text{ moles A}}\right\} \; \times \; \left[\frac{g\,B}{1 \text{ mole B}}\right] \quad (\rightarrow g\,B)$$

4.7 Structure, Stability, and Spontaneity
   A. Energy changes also take place during chemical reactions. *Exothermic* reactions result in the release of heat; *endothermic* reactions require energy to be supplied to convert reactants to products. The heat released or absorbed during a reaction is called the *heat of reaction*. The *enthalpy change* is the heat of reaction for processes at constant pressure.

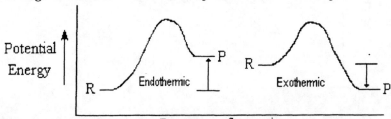

   B. In exothermic reactions, the electronic arrangement of the products represents a more stable (lower potential energy) arrangement than the original arrangement in the reactants. Therefore, exothermic reactions are energetically favorable, or "downhill."
   C. The conversion of reactants to products always proceeds through an intermediate *activated complex*. Because a reaction is energetically favorable does not mean that it is instantaneous. It may or may not proceed rapidly, depending on the *energy of activation* needed to form the activated complex. The energy of activation is the energy needed to get to the "top of the energy hill" (see 4.11).

## Chapter 4 - Chemical Reactions

4.8 Forward and Reverse Reactions
   A. Many reactions involve the interconversion of reactants and products; that is, the products can react to re-form the original reactants. Such reactions are said to be *reversible* reactions.
   B. Opposing reactions are common in nature. Respiring cells "burn" glucose to $CO_2$ and $H_2O$ with the release of energy (exothermic), while other cells can carry out the endothermic process of *photosynthesis*, using the energy of sunlight to convert $CO_2$ and $H_2O$ into glucose.
   C. For reversible reactions, the energy barrier may be approached from either side.

4.9 Reaction Rates: Collisions and Orientation
   A. The *reaction rate* refers to how rapidly the product is produced per given unit of time, $\Delta[P]/s$.
   B. In order for reactant molecules to produce a product, the reactant species must come together via molecular *collisions* with the proper *orientation* and with sufficient *energy* to overcome the *energy of activation* barriers. "Proper orientation" simply means that the reactive surfaces on the reactant molecules must approach one another with proper juxtaposition to permit required bonding rearrangements to take place.
   C. Reaction rates are affected by temperature, presence of catalysts, and by concentration.
   D. Reaction rates: the effect of temperature
      1. Increasing the temperature of a reaction by 10 °C often doubles the rate of the reaction. Changes in temperature affect rates of chemical reactions in two ways:
         a. At higher temperatures the molecules move more rapidly, so they collide more frequently. This is a minor effect.
         b. More importantly, a small increase in temperature, which might bring about only a 2% increase in average molecular velocity, can cause a doubling of the number of molecules with energy above the threshold needed to overcome the *energy of activation* for the reaction.
      2. It is noteworthy that chemical reactions in our bodies take place at essentially constant temperature, 37 °C, yet occur very rapidly due to the presence of enzymes, which are biocatalysts. The activity of these enzymes can be "killed" by temperatures that are either too high or too low.
   E. Reaction rates: catalysis
      1. A *catalyst* is a substance that increases the rate of a chemical reaction without itself being changed.
      2. Most catalysts act by providing an alternate pathway, or different *reaction mechanism*, for the reaction to occur, one that has a lower activation energy requirement.

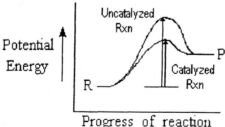

      3. Biological catalysts (*enzymes*) mediate nearly all reactions that take place in living systems.
   F. Reaction rates: the effect of concentration
      1. The concentration of reactants affects the *rate* of chemical reactions by increasing the number of collisions that will occur.
      2. The dependence of reaction rate on the concentration of reactants must be determined experimentally and is expressed as a *rate law*.
      3. Information about rates of chemical reactions can be used to postulate a *mechanism* for the chemical reaction, i.e., a step-by-step description of the chemical changes that occur in the overall process.

# Chapter 4 - Chemical Reactions

4.10 Equilibrium in Chemical Reactions
   A. A condition of *dynamic equilibrium* is established for a reversible reaction when the rate of the forward reaction equals the rate of the reverse reaction.
   B. Equilibrium is established for reversible reactions in isolated systems.
   C. When a system at equilibrium is subjected to a "stress" (such as the addition of reactants or products, or a change in temperature or pressure), the system rearranges in such a way as to minimize that stress. This is referred to as *Le Chatelier's principle* (1884).

## DISCUSSION

The subject of this chapter is the chemical equation and the information it contains. You probably never realized how much information is contained in a chemical equation. Much of the chapter is devoted to a discussion of the concepts and terminology needed to extract every last bit of data from an equation. Therefore, one of the first things you should do is make sure you understand the terms listed in the Key Terms section.

Molar mass, formula mass, mole, and Avogadro's number are interrelated terms. The formula mass of a compound expressed in grams (the gram formula mass) is a mole of the compound; a mole of the compound contains Avogadro's number of the compound units, which is equal to $6.02 \times 10^{23}$ units.

**1 gram formula mass = 1 mole = Avogadro's number = $6.02 \times 10^{23}$ units**

You can treat this relationship as a multiple conversion factor. If you calculate the formula mass of a compound and remember the above relationships, you can interconvert units expressed in grams to moles or to molecules and vice versa.

## Example. Interconversions involving methane, $CH_4$

Formula mass:  atomic mass of C = 12      $1 \times 12 = 12$
               atomic mass of H = 1       $4 \times 1 = \underline{\ 4}$
                                          formula mass = 16

Once we have the gram formula mass or molar mass, we also know that there are:
   16 g of $CH_4$/1 mole of $CH_4$; Avogadro's number of $CH_4$ molecules
         or  $6.02 \times 10^{23}$ $CH_4$ molecules/16 g of $CH_4$

How many moles are in 4 g of $CH_4$?  4 g × [1 mole] = 0.25 mole
                                              16 g

How many grams are in 4 moles of $CH_4$?  4 moles × [16 g] = 64 g
                                                    1 mole

How many molecules are in 4 g of $CH_4$?  4 g × [$6.02 \times 10^{23}$ molecules] = $1.5 \times 10^{23}$ molecules
                                                      16 g

How many molecules are in 4 moles of $CH_4$?  4 mol × [$6.02 \times 10^{23}$ molec] = $24 \times 10^{23}$ molecules
                                                         1 mole

Let's suppose you are now totally at ease with moles and formula masss and such. That brings us to equations. **The first thing you should check in an equation is whether or not it is balanced.** (We'll supply the correct reactants and products.) If the equation isn't balanced, the quantitative information derived from it may be incorrect. As we indicated in this chapter, you won't be balancing extremely

## Chapter 4 - Chemical Reactions

complex equations, but you should be able to handle those shown at the end of the chapter. Once you have a balanced equation, the coefficients in that equation give you the following information directly:
  a. the combining ratio of molecules (or other formula units like ion pairs)
  b. the combining volume ratios of gaseous reactants and products, assuming temperature and pressure remain constant
  c. the combining ratio of moles of molecules (or other formula units)

The coefficients **do not** give you directly the combining weight ratios. Thus, from the equation:

$$CH_4 + 2\,O_2 \rightarrow CO_2 + 2\,H_2O \quad \text{(all gases)} \text{ you know that:}$$

  a. 1 molecule of methane ($CH_4$) reacts with 2 molecules of oxygen ($O_2$) to give 1 molecule of carbon dioxide ($CO_2$) and 2 molecules of water ($H_2O$).
  b. 1 volume of methane gas reacts with 2 volumes of oxygen gas to produce 1 volume of carbon dioxide gas and 2 volumes of water vapor.
  c. 1 mole of methane reacts with 2 moles of oxygen to give 1 mole of carbon dioxide and 2 moles of water.

The equation does **not** say that 1 gram of methane reacts with 2 grams of oxygen to produce 1 gram of carbon dioxide and 2 grams of water. If you want to find how many grams of oxygen react with 1 gram of methane, you must first convert grams to moles and only then use the equation to determine the combining ratio. After using the equation, you'll have the answer in moles and must convert to grams.

The examples in the chapter demonstrate the use of equations to obtain information about combining ratios. Review those examples, then the problems at the end of the chapter. For more practice, try problems 5 and 6 below.

## Problems

1. Calculate the formula mass. We're using relatively complicated formulas just to make sure you understand when a subscript applies to a particular atom in the formula and when it applies to a polyatomic ionic group of atoms. You'll require a periodic table or a list of atomic masss.
    a. $(NH_4)_2SO_4$     c. $Be(NO_3)_2$     e. $Al_2(C_2O_4)_3$     g. $SrSO_4$
    b. $Ca(NO_3)_2$       d. $(NH_4)_2C_2O_4$  f. $Ca(C_2H_3O_2)_2$

2. The following questions refer to the compound $C_5H_8O_2$.
    a. How many moles of $C_5H_8O_2$ are in 100 g? In 200 g? In 25 g? In 3.687 g?
    b. How many grams of $C_5H_8O_2$ are in 1 mole? In 8 moles? In 0.8 mole? In 0.01 mole?
    c. How many molecules of $C_5H_8O_2$ are in 1 mole of the compound? In 0.5 mole? In 3 moles? In 100 g? In 50 g? In 300 g?
    d. How many carbon atoms are in one $C_5H_8O_2$ molecule? In one mole of $C_5H_8O_2$? In 100 g of $C_5H_8O_2$?
    e. How many hydrogen atoms are in one $C_5H_8O_2$ molecule? In one mole of $C_5H_8O_2$? In 100 g of $C_5H_8O_2$?

3. For additional practice interconverting these units, answer these questions for the compound $PbCrO_4$. A calculator would be useful because these numbers will not be as easy.
    a. How many moles of $PbCrO_4$ are in 100 g? In 200 g? In 25 g? In 3.687 g?
    b. How many grams of $PbCrO_4$ are in 1 mole? In 8 moles? In 0.8 mole? In 0.01 mole?
    c. How many molecules of $PbCrO_4$ are in 1 mole of the compound? In 0.5 mole? In 3 moles? In 100 g? In 50 g? In 300 g?
    d. How many oxygen atoms are in one $PbCrO_4$ molecular unit? In one mole of $PbCrO_4$? In 100 g of $PbCrO_4$?

# Chapter 4 - Chemical Reactions

4. Here are some additional equations to balance.
    a. $Zn + KOH \rightarrow K_2ZnO_2 + H_2$
    b. $HF + Si \rightarrow SiF_4 + H_2$
    c. $B_2O_3 + H_2O \rightarrow H_6B_4O_9$
    d. $SiCl_4 + H_2O \rightarrow SiO_2 + HCl$
    e. $SnO_2 + C \rightarrow Sn + CO$
    f. $Fe_2O_3 + C \rightarrow FeO + CO_2$
    g. $Fe_3O_4 + C \rightarrow Fe + CO$
    h. $Fe(OH)_3 + H_2S \rightarrow Fe_2S_3 + H_2O$
    i. $Ca_3P_2 + H_2O \rightarrow PH_3 + Ca(OH)_2$
    j. $Bi_2O_3 + C \rightarrow Bi + CO$

5. Refer to the equation: $CS_2 + 2\,CaO \rightarrow CO_2 + 2\,CaS$.
    a. How many moles of $CO_2$ are obtained from the reaction of 2 moles of $CS_2$?
       From the reaction of 2 moles of CaO?
    b. How many moles of CaO are consumed if 0.3 mole of $CS_2$ react?
       If 0.3 mole of CaS is produced?
    c. How many grams of CaS are obtained if 152 g of $CS_2$ are consumed in the reaction?
       If 7.6 g of $CS_2$ are consumed?
       If 22 g of $CO_2$ are produced?
       If 44 g of $CO_2$ are produced?
    d. How many grams of CaO are required to react completely with 38 g of $CS_2$?
       With 152 g of $CS_2$?   To produce 36 g of CaS?

6. Refer to the equation: $C_3H_8 + 5\,O_2 \rightarrow 3\,CO_2 + 4\,H_2O + 500$ kcal
       (All compounds are gases; temperature and pressure are held constant.)
    a. If 5 L of $C_3H_8$ react, what volume of $CO_2$ will be produced?
       What volume of $O_2$ will react?
    b. Answer the same questions for the reaction of 5 mL of $C_3H_8$.
       For the reaction of 5 m$^3$ of $C_3H_8$.
    c. If 22 g of $C_3H_8$ are burned, how many grams of $CO_2$ are produced?
       How many grams of $O_2$ are consumed? How many kilocalories of heat are produced?

The last question in Problem 6 brings energy into the discussion of chemical reactions. That question implies correctly that you treat energy like any other product or reactant, using the balanced equation to establish the ratio of energy produced (or consumed) to moles of chemical species involved. And in case you're wondering about this point, note that you are not able to supply the energy portion of an equation. The amount of energy required for the balanced equation would simply have to be given in a problem such as that in 6 above.

Terms and concepts associated with the energy changes accompanying chemical reactions were reviewed at the end of the chapter. Your understanding of this terminology, plus the concepts of equilibria, Le Chatelier's principle, and other material treated in this chapter are reviewed in the **SELF-TEST** below.

## SELF-TEST

1. How many moles are there in 120 g of glucose ($C_6H_{12}O_6$)?
    a. 1.0    b. 0.5    c. 0.67    d. 1.2    e. 1.5
2. How much does Avogadro's number of sulfur atoms weigh?
    a. 16 amu    b. 32 amu    c. 16 g    d. 32 g    e. $6.02 \times 10^{23}$ g

## Chapter 4 - Chemical Reactions

3. How many atoms are there in 2 moles of helium?
   a. $2 \times 10^{23}$   b. $6.02 \times 10^{23}$   c. $12.04 \times 10^{23}$   d. $6.02 \times 10^{46}$

4. One-half mole of $CO_2$ weighs:
   a. 8 g   b. 12 g   c. 22 g   d. 32 g   e. 48 g   f. 64 g

5. Avogadro's number of hydrogen molecules ($H_2$) weighs:
   a. 1 g   b. 2 g   c. 3 g   d. 4 g   e. $6.02 \times 10^{23}$ g

6. How many molecules of water are in 9 g of $H_2O$?
   a. $23 \times 10^{23}$   b. $6 \times 10^{23}$   c. $2 \times 10^{23}$   d. $3 \times 10^{23}$   e. $18 \times 10^{23}$

7. How many hydrogen atoms are present in 0.5 mole of $H_2O$?
   a. Avogadro's number   b. $0.5 \times$ Avogadro's number   c. $2 \times$ Avogadro's number

8. What is the molar mass of $CO_2$?
   a. 22   b. 16   c. 12   d. 44   e. 60

9. Avogadro's number is <u>not</u>:
   a. the number of atoms in one mole of He.
   b. the number of molecules in one mole of $H_2$.
   c. the number of atoms in one mole of $Br_2$.

10. If one mole of A weighs 40 g and one mole of B weighs 20 g, then:
    a. T / F  each A atom weighs twice as much as each B atom.
    b. T / F  40 g of A contains twice as many atoms as 20 g of B.
    c. T / F  1 mole of A weighs as much as two moles of B.

11. Consider the following reaction diagram. Identify the letter corresponding to the activation energy for the reverse reaction ($P \rightarrow R$).

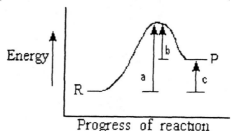

   a. a   b. b   c. c

12. Is the reaction depicted ($R \rightarrow P$) in Question 11 exothermic or endothermic?
    a. exothermic   b. endothermic

13. According to this energy diagram:

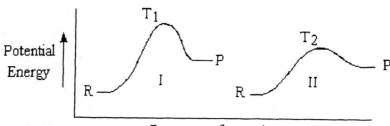

   a. Reaction ($R \rightarrow P$) is endothermic or exothermic?
   b. For which reaction was a catalyst used? Reaction I or reaction II?
   c. For reaction I the activation energy is the difference between $T_1$ and R,  $T_1$ and P,  R and P?
   d. The net energy change for the reaction ($R \rightarrow P$) is the difference in energy between
      R and $T_2$   $T_2$ and P   R and P?
   e. For the reverse reaction ($P \rightarrow R$), the net energy change for the reaction results in energy being _____ by the system.  ( absorbed / released / unused )

## Chapter 4 - Chemical Reactions

14. An increase in reaction rate is not expected to accompany an increase in:
    a. concentration of reactants   b. temperature   c. activation energy

15. In general, which reaction would be expected to have the higher activation energy?
    a. a strongly exothermic reaction   b. a strongly endothermic reaction

16. An increase in temperature generally results in an increase in the rate of a reaction. Which of the following does not account, at least in part, for this phenomenon?
    a. Reacting particles collide more frequently.
    b. Collisions of faster-moving particles supply the activation energy for the reaction.
    c. At higher temperature, reactions change from endothermic to exothermic.

17. The decomposition of potassium chlorate, $KClO_3$, produces oxygen and potassium chloride. What is the coefficient for oxygen in a properly balanced equation?
    $$KClO_3 \rightarrow KCl + O_2$$
    a. 1   b. 2   c. 3   d. 4   e. 6

18. How many moles of oxygen can be produced from the decomposition of one mole of $KClO_3$?
    a. 1   b. 1.5   c. 2   d. 0.5   e. 3

19. How many grams of oxygen can be produced from the decomposition of 100 g of $KClO_3$?
    a. 39   b. 28   c. 53   d. 84   e. 93

20. What is the sum of the coefficients in a properly balanced equation describing the combustion of propane gas to produce carbon dioxide and water?
    $$C_3H_8 + O_2 \rightarrow CO_2 + H_2O$$
    a. 6   b. 8   c. 4   d. 13   e. 15

21. How many moles of oxygen are required to react with 3 moles of propane?
    a. 4   b. 12   c. 5   d. 15   e. 18

22. How many grams of $CO_2$ are produced from the complete combustion of 100 g of propane?
    a. 100   b. 300   c. 88   d. 132   e. 200

23. A catalyst
    a. increases the activation energy and increases the rate of a reaction.
    b. decreases the activation energy and decreases the rate of the reaction.
    c. increases the activation energy and decreases the rate of the reaction.
    d. decreases the activation energy and increases the rate of the reaction.

24. Refer to this reaction in which all compounds are gases:
    $$2A + 3B \rightarrow 3X + Y + heat$$
    The reaction is originally at equilibrium. Predict the direction of the shift in equilibrium for:

    | Stress | Shift | | |
    |---|---|---|---|
    | a. addition of compound A | right | left | no change |
    | b. removal of compound X | right | left | no change |
    | c. an increase in pressure | right | left | no change |
    | d. an increase in temperature | right | left | no change |
    | e. addition of a catalyst | right | left | no change |

25. Ammonia is produced by the reaction of nitrogen and hydrogen. The production of ammonia would be favored by _____ pressure.
    $$N_2 + 3H_2 \rightarrow 2NH_3$$
    a. high   b. low   c. neither high nor low

# Chapter 4 - Chemical Reactions

## ANSWERS

### Problems

1.     a. 132   b. 164   c. 133   d. 124   e. 318   f. 158   g. 184

2. The formula mass of $C_5H_8O_2$ = 100
   - a. 1 mole;   2 moles;   0.25 mole;   0.03687 mole
   - b. 100 g;   800 g;   80 g;   1 g
   - c. $6.02 \times 10^{23}$;   $3.01 \times 10^{23}$;   $18.06 \times 10^{23}$;   $6.02 \times 10^{23}$;   $3.01 \times 10^{23}$;   $18.06 \times 10^{23}$
   - d. 5;   5 moles or $30 \times 10^{23}$;   5 moles or $30 \times 10^{23}$
   - e. 8;   8 moles or $48 \times 10^{23}$;   8 moles or $48 \times 10^{23}$

3. The formula mass of $PbCrO_4$ = 323
   - a. 0.31 mole; 0.62 mole; 0.077 mole; 0.0114 mole
   - b. 323 g; 2584 g; 258 g; 3.23 g
   - c. $6.02 \times 10^{23}$; $3.01 \times 10^{23}$; $18.06 \times 10^{23}$; $1.86 \times 10^{23}$; $0.93 \times 10^{23}$; $5.6 \times 10^{23}$
   - d. 4; 4 moles or $2.4 \times 10^{24}$; 0.309 mole or $1.86 \times 10^{23}$

4. 
   - a. $Zn + 2\,KOH \rightarrow K_2ZnO_2 + H_2$
   - b. $4\,HF + Si \rightarrow SiF_4 + 2\,H_2$
   - c. $2\,B_2O_3 + 3\,H_2O \rightarrow H_6B_4O_9$
   - d. $SiCl_4 + 2\,H_2O \rightarrow SiO_2 + 4\,HCl$
   - e. $SnO_2 + 2\,C \rightarrow Sn + 2\,CO$
   - f. $Fe_2O_3 + CO \rightarrow 2\,FeO + CO_2$
   - g. $Fe_3O_4 + 4\,C \rightarrow 3\,Fe + 4\,CO$
   - h. $2\,Fe(OH)_3 + 3\,H_2S \rightarrow Fe_2S_3 + 6\,H_2O$
   - i. $Ca_3P_2 + 6\,H_2O \rightarrow 2\,PH_3 + 3\,Ca(OH)_2$
   - j. $Bi_2O_3 + 3\,C \rightarrow 2\,Bi + 3\,CO$

5. 
   - a. 2 moles of $CO_2$; 1 mole of $CO_2$
   - b. 0.6 mole of CaO; 0.3 mole of CaO
   - c. 288 g of CaS; 14.4 g of CaS; 72 g of CaS; 144 g of CaS
   - d. 56 g of CaO; 224 g of CaO; 28 g of CaO

6. 
   - a. 15 L of $CO_2$; 25 L of $O_2$
   - b. for 5 mL of $C_3H_8$: 15 mL of $CO_2$; 25 mL of $O_2$
     for 5 m$^3$ of $C_3H_8$: 15 m$^3$ of $CO_2$; 25 m$^3$ of $O_2$
   - c. 66 g of $CO_2$; 80 g of $O_2$; 250 kcal

### Self-Test

| | | | |
|---|---|---|---|
| 1. c | 10. a. T, b. F, c. T | 14. c | 23. d |
| 2. d | 11. b | 15. b | |
| 3. c | 12. b | 16. c | 24. a. right |
| 4. c | 13. a. endothermic | 17. c |      b. right |
| 5. b |      b. reaction II | 18. b |      c. right |
| 6. d |      c. $T_1$ and R | 19. a |      d. left |
| 7. a |      d. R and P | 20. d |      e. no change |
| 8. d |      e. released | 21. d | 25. a |
| 9. c | | 22. b | |

# 5 Oxidation and Reduction

**KEY WORDS**

| | | | |
|---|---|---|---|
| *oxide* | *redox* | *reducing agent* | *disinfectants* |
| *oxidation* | *reduced* | *oxidizing agent* | *bleaching* |
| *oxidation number* | *oxidation state* | *antiseptics* | *photosynthesis* |
| *half-reaction* | *reduction* | *antioxidants* | |

**SUMMARY**

Chemical reactions can be classified in many ways; in this chapter we study oxidation-reduction reactions. Reduced forms of matter (methane) are high in energy, while oxidized forms of matter ($CO_2$) are low in energy.

5.1 Oxygen and Hydrogen: An Overview of Oxidation and Reduction
 A. Oxygen: abundant and essential
  1. Oxygen is the most abundant element on this planet.
   a. Air         –   21% oxygen (free)
   b. Water       –   89% oxygen (combined)
   c. Earth's crust –   45% oxygen (combined)
   d. People      –   60% oxygen (combined)
  2. Oxygen is a gas at room temperature. The normal boiling point of $O_2$ is $-183$ °C.
  3. Fuels such as natural gas, gasoline, coal, and the foods we eat all need oxygen for combustion to release their stored chemical energy.
 B. Chemical properties of oxygen: oxidation
  1. Materials react with oxygen to form *oxides* by the process called ***oxidation.***
  2. The substances that combine with oxygen are said to be oxidized.
   a. $4\,Fe + 3\,O_2 \rightarrow 2\,Fe_2O_3$     iron (III) oxide (rust)
   b. $CH_4 + 2\,O_2 \rightarrow 2\,H_2O + CO_2$   carbon dioxide
 C. Hydrogen: a reactive lightweight element
  1. Hydrogen represents only 0.9% of the mass of the Earth's crust but 15.1% of the atoms. Hydrogen is the most abundant element (90% of all atoms) in the universe. Free or uncombined hydrogen is very rarely found on Earth, but combined hydrogen is found in water, natural gas, petroleum products, and in all foodstuffs.
  2. Preparation and properties of hydrogen
   a.   $Zn + HCl \rightarrow ZnCl_2 + H_2 \uparrow$
       (active metal) (acid)         (gas)
   b. Hydrogen is a colorless and odorless gas. It is very light but highly flammable.
   c. Absorbed $H_2$ on metal surfaces, such as Pt or Ni, is unusually reactive and used in surface catalysis.
 D. Chemical properties of hydrogen: reduction
  1. Hydrogen likes to combine with oxygen to form water. It will react with many metal oxides to remove the oxygen and form the free metal.
       $CuO + H_2 \rightarrow Cu + H_2O$
  2. The substance that reacts with hydrogen is said to be ***reduced***, and this process is called reduction.

## Chapter 5 - Oxidation and Reduction

E. *Oxidation*: simple definitions
    1. Addition of oxygen:      $2\ Ba + O_2 \rightarrow 2\ BaO$
    2. Loss of hydrogen:      $CH_3CH_2OH + X \rightarrow CH_3COH + XH_2$
    3. Loss of electrons:      $Fe^{2+} \rightarrow Fe^{3+} + e^-$

F. *Reduction*: simple definitions
    1. Loss of oxygen      $CuO + H_2 \rightarrow Cu + H_2O$
    2. Addition of hydrogen      $N_2 + 3\ H_2 \rightarrow 2\ NH_3$
    3. Addition of electrons      $Fe^{3+} + e^- \rightarrow Fe^{2+}$

5.2 Oxidation Numbers and Oxidation-Reduction (Redox) Equations

A. *Oxidation* can also be characterized by an increase in *oxidation state* or *oxidation number*. Oxidation numbers are a form of bookkeeping to help us understand the changes that take place in "formal charge" during oxidation-reduction (*redox*) reactions. Rules for assigning oxidation numbers:

| Class | Oxidation Number | Example |
|---|---|---|
| 1. Free elements | 0 | $Al^{(0)}$ |
| 2. Metal compounds | | |
|     Group 1A elements | +1 | $Na^{(+1)}Cl$ |
|     Group 2A elements | +2 | $Ba^{(+2)}Cl_2$ |
| 3. Hydrogen | | |
|     In most compounds | +1 | $H^{(+1)}_2O$, $CH^{(+1)}_4$ |
|     Metal hydrides | -1 | $NaH^{(-1)}$ |
| 4. Oxygen | | |
|     In most compounds | -2 | $H^{(+1)}_2O^{(-2)}$ |
|     In peroxides | -1 | $H^{(+1)}_2O^{(-1)}_2$ |
| 5. Uncharged compounds | | |
|     Sum of oxidation numbers = 0 | | $C^{(-4)}H^{(+1)}_4$ |
| 6. Polyatomic ions | | |
|     Sum of oxidation numbers = charge | | $H_3PO_4$: $O(-2)$, $H(+1)$, $P(+5)$ |

B. Identifying redox reactions
    1. In a redox reaction the oxidation number of one or more elements increases – the oxidation process – and the oxidation number of one or more elements decreases – the reduction process.
         $Mg(s) + Cu^{2+}(aq) \rightarrow Mg^{2+}(aq) + Cu(s)$
    2. A redox reaction can be thought of as two *half-reactions*.
         Oxidation:    $Mg(s) \rightarrow Mg^{2+}(aq) + 2\ e^-$
         Reduction:    $Cu^{2+}(aq) + 2\ e^- \rightarrow Cu(s)$

5.3 Oxidizing and Reducing Agents

A. The substance being reduced is the *oxidizing agent*. Oxygen is the most common oxidizing agent and is used to "burn" all sorts of fuels and foodstuffs.
         $C_6H_{12}O_6 + 6\ O_2 \rightarrow 6\ CO_2 + 6\ H_2O$
         Glucose     (Oxygen is being reduced and is the oxidizing agent.)

B. In a redox reaction the substance being oxidized is called the *reducing agent*.
         $CuO + H_2 \rightarrow Cu + H_2O$
         (Hydrogen is being oxidized and is the reducing agent.)

C. Some common oxidizing agents
1. The dichromate ion ($Cr_2O_7^{2-}$) is another common laboratory oxidizing agent as Cr goes from $Cr^{+6}$ to $Cr^{+3}$.

$$Cr_2O_7^{2-} + 3\,C_2H_5OH + 8\,H^+ \rightarrow 2\,Cr^{3+} + 3\,C_2H_4O + 7\,H_2O$$

2. Hydrogen peroxide is a syrupy, colorless liquid usually used as a 3% or 30% aqueous solution. It forms water when used as an oxidizing agent.

$$PbS + 4\,H_2O_2 \rightarrow PbSO_4 + 4\,H_2O$$

3. Potassium permanganate ($KMnO_4$) is a black solid that forms deep purple solutions. The permanganate ion ($Mn^{+7}$) is a common laboratory oxidizing agent, as it is reduced to $Mn^{+2}$.

$$MnO_4^- + 5\,Fe^{2+} + 8\,H^+ \rightarrow Mn^{2+} + 5\,Fe^{3+} + 4\,H_2O$$
(purple)                              (colorless)

4. The halogens are common oxidizing agents.

$$Cl_2 + Mg \rightarrow Mg^{2+} + 2\,Cl^-$$

5. Oxidation and antiseptics
   a. Many common *antiseptics* are mild oxidizing agents: 3% $H_2O_2$, 0.01% $KMnO_4$, $KClO_3$, NaOCl, solutions of iodine.
   b. Common *disinfectants* include bleaching powder [$Ca(OCl)_2$], chlorine ($Cl_2$), and ozone ($O_3$). Ozone is used in place of chlorine in many European cities to treat drinking water.
6. Oxidation: bleaching and stain removal
   a. Common *bleaching* agents act by removing mobile or highenergy electrons whose absorption properties account for the unwanted colors.
   Laundry bleaches (Clorox) – 5.25% NaOCl (sodium hypochlorite)
   Bleaching powder – $Ca(OCl)_2$
   Hydrogen peroxide – bleaches hair by oxidizing the melanin pigments to colorless compounds.
   b. Stain removal can involve solubilization or conversion to colorless materials by either oxidizing or reducing agents.
   Hydrogen peroxide – blood stains removed from cotton
   Oxalic acid – rust spots removed through complex formation
   Cornstarch – used as an absorbent
   Acetone – used as a solvent

D. Some reducing agents of interest
1. Hydrogen gas and elemental carbon are used as reducing agents.

$$2\,Fe_2O_3 + 3\,C \rightarrow 4\,Fe + 3\,CO_2$$

2. Hydrogen gas serves as a reducing agent.

$$WO_3 + 3\,H_2 \rightarrow W + 3\,H_2O$$

3. Hydrogen peroxide can also serve as a reducing agent as $O(-1)$ is oxidized to free oxygen.

$$2\,MnO_4^- + 5\,H_2O_2 + 6\,H^+ \rightarrow 2\,Mn^{2+} + 5\,O_2 + 8\,H_2O$$

4. Organic hydroquinones are used to "fix" silver in photographic film by reducing $Ag^+$ ions that have been exposed to light to free Ag metal.
5. Vitamin C (ascorbic acid) and Vitamin E are *antioxidants*, a word commonly used for reducing agents in the food industry.

## 5.4 Oxidation, Reduction, and Living Things
A. Reduced compounds represent a form of stored potential energy. The driving force to produce reduced compounds is ultimately derived from the sun in the processes of *photosynthesis*.

$$6\,CO_2 + 6\,H_2O + \text{energy} \rightarrow 6\,O_2 + C_6H_{12}O_6$$

## Chapter 5 - Oxidation and Reduction

Plants and the animals that feed on plants use these materials to make other reduced compounds such as the carbohydrates, fats, and proteins that constitute our major food materials.

B. We produce energy to meet our metabolic and body needs by releasing the energy stored in these reduced compounds as we combust them back to $CO_2$ and water.

$$Carbohydrates + O_2 \rightarrow CO_2 + H_2O + energy$$
$$Fats + O_2 \rightarrow CO_2 + H_2O + energy$$
$$Proteins + O_2 \rightarrow CO_2 + H_2O + energy + urea$$

## DISCUSSION

It is impossible to overemphasize the importance of oxidation-reduction processes. Think of it this way: You are powered by the energy of sunlight, but you can't simply unfold solar panels as artificial satellites do and convert sunlight to stored electrical energy to be tapped as necessary. You are a chemical factory and not a satellite, solar-powered or otherwise. So somehow you have to tap that solar energy in a chemical way. This is precisely the role of oxidation-reduction reactions in life processes – they plug you into the sun. The later chapters of the text will detail the sequence of reactions that accomplishes this objective. Right now we are concerned with familiarizing ourselves with the general features of oxidation-reduction reactions.

We've spent much time in this chapter developing your ability to recognize when a compound is oxidized or reduced. In addition to the three definitions used in the text, we have included the concept of *oxidation number*. On first encounter, oxidation numbers may seem to be mysterious and magical. They were conjured up by chemists to make it easier to deal with oxidation-reduction reactions. It is the mystery of oxidation numbers that we have tried to dispel. They simply represent a means of assessing oxidation or reduction. You should be able to look at an equation and recognize when oxidation (and reduction) has occurred. (Always remember that one process is impossible without the other.) If you can do that without referring to oxidation numbers, fine. But we think you'll find, just as chemists already have, that occasionally a consideration of oxidation numbers makes the evaluation process a little easier.

Problems 1 and 2 below offer practice in oxidation states and in evaluating the components of oxidation–reduction equations. These problems supplement those at the end of the chapter.

## Problems

1. Determine the oxidation states of the underlined atoms.
   a. $\underline{O}_3$
   b. $\underline{N}_2$
   c. $\underline{N}H_3$
   d. $\underline{Cr}O_3$
   e. $\underline{Cr}_2O_7^{2-}$
   f. $NaO\underline{Cl}$
   g. $H\underline{Cl}O_4$
   h. $H\underline{Cl}O_2$
   i. $K\underline{I}O_3$
   j. $H_5\underline{I}O_6$
   k. $H_3\underline{P}O_4$
   l. $H_4\underline{P}_2O_7$
   m. $CaH\underline{P}O_4$
   n. $H\underline{S}_2O_7^-$
   o. $\underline{N}_2O_3$
   p. $\underline{N}_2O_4$
   q. $\underline{N}_2O_5$
   r. $\underline{N}O_2^-$
   s. $\underline{C}_2H_6$
   t. $\underline{C}H_2O_2$
   u. $\underline{C}_2H_4O_2$
   v. $\underline{C}H_4O$
   w. $\underline{C}O_2$
   x. $\underline{C}_3H_6O_3$
   y. $\underline{C}_3H_4O_5$
   z. $\underline{C}_{12}H_{22}O_{11}$

2. Identify the element being oxidized, the element being reduced, the oxidizing agent, and the reducing agent for each equation.
   a. $4\,Na + CO_2 \rightarrow 2\,Na_2O + C$
   b. $C_2H_4 + 3\,O_2 \rightarrow 2\,CO_2 + 2\,H_2O$
   c. $2\,Ag^+ + Cu \rightarrow 2\,Ag + Cu^{2+}$
   d. $5\,CO + I_2O_5 \rightarrow I_2 + 5\,CO_2$
   e. $3\,SO_2 + 2\,CrO_3 + 3\,H_2O \rightarrow Cr_2O_3 + 3\,H_2SO_4$

# Chapter 5 - Oxidation and Reduction

## SELF-TEST

1. The most common element in the universe is:
   a. $O_2$   b. $H_2$   c. $N_2$   d. He   e. C
2. On Earth we do <u>not</u> find a large abundance of oxygen in the form of:
   a. $O_2$   b. $H_2O$   c. $H_2O_2$   d. $SiO_2$
3. A substance is oxidized if it:
   a. gains oxygen atoms       b. gains hydrogen atoms
   c. gains electrons          d. all of these          e. none of these
4. In the reaction, $C_2H_4 + H_2O \rightarrow C_2H_6O$, carbon is:
   a. oxidized   b. reduced   c. neither oxidized nor reduced
5. Hydrogen is no longer used in lighter-than-air ships because:
   a. It is more dense than helium.
   b. It is a strong oxidizing agent.
   c. Under certain conditions it reacts explosively to form water.
   d. It is too difficult to obtain from petroleum products.
6. Platinum is a useful catalyst in reactions involving hydrogen because:
   a. The hydrogen absorbed on the catalyst's surface is more reactive than molecular hydrogen.
   b. The activation energy of the reaction using platinum is lower.
   c. Both of the above statements are correct.
   d. Neither statement a nor b above is correct.
7. Two common oxidizing agents change color when they are reduced. They are:
   a. $H_2$   b. $KMnO_4$   c. $Na_2Cr_2O_7$   d. $O_2$   e. NaOCl
8. Which compound can be an oxidizing agent or a reducing agent?
   a. $H_2O_2$   b. $KMnO_4$   c. $H_2$
9. Hydrogen peroxide used as an oxidizing agent is reduced to:
   a. $O_2$   b. $H_2O$   c. $H_2O_2$   d. $H_2$
10. Disinfectants, antiseptics and bleaches are frequently:
    a. oxidizing agents    b. reducing agents
11. In photosynthesis, carbon dioxide is _____ to a sugar.
    a. oxidized    b. reduced
12. Which is <u>not</u> a reason that ozone is preferred over chlorine as a disinfectant for water?
    a. Ozone is less expensive than chlorine.
    b. Chlorine has been shown to form toxic by-products.
    c. Ozone is more effective against some viruses.
    d. Chlorine imparts a distinctive taste to water.
13. In general, animals, including humans, are:
    a. oxidizing agents    b. reducing agents
14. The oxidation number of Cl in $HClO_3$ is:
    +1  +2  +3  +5  +7  +8  −1  −2  −3  −4  −6  −7  −8
15. The oxidation number of Sn in $SnO_3^{2-}$ is:
    +1  +2  +3  +4  +5  +6  +7  8  −1  −2  −3  −6  −7  −8
16. The oxidation number of P in $H_2PO_4^-$ is:
    +1  +2  +3  +4  +5  +6  +7  +8  −1  −2  −3  −4  −5  −8
17. In the equation:   $FeO + C \rightarrow Fe + CO$
    a. FeO is an(a):      oxidizing agent       reducing agent
    b. C is:              oxidized              reduced

## Chapter 5 - Oxidation and Reduction

18. According to the equation: $SO_2 + HNO_3 + H_2O \rightarrow H_2SO_4 + NO$
    a. sulfur is being:     oxidized            reduced
    b. $HNO_3$ is the:        oxidizing agent      reducing agent
19. In the reaction:       $N_2 + 2 H_2O \rightarrow NH_4^+ + NO_2^-$, nitrogen is:
    a. oxidized     b. reduced     c. both     d. neither
20. Consider: $3 C_2H_4 + 2 MnO_4^- + 4 H_2O \rightarrow 3 C_2H_6O_2 + 2 MnO_2 + 2 OH^-$.
    a. What is the oxidation number of C in $C_2H_4$?
       +1   +2   +3   +4   +5   −1   −2   −3   −4   −5   0
    b. What is the oxidation number of C in $C_2H_6O_2$?
       +1   +2   +3   +4   +5   −1   −2   −3   −4   −5   0
    c. Is carbon being oxidized or reduced?
21. Balance each of the following redox reactions, and underline the oxidizing agents.
    a. $KCl + MnO_2 + H_2SO_4 \rightarrow K_2SO_4 + MnSO_4 + Cl_2 + H_2O$
    b. $KMnO_2 + FeSO_4 + H_2SO_4 \rightarrow K_2SO_4 + MnSO_4 + Fe_2(SO_4)_3 + H_2O$
    c. $Cu + HNO_3 \rightarrow Cu(NO_3)_2 + NO + H_2O$
22. Butane lighters rely on the ready combustion of butane ($C_4H_{10}$) to form carbon dioxide and water.
    a. Write a balanced reaction for the combustion of butane with oxygen.
    b. How many moles of carbon dioxide are produced from 6 moles of butane?
    c. How many grams of oxygen are required to completely combust 100 g of butane?
    d. If 200 g of butane are reacted with 200 g of oxygen, which reactant will be present in excess?

## ANSWERS

**Problems**

1. a. 0   b. 0   c. −3   d. +6   e. +6   f. +1   g. +7   h. +3   i. +5   j. +7   k. +5   l. +5   m. +5
   n. +6   o. +3   p. +4   q. +5   r. +3   s. −3   t. +2   u. 0   v. −2   w. +4   x. 0   y. +2   z. 0
2. a. Na is oxidized, and C is reduced. Na is the reducing agent, and $CO_2$ is the oxidizing agent.
   b. C is being oxidized, and O is being reduced.
      $C_2H_4$ is the reducing agent, and $O_2$ is the oxidizing agent.
   c. Cu is being oxidized, and Ag+ is reduced. Cu is the reducing agent, and Ag+ is the oxidizing agent.
   d. C is being oxidized, and I is being reduced.
      CO is the reducing agent, and $I_2O_5$ is the oxidizing agent.
   e. S is being oxidized, Cr is being reduced. $SO_2$ is a reducing agent, and $CrO_3$ is an oxidizing agent.

**Self-Test**

| | | | | | |
|---|---|---|---|---|---|
| 1. b | 5. c | 9. b | 13. a | 17. a. oxidizing agent | 19. c |
| 2. c | 6. c | 10. a | 14. +5 |      b. oxidized | 20. a. −2 |
| 3. a | 7. b, c | 11. b | 15. +4 | 18. a. oxidized |      b. −1 |
| 4. c | 8. a | 12. a | 16. +5 |      b. oxidizing agent |      c. oxidized |

21.    a. $2 KCl + \underline{MnO_2} + 2 H_2SO_4 \rightarrow K_2SO_4 + MnSO_4 + Cl_2 + 2 H_2O$
         b. $2 K\underline{Mn}O_2 + 10 FeSO_4 + 8 H_2SO_4 \rightarrow K_2SO_4 + 2 MnSO_4 + 5 Fe_2(SO_4)_3 + 8 H_2O$
         c. $3 Cu + 8 H\underline{N}O_3 \rightarrow 3 Cu(NO_3)_2 + 2 NO + 4 H_2O$
22.    a. $2 C_4H_{10} + 13 O_2 \rightarrow 8 CO_2 + 10 H_2O$
         b. 24 m $CO_2$      c. 1.72m $C_4H_{10}$ = 11.2m or 359g $O_2$    d. $C_4H_{10}$

# 6 Gases

**KEY WORDS**

| | | | |
|---|---|---|---|
| matter | kinetic-molecular theory | Henry's law | Gay-Lussac's law |
| solid | Boyle's law | diffusion | ideal gas law |
| Charles' law | Dalton's law | STP | gaseous |
| partial pressure | vapor pressure | molar volume | hemoglobin |
| liquid | pressure | temperature | barometer |
| atmosphere | Avogadro's Law | absolute temperature | relative humidity |
| mmHg | heat index | Kelvin scale | pascal |
| | | universal gas constant | combined gas law |

**SUMMARY**

Earth's atmosphere is composed of gases weighing about $5.2 \times 10^{15}$ metric tons (14.7 lb/in.$^2$). It is estimated that 99% of Earth's atmosphere lies within 30km of Earth's surface.

6.1 Air: A Mixture of Gases
  A. *Matter* can exist in a *solid*, *liquid*, or *gaseous* state. The theory of gases will help us to understand all three states of matter.
  B. Dry air: 78% $N_2$, 21% $O_2$, 1% Ar. Humid air can be up to 4% water vapor.

6.2 The Kinetic-Molecular Theory
  A. Experimentally, gases have low densities, are readily compressed, fill their containers, and expand on heating.
  B. *Kinetic-molecular theory* of gases
    1. All matter is composed of tiny, discrete particles called molecules.
    2. Molecules are in rapid, constant, random motions.
    3. Distances between particles are large compared to the diameter of the particles.
    4. Forces between gaseous particles are small.
    5. Gaseous particles undergo elastic collisions with themselves and with their containers.
    6. The average kinetic energy of a gas is proportional to its absolute *temperature*.
  C. These collisions are responsible for the "*pressure*" of gases.
  D. *Temperature* is just a reflection of the average kinetic energy of the gas particles.

6.3 Atmospheric Pressure
  A. **Pressure** is *force per unit of area* ($P = F/A$). The pressure of the *atmosphere* can be measured by how high a column of liquid it will support. Such a device is called a *barometer*. The mercury barometer was invented by Torricelli in 1643.
  B. Units of pressure
    1. Common: 1 atm = 760 *mmHg* = 760 Torr = 29.92 in.Hg = 14.7 psi (lb/in.$^2$)
    2. SI: The SI unit for pressure is the *pascal* (1 Pa = 1 N/m$^2$) 1 atm = 101.3 kPa

6.4 Boyle's Law: The Pressure-Volume Relationship
  A. Boyle (1662) determined that for a given mass of gas at constant temperature, the volume varies inversely with the pressure.
     **Boyle's Law**:  $PV = k$;  $P \propto 1/V$;  $V \propto 1/P$  (where $k$ is a constant)

## Chapter 6 - Gases

    B. Boyle's law can be explained by kinetic-molecular theory, since the number of collisions against the container walls (pressure) will go down as the distance between the walls (volume) increases.
    C. Artificial respirators, such as the iron lung, are based on Boyle's law. As the bellows move out the pressure surrounding the chest is reduced, allowing air to flow into the patient's lungs.

6.5 Charles's Law: The Temperature-Volume Relationship
    A. Charles (1787) observed that the volume of a gas at constant pressure varies directly with its absolute or Kelvin temperature.
        ***Charles's Law***: $V \propto T$; $V = kT$; $V/T = k$ (T expressed in Kelvin, $k$ is a constant)
    B. A plot of V vs. T (volume vs. Temperature) gives a straight line. By extrapolating this line, it would appear that the volume of a gas would become zero at –273.15 °C. In 1848 Kelvin assigned this temperature the zero point on an *absolute temperature* scale or ***Kelvin scale***.
    C. Charles's law can also be explained by kinetic-molecular theory. Since the average kinetic energy increases with increasing temperature, the particles move further per unit of time and therefore can maintain the same pressure in a larger volume at the higher temperature.
    D. Gay-Lussac's Law: Pressure-Temperature Relationship
        1. Joseph Gay-Lussac conducted experiments similar to those of Charles to determine that the pressure of a gas at constant volume varies directly with its absolute temperature.
            ***Gay-Lussac's Law***: $P \propto T$; $P = kT$; $P/T = k$ (T expressed in Kelvin, $k$ is a constant)
        2. Gay-Lussac's law can be explained by kinetic-molecular theory by the same reasoning used for Charles's law.

6.6 Avogadro's Law: The Molar Volume of a Gas
    A. ***Avogadro's Law*** – at a fixed temperature and pressure the volume of a gas is directly proportional to the number of molecules or moles of a gas present.
        $V \propto n$;         $V = kn$ ($k$ is a constant)
    B. Standard temperature and pressure (***STP***) conditions are defined as 0 °C and 1 atmosphere (atm).
    C. The volume of 1 mole of most gases at STP is 22.4 L. This quantity is known as the ***molar volume*** of a gas.

6.7 The Combined Gas Law
    A. Boyle's and Charles's laws can be combined to express the pressure, volume, and temperature relationship of an ideal gas.
    B. ***Combined gas law***.     $PV \propto T$   or   $P_1V_1/T_1 = P_2V_2/T_2$

6.8 The Ideal Gas Law ($PV = nRT$)
    A. Avogadro's hypothesis that equal volumes of gases at the same temperature and pressure contain equal numbers of molecules allows us to relate $P$, $V$, $T$ and moles ($n$).
        $PV/T \propto n$   or   $PV \propto nT$   or   $PV/nT$ = constant = $R$ (Ideal Gas Constant)
    B. The ideal gas law can be expressed in equation form as follows:
        ***Ideal Gas Law***:     $PV = nRT$   ($R$ = ***universal gas constant*** = 0.0821 L-atm/mol-K)

6.9 Henry's Law: The Pressure-Solubility Relationship
    A. Henry (1801) reported that the solubility of a gas in a liquid at a given temperature is directly proportional to the pressure of the gas at the surface of the liquid, ***Henry's Law***.
    B. Henry's law has important applications to the manufacture of carbonated beverages, to deep-sea diving (diver's bends), and to several areas of therapy.

# Chapter 6 - Gases

6.10 Dalton's Law of Partial Pressures

A. Dalton (1803) discovered that the total pressure of a mixture of gases is equal to the sum of the *partial pressures* exerted by the separate gases. This finding is known as Dalton's law.
   **Dalton's law** of partial pressures: $P_{total} = P_1 + P_2 + P_3 + \cdots$

B. The *vapor pressure* of a liquid refers to the partial pressure of the gas phase above the liquid phase.

C. The vapor pressure of water varies with temperature. Cold air holds less water vapor than warm air does.

| T(°C): | 0 | 20 | 30 | 40 | 60 | 80 | 100 |
|---|---|---|---|---|---|---|---|
| Vapor pressure in Torr: | 5 | 18 | 32 | 55 | 149 | 355 | 760 |

D. *Relative humidity* compares the actual amount of water vapor in the air to the maximum amount the air could hold if it were saturated at that temperature. For example, if the water pressure in air is 9 mmHg, it is at 50% relative humidity (relative to the maxumum value), since the vapor pressure of water at 20 °C is 18 mmHg. Higher humidity makes you feel hotter by inhibiting evaporation of sweat. The *heat index* relates temperature to relative humidity.

6.11 Partial Pressures and Respiration

A. *Diffusion* refers to the process whereby a substance "moves" from a region of higher concentration to regions of lower concentration.

B. Inspired air is rich in $O_2$ and poor in $CO_2$, while metabolizing cells are rich in $CO_2$ and poor in $O_2$. Diffusion facilitates the transfer of needed $O_2$ from the lungs to the cells and the removal of waste $CO_2$ from the cells out via the lungs.

C. Our bodies need more oxygen than can be dissolved in blood plasma alone. Most of the oxygen in the bloodstream combines with a protein molecule called *hemoglobin*, which picks up $O_2$ at the lungs and releases it at the respiring cells. Hemoglobin also helps with $CO_2$ transport, but most $CO_2$ is transported as the bicarbonate ion ($HCO_3^-$). Carbon monoxide poisoning is due in part to tight CO binding to hemoglobin, so the hemoglobin is no longer available for $O_2$ transport.

## DISCUSSION

As always, we assume that you've worked through the problems at the end of the chapter in the text. You should, therefore, have a pretty good idea of how well you understand the gas laws of Boyle, Charles, and Gay-Lussac.

The following diagrams present the briefest summary of these laws:

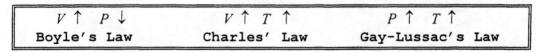

The diagrams should call to mind which of the variables vary directly and which vary inversely. Thus, volume and pressure move in opposite directions. Volume and temperature move together, as do pressure and temperature. Also remember that any units of volume and pressure can be used in these gas laws, as long as you are consistent, but temperature variations must be calculated using the absolute (Kelvin) scale. Beyond these few reminders, the best way for you learn these gas laws is to work more practice problems to sharpen your skills.

Chapter 6 - Gases

**Problems**

1. A gas with a pressure of 500 mmHg at 0 °C occupies a volume of 500 mL.
    a. If the temperature is changed to 273 °C while the pressure is held constant, what will the new volume be?
    b. If the volume of the original gas was compressed to 100 mL and the temperature was maintained constant, what would the pressure be?
    c. What would the volume be at STP?
2. If a gas occupies 2 L at 27 °C and 1 atm:
    a. What would the volume be if the temperature remained constant but the pressure decreased to 0.25 atm?
    b. What would the volume be if the pressure remained constant but the temperature decreased to −123 °C?
    c. What would the volume be at STP?
3. A balloon was filled with gas to a volume of 60 L at a temperature of 30 °C and a pressure of 750 Torr.
    a. What will the volume of the balloon be if it drifts to a point where the temperature is still 30 °C but the pressure is only 250 Torr?
    b. What will the volume of the balloon be if the pressure remains constant but the temperature drops to −71 °C?
    c. What would the volume be at STP?
4. A rigid container is filled with gas at 86 °F and 20 psi and then sealed. What will the pressure inside the container be if it is heated to 212 °F?
5. What is the pressure of 0.25 mol of He that occupies a volume of 4.0 L at 300 K?

Use the **SELF-TEST** to check your understanding of the remaining material in Chapter 6. Remember that some of the **SELF-TEST** questions do require calculations like the problems above; that is, they are asking for more than simple recall of factual material. You may need to refer to a periodic table for atomic weights.

**SELF-TEST**

1. One atmosphere is approximately equal to _____ pascals.
    a. 1.0    b. 100    c. 10,000    d. 100,000    e. 1000
2. Which gas has the highest partial pressure in the atmosphere?
    a. $CO_2$    b. $H_2O$    c. $O_2$    d. $N_2$
3. If a pressure is reported as 1 Torr, it is also:
    a. 760 mmHg    b. 760 atm    c. 1 mmHg    d. 1 atm
4. The conditions expressed in the abbreviation STP do not include:
    a. 0 °C    b. 273 K    c. 760 mmHg    d. 1 atm    e. 1 Torr
5. Expressed in inches of mercury, one standard atmosphere is 29.92" of Hg. What is the atmospheric pressure in mmHg when the weatherman reports that the "barometer is holding steady at 27.8 in.Hg?"
    a. 740    b. 706    c. 930    d. 785    e. 721
6. According to the kinetic molecular theory, if you increase the temperature of a gas without changing the volume of the gas, the particles of the gas will:
    a. strike the container walls more often    b. strike the walls with less force
    c. lose kinetic energy    d. increase in size

Chapter 6 - Gases

7. Boyle's law would be explained by the kinetic molecular theory in the following way:
   a. If the volume of a gas sample decreases, the particles move faster and strike the walls of the container more often, and this is measured as an increase in pressure.
   b. If the volume of a gas is decreased, the distance between the walls of the container decreases and the particles strike the walls more often, resulting in an increase in pressure.
   c. Both of the above explanations are incorrect because a decrease in volume results in decrease in the pressure of a gas.
8. An increase in the pressure of a gas at constant temperature corresponds to:
   a. an increase in the concentration of the gas
   b. a decrease in the concentration of the gas
   c. an increase in the size of the particles of the gas
9. In applying each of the following gas laws, which variable is held constant?
   a. Boyle's law         (pressure, temperature, or volume)
   b. Gay-Lussac's law    (pressure, temperature, or volume)
   c. Charles's law       (pressure, temperature, or volume)
10. Henry's law can be used to explain the
    a. operation of an iron lung
    b. bends experienced by deep-sea divers
    c. process of respiration
11. If a gas occupies 2 L at 3 atm, how many liters will it occupy at 6 atm if the temperature isn't changed?
    a. 1    b. 2    c. 4    d. 8    e. 16    f. 32
12. If a gas occupies 2 L at 273 °C, what is the volume at 0 °C if the pressure does not change?
    a. 1/273    b. 0.5    c. 1    d. 2    e. 4    f. 273
13. If a mixture of gases has a total pressure of 200 mmHg and the partial pressure of nitrogen in the mixture is 60 mmHg, the percent of nitrogen in the mixture is:
    a. 60%    b. 50%    c. 200%    d. 15%    e. 30%
14. If the concentration of water vapor in the air is 1% and the total atmospheric pressure equals 1 atm, the partial pressure of water vapor is:
    a. 0.1 atm    b. 1 mmHg    c. 7.6 mmHg    d. 100 atm    e. 760 mmHg
15. If a sample of gas at STP contains 10% helium, what is the partial pressure of helium in the sample?
    a. 0.1 atm    b. 0.1 mmHg    c. 10 atm    d. 10 mmHg    e. 7.6 atm
16. If the relative humidity is 50% on a day when the temperature is 30 °C, the partial pressure of water vapor in the air is:
    a. 16 mmHg    b. 30 mmHg    c. 50 mmHg    d. 32 mmHg    e. 64 mmHg
17. The molar volume of argon gas (atomic number 18) is:
    a. 4 g/L    b. 4 moles    c. 4 L    d. $6.02 \times 10^{23}$    e. 22.4 L
18. A flask contains Avogadro's number of hydrogen atoms at STP. The volume of the flask is how many liters?
    a. 1    b. 11.2    c. 20    d. 22.4    e. $6.02 \times 10^{23}$
19. If a container holds 0.5 mole of $N_2$ gas at STP, the container holds how much $N_2$?
    a. $6.02 \times 10^{23}$ molecules    b. $0.5 \times 10^{23}$ molecules    c. 7 g    d. 14 g    e. 28 g
20. If a container holds 2 g of helium at STP, how big is the container?
    a. 0.5 L    b. 11.2 L    c. 20 L    d. 22.4 L    e. 44.8 L
21. The approximate density of $CO_2$ gas at STP is:
    a. 1 g/mL    b. 2 g/mL    c. 2 g/L    d. 44 g/ml    e. 44 g/L    f. 22.4 g/L
22. What is the approximate mass of 2 L of butane or $C_4H_{10}$ gas at STP?
    a. 10 g    b. 20 g    c. 22.4    d. 58 g    e. 5 g

## Chapter 6 - Gases

23. How many moles of Ne occupy 2.0 L at STP?
    a. 20    b. 11.2    c. 0.05    d. 0.09    e. 1.0
24. What is the approximate temperature of 1.0 mol of an ideal gas that occupies 22.4 L at 0.50 atm pressure?
    a. 0 °C    b. 273 K    c. 100 °C    d. 200 °C    e. 273 °C
25. Nitrogen is collected over water at 30 °C. If the total pressure within the collection vessel is 1 atm, the partial pressure of nitrogen is:
    a. 0.68 atm    b. 32 Torr    c. 728 Torr    d. 760 mmHg    e. 792 mmHg
26. Consider an autoclave where we want a steam temperature of 160 °C. For the fixed volume of the autoclave, steam is at 100 °C when the pressure is 1 atm. What pressure is required for steam at this higher temperature?
    a. 1.60 atm    b. 1.16 atm    c. 1.10 atm    d. 5.40 atm    e. 792 mmHg
27. Consider an automobile tire that is initially at 32 psi (2.17 atm) at a temperature of 25 °C. What is the pressure in the tire after 100 miles of driving if the tire temperature is now 50 °C? (Assume the volume remains constant.)
    a. 4.34 atm    b. 64 psi    c. 3.06 atm    d. 28.9 psi    e. 34.7 psi
28. The bends result from which dissolved gas being released as a deep-sea diver ascends?
    a. $H_2O$    b. $O_2$    c. $N_2$    d. $CO_2$    e. He
29. The mechanism by which we inspire air depends on our ability to _____ the size of our chest cavity.
    a. increase    b. decrease
30. In respiratory therapy, the highest humidity is imparted to a gas mixture to be administered through the:
    a. mouth    b. nose    c. trachea
31. The formation of oxyhemoglobin is favored by _____ $P_{O_2}$.
    a. high    b. low
32. If the partial pressure of a gas in the alveoli is 104 Torr and the gas tension of venous blood is 40 Torr:
    a. The net diffusion of the gas will be from alveoli to blood.
    b. The net diffusion of the gas will be from blood to alveoli.
    c. Nothing happens because gas tension and partial pressure have nothing to do with one another.
33. Where would you expect the $P_{CO_2}$ to be highest?
    a. in the alveoli immediately after inspiration of fresh air
    b. in the venous blood
    c. in the arterial blood
    d. in a cell that is metabolically active

# Chapter 6 - Gases

## ANSWERS

### Problems

1. a. 1.00 L     b. 2500 mmHg;    c. 329 mL
2. a. 8 L        b. 1 L           c. 1.82 L
3. a. 180 L      b. 40 L          c. 53 L
4. 24.6 psi
5. 1.5 atm

### Self-Test

| | | | | |
|---|---|---|---|---|
| 1. d | 8. a | 13. e | 20. b | 27. e |
| 2. d | 9. temperature | 14. c | 21. c | 28. c |
| 3. c |     volume | 15. a | 22. e | 29. a |
| 4. e |     pressure | 16. a | 23. d | 30. c |
| 5. b | 10. b | 17. e | 24. e | 31. a |
| 6. a | 11. a | 18. b | 25. c | 32. a |
| 7. b | 12. c | 19. d | 26. b | 33. d |

# 7 Liquids and Solids

**KEY WORDS**

| | | | |
|---|---|---|---|
| *interionic forces* | *dispersion force* | *condensation* | *unit cell* |
| *dipole forces* | *liquids* | *dynamic equilibrium* | *metallic solids* |
| *hydrogen bond* | *viscosity* | *normal boiling point* | *ionic solids* |
| *donor* | *surface tension* | *distillation* | *molar heat* |
| *acceptor* | *gas* | *solids* | *heat of fusion* |
| *intermolecular forces* | *London forces* | *crystal lattice* | *molecular solids* |
| *melting point* | *specific heat* | *density* | *ion pairs* |
| *sublimation* | *covalent network* | *body-centered cubic* | *ionic bonds* |
| *evaporation* | *vaporization* | *face-centered cubic* | *crystalline* |

**SUMMARY**

7.1 Intermolecular Forces: Some General Considerations
   A. *Intermolecular forces* determine the physical properties of fluids and solids. Strong intermolecular forces favor solids and require higher temperatures to separate the molecules. Gaseous molecules tend to have weak *intermolecular forces*, allowing the distances between gas particles to be very large. The particles in liquids and solids are held closer together because the intermolecular forces between their particles are much larger.
      1. Types of intermolecular forces: *interionic > hydrogen bonds > dipole > dispersion*
      2. Strength of intermolecular forces: *solid > liquid > gas*
   B. The physical state of a substance depends on its molecular weight and the type of forces operating between particles.
      1. All ionic substances are solids at room temperature.
      2. Most metals are solids at room temperature.
      3. Molecular substances exist as solids, liquids, or gases at room temperature. For a series of related molecules with the same type of intermolecular forces, the heavier molecules in the series will be solids and the lighter molecules in the series will be gases.

|  | $F_2$ | $Cl_2$ | $Br_2$ | $I_2$ |
|---|---|---|---|---|
| Molecular weight: | 38 | 71 | 160 | 254 |
| Physical state: | gas | gas | liquid | solid |

7.2 Ionic Bonds as Forces Between Particles
   A. *Interionic forces* are the strongest of the intermolecular forces. Most ionic compounds are solids at room temperature. There are no "molecules" in ionic solids, just a lattice of *"ion pairs."*
   B. The strength of the *"ionic bonds"* increases with increasing charge on the ion and decreases as the distance between charges increases. Thus, ionic forces decrease, and *dispersion* forces become more important for very large, singly charged ions such as $Ag^+$, $I^-$.

7.3 Dipole Forces
   A. Unsymmetrical molecules containing polar bonds are dipoles with centers of partially negative and positive charges.

## Chapter 7 - Liquids and Solids

B. Polar molecules attract one another as the positive end of one molecule interacts with the negative end of another molecule. Dipole forces are weaker than ionic bonds.

$N_2$ (b.p. = –196 °C); $^{\delta+}C = O^{\delta-}$ (b.p. = –192 °C) both with molecular wt. 28.0

### 7.4 Hydrogen Bonds
A. Compounds containing H attached to small, electronegative elements such as N, O, or F exhibit stronger attractive forces than would be expected on the basis of dipolar forces alone. These forces are called hydrogen bonds. Note the anomalously high boiling point of water.

|         | $H_2O$ | : | $H_2S$ | $H_2Se$ | $H_2Te$ |
|---------|--------|---|--------|---------|---------|
| Mol.wt. | 18     | : | 34     | 84      | 130     |
| b.p.    | 100 °C | : | –60 °C | –41 °C  | –2 °C   |

B. A *hydrogen bond* is not a covalent bond, but is rather an interaction of the partially positive hydrogen of the *donor* molecule with the lone pair of nonbonded electrons of a N, O, or F of the *acceptor* molecule. Hydrogen bonds are usually represented by a dashed line.

C. Hydrogen bonds are limited primarily to those molecules containing N, O, or F, and are very important in stabilizing the structures of important biomacromolecules such as proteins, nucleic acids, and polysaccharides.

### 7.5 Dispersion Forces
A. Nonpolar compounds also have attractive intermolecular forces due to momentary induced dipoles that arise from the motions of electrons about the nuclei of all compounds.
B. These transient attractive forces, called *dispersion forces* or *London forces*, are fairly weak but are present in all molecules and increase as the size and number of electrons of the molecule increase.

|                    | $F_2$ | $Cl_2$ | $Br_2$ | $I_2$ |
|--------------------|-------|--------|--------|-------|
| Molecular weight:  | 38    | 71     | 160    | 254   |
| Physical state:    | gas   | gas    | liquid | solid |

### 7.6 The Liquid State
A. Molecules of a liquid are in constant motion but close together; thus they can be compressed only slightly and diffuse much more slowly than gases.
B. Properties of liquids
 1. *Viscosity* – resistance to flow; viscosity is less for small symmetrical molecules and also decreases with increasing temperature.
 2. *Surface tension* – Surface molecules of liquids with strong intermolecular forces tend to form "beads" to minimize the surface area.

### 7.7 From Liquid to Gas: Vaporization
A. Terms
 1. *Vaporization* – the conversion of a liquid to vapor; heat is required to vaporize a liquid to vapor.
 2. *Evaporation* – the conversion of a liquid to a vapor upon exposure to air.
 3. *Condensation* – the conversion of vapor to liquid; heat is released during condensation.

# Chapter 7 - Liquids and Solids

  4. *Equilibrium* – a dynamic state of a system in which opposing processes are occurring at the same rate.
  5. *Boiling point* – the temperature at which the vapor pressure of a liquid equals atmospheric pressure.
  6. *Normal boiling point* – the temperature at which a liquid boils under standard pressure (1 atm).
  7. *Distillation* – the separation of volatile components by boiling off, and condensing the vapor.
  8. *Molar heat of vaporization* – the quantity of heat required to vaporize 1 mole of a liquid at constant pressure; for water, this quantity is 540 cal/g or 9.7 kcal/mol.
 B. A liquid placed in a closed container will evaporate until it reaches its vapor pressure at that temperature. No further net evaporation takes place because the rate of vaporization equals the rate of condensation. The system is in *dynamic equilibrium*.

7.8 The Solid State
 A. Molecules of a solid are close together but vibrate in fixed positions.
 B. Particles in *crystalline* solids are arranged in regular patterns. The *unit cell* is the small, repeating segment of a *crystal lattice*. Simple cubic, *body-centered cubic*, and *face-centered cubic* are three common unit cell types found in crystalline solids.
 C. Classes of solids
  1. *Ionic solids* – high melting points, hard; e.g., NaCl
  2. *Molecular solids* – low melting points; e.g., paraffin
  3. *Covalent network* – atoms linked by covalent bonds in three dimensions, forming one gigantic "molecule"; e.g., diamond
  4. *Metallic solids* – malleable, good conductors, in metallic bonding a sea of electrons surround the metal ions; e.g., copper

7.9 From Solid to Liquid: Melting (Fusion)
 A. As the temperature is increased, most solids attain sufficient energy to overcome the forces that hold them in the solid lattice and they melt to form a liquid. The temperature at which this takes place is called the *melting point*.
 B. The molar *heat of fusion* is the heat energy input required to convert 1 mole of a solid to liquid at its normal melting point. This quantity is generally less than the molar heat of vaporization because the difference in intermolecular forces between solid and liquid is less than it is between liquid and gas. The heat of fusion for water is 80 cal/g or 1.44 kcal/mol.
 C. A few volatile solids pass directly from the solid state to the vapor state ($CO_2$, or "Dry Ice"). This process is called *sublimation*.

7.10 Water: A Most Unusual Liquid
 A. Physical state – Most substances of low molecular weight (~18) are gases at room temperature. Water is a liquid because of its strong intermolecular forces.
 B. *Density* – The density of ice at 4 °C is 0.917 g/cm$^3$, less than the 1.00 g/cm$^3$ of water at 4 °C. Thus ice floats and northern lakes do not freeze solid. This same property of ice expansion can cause car radiators and living cells to rupture upon freezing. Flash-freezing avoids this growth of large ice crystals.
 C. *Specific heat* – Water has a high specific heat, 1.0 cal/(g-°C), and thus resists rapid changes in temperature. The water in our bodies and the oceans of Earth act as thermostats to moderate temperature variations.

# Chapter 7 - Liquids and Solids

D. Heat of vaporization – Water has a very high heat of vaporization. Thus, large amounts of body heat can be dissipated by small amounts of perspiration.

E. Polar solvent – Water is a polar molecule capable of dissolving polar and ionic substances. A typical cell is nearly 65% water. Water is the "solvent of life."

## DISCUSSION

Many new terms were introduced in Chapter 7. A knowledge of these is essential to an understanding of the concepts introduced in the chapter. Take time now to review the list of terms in the KEY WORDS section. Each key word is identified in the SUMMARY. Numerical problems in this chapter deal with energy changes associated with changes of state and changes of temperature. Energy must be supplied to a system to convert a solid to a liquid, to convert a liquid to a gas, or to raise the temperature of a gas or a liquid or a solid. The reverse changes (liquid to solid, gas to liquid, or decreases in temperature) occur if energy is removed from the system.

The system gains or loses energy in definite stages. You can raise the temperature of a solid by supplying it with energy only until it reaches the melting point. Then any energy you add will be used up in converting solid to liquid. While this change is occurring, the temperature stays the same (at the melting point). As soon as the solid is completely converted to a liquid, additional energy will be used to increase the temperature of the liquid. Eventually the boiling point of the liquid is reached. Then the temperature remains constant while additional energy is used to convert liquid to gas. When this conversion is complete, the temperature of the gas can be raised by supplying more energy.

To calculate the energy needed for each of these changes, you need the proper heat constant.

| Change Effected | Constant Used |
|---|---|
| Increase or decrease temperature of solid | Specific heat of solid |
| Conversion of solid to liquid or vice versa | Heat of fusion |
| Increase or decrease in temperature of liquid | Specific heat of liquid |
| Conversion of liquid to gas or vice versa | Heat of vaporization |
| Increase or decrease in temperature of gas | Specific heat of gas |

A change that starts with a solid and ends with a gas should simply be treated as a multistep calculation (see Example 7.8 in the text). Problems at the end of the chapter offer practice in dealing with these energy changes. A few more are offered for additional practice. Since the energy changes involving water are so important to life on this planet, all of the additional problems deal with this substance. The pertinent constants are:

| | | |
|---|---|---|
| **Specific heat** of ice | = | 0.50 cal/g/°C |
| **Specific heat** of water | = | 1.0 cal/g/°C |
| **Specific heat** of steam | = | ~0.5 cal/g/°C |
| **Heat of fusion** of water | = | 80 cal/g |
| **Heat of vaporization** of water | = | 540 cal/g |

## Chapter 7 - Liquids and Solids

**Problems**

1. 200 g of water at 20 °C was heated to 40 °C. How much energy was required?
2. A sample of water weighing 80 g cooled from 100 °C to 25 °C. How much heat was released to the surroundings?
3. How many calories are required to convert 15 g of water to steam at 100 °C?
4. How many calories are required to convert 15 mL of water, originally at 20 °C, to steam at 100 °C?
5. Approximately how much heat is removed from the body in one day by the evaporation of 500 mL of perspiration?
6. How many calories are required to convert 1 mol of ice to water at 0 °C?
7. If 1 g of steam at 120 °C is condensed, cooled, and converted to ice at −20 °C, how much energy is released to the surroundings?
8. The types of bonding in column B are responsible for maintaining the lattice structure of the materials in column A. Choose from column B the type of bonding that plays the <u>most</u> significant role for each item in column A.

| Column A | Column B |
|---|---|
| a. ___ Fe | 1. ionic bonding |
| b. ___ $H_2$ | 2. dipole interactions |
| c. ___ $CO_2$ | 3. hydrogen bonding |
| d. ___ HCN | 4. dispersion forces |
| e. ___ $NH_3$ | 5. metallic bonding |
| f. ___ KCl | |

## SELF-TEST

1. Predict which of these compounds would be a gas at STP.
   a. Au    b. $CH_3-CH_2-CH_3$    c. Mg    d. NaCl    e. $I_2$
2. Which of the following intermolecular forces are weakest?
   a. ion-ion    b. dipole-dipole    c. hydrogen bonding    d. London/dispersion
3. Dipole-dipole interactions are _____ hydrogen bonds.
   a. stronger than    b. weaker than    c. equal in strength to
4. In general, which is the strongest type of intermolecular bonding?
   a. ionic bonding    b. dipole interactions    c. hydrogen bonding
5. At room temperature ionic compounds exist as:
   a. solids    b. liquids    c. gases
6. For which compound would you expect the interionic forces to be strongest?
   a. LiF    b. BeO    c. BN
7. For which compound would the dispersion forces be greatest?
   a. $Cl_2$    b. $Br_2$    c. $I_2$
8. Which type of force would not be operating in a solid sample of HCl?
   a. dipolar forces    b. interionic forces    c. dispersion forces
9. Which represents the best arrangement for a pair of dipoles?

   a. (+  −) (+  −)   b. (+  −) (−  +)   c. (+  −) / (−  +)   d. (+  −) / (+  −)

10. For which compound is hydrogen bonding a significant attractive force?
    a. F−N=O    b. Cl−N=O    c. H−N=O    d. H−C≡C−H

11. Which of the following correctly illustrates hydrogen bonding?

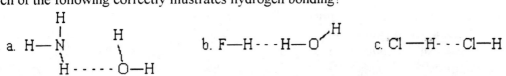

12. Which compound would you expect to have the highest melting point?
    a. $C_3H_8$    b. $C_8H_{18}$    c. $C_{18}H_{38}$
13. Which compound would you expect to have the strongest hydrogen bonding in the liquid state?
    a. $CH_3$–O–$CH_3$    b. $CH_3$–OH    c. $CH_3$–$CH_3$    d. $CH_3$–CO–$CH_3$ (acetone)
14. If compound A has a lower vapor pressure than compound B at a given temperature, which compound would be expected to have the lower boiling point?
    a. A    b. B    c. The boiling point would be the same at 760 mmHg.
15. Which noble gas would have the highest boiling point?
    a. He    b. Ne    c. Ar    d. Kr    e. Xe
16. The atmospheric pressure is 670 Torr. Water will boil at:
    a. less than 100 °C    b. exactly 100 °C    c. more than 100 °C
17. The vapor pressures of liquids X, Y, and Z at 75 °C are 300, 644, and 1126 Torr, respectively. Which of these, if any, has a normal boiling point below 75 °C?
    a. X    b. Y    c. Z    d. all of these    e. none of these    f. X and Y
18. All of the following are gases at room temperature. Which one will liquefy most easily when pressurized at a fixed temperature?
    a. $O_2$    b. $NH_3$    c. $SiH_4$    d. $F_2$    e. $H_2$
19. Which of the following processes is endothermic?
    a. cooling    b. condensation    c. solidification    d. melting
20. The boiling points of water, ethanol, diethylether, and chloroform are 100 °C, 78.5 °C, 28 °C, and 119 °C. Which liquid would you predict to have the lowest vapor pressure at room temperature?
    a. water    b. ethanol    c. diethyl ether    d. chloroform
21. If a liquid and its vapor are at equilibrium in a closed container and the temperature is increased, which rate increases?
    a. rate of vaporization    b. rate of condensation    c. both    d. neither
22. In general, which is expected to be larger for a given substance?
    a. molar heat fusion    b. molar heat vaporization    c. specific heat
23. Which would be expected to have the highest heat of vaporization?
    a. $CH_4$    b. $H_2O$    c. LiF
24. When is the viscosity of a lubricating oil highest?
    a. at room temperature    b. at 150 °C    c. at –15 °C
25. If you start with 1 g of ice at 0 °C and add 120 calories of heat energy, the temperature of the water will be:
    a. 0 °C    b. 10 °C    c. 20 °C    d. 30 °C    e. 40 °C    f. 80 °C    g. 100 °C
26. The molar heat capacity of some molecule (X, liquid) is 10 cal/K-mol, its heat of vaporization is 5000 cal/mol, and its normal boiling point is 75 °C. How many calories are required to convert 1 mol of X at 60 °C to 1 mol of X vapor at 75 °C at 1 atm pressure?
    a. 5150    b. 4850    c. 150    d. 5650    e. 4750
27. Refer to the following specific heats:
    For X = 0.1 cal/g-°C    For Y = 0.4 cal/g-°C    For Z = 0.8 cal/g-°C
    a. Which material has the highest heat capacity?    X    Y    Z
    b. If all are supplied with the same amount of heat, which will reach the highest temperature?
        X    Y    Z

## Chapter 7 - Liquids and Solids

28. Which property of water is regarded as unusual?
    a. physical state at room temperature
    b. the relative density of solid and liquid
    c. heat of vaporization
    d. specific heat
    e. all of these
    f. none of these

## ANSWERS

**Problems**

1. 4 kcal    2. 6.0 kcal    3. 8.1 kcal    4. 9.3 kcal;
5. 270 kcal    6. 1.44 kcal    7. 0.74 kcal
8. a. 5    b. 4    c. 4    d. 2    e. 3    f. 1

**Self-Test**

| | | | |
|---|---|---|---|
| 1. b | 8. b | 15. e | 22. b |
| 2. d | 9. c | 16. a | 23. c |
| 3. b | 10. c | 17. c | 24. c |
| 4. a | 11. a | 18. b | 25. e |
| 5. a | 12. c | 19. d | 26. a |
| 6. c | 13. b | 20. d | 27. Z, X |
| 7. c | 14. b | 21. c | 28. e |

# 8 Solutions

## KEY WORDS

| | | | |
|---|---|---|---|
| solution | osmosis | deliquescent | hypotonic |
| homogeneous | dilute | dynamic | plasmolysis |
| solute | concentrated | equilibrium | hemolysis |
| solvent | concentration | molarity | hypertonic |
| hydration | saturated | colligative property | crenation |
| hydrate | supersaturated | unsaturated | colloid |
| aqueous solution | anhydrous | precipitate | Tyndall effect |
| miscible | efflorescent | insoluble | emulsion |
| soluble | hygroscopic | isotonic | dialysis |
| osmolality | osmotic pressure | boiling pt. elevation | freezing pt. depression |
| osmol | mass/volume % | colloidal dispersions | emulsifying agent |
| reverse osmosis | % by mass | % by volume | semipermeable membrane |

## SUMMARY

8.1 Some Types of Solutions
  A. *Solution* – a *homogeneous* mixture of two or more substances
  B. *Solute* – the substance being dissolved; minor component
  C. *Solvent* – the substance doing the dissolving; major component
  D. *Aqueous solution* – solution in which the solvent is water

8.2 Qualitative Aspects of Solubility
  A. Miscible vs. Insoluble
    1. Substances are *miscible* when they can be mixed in all proportions.
    2. Substances are said to be *insoluble* when their solubility is very low, near zero.
  B. Dilute vs. Concentrated
    1. *Dilute* solutions contain relatively little solute.
    2. *Concentrated* solutions contain a great deal of solute.

8.3 Solubility of Ionic Compounds
  A. The dissolution of salts requires energy for the breakdown of the crystal lattice and separation of solvent molecules while energy is released by the solvation of ions. Substances are most *soluble* when the energy released exceeds the energy absorbed.
  B. *Hydration* is the process in which water molecules surround the solute. The unique structure of water, being a polar molecule with hydrogen-bonding capabilities, enables water to dissolve many ionic compounds. Most compounds of $Na^+$, $K^+$, $NH_4^+$, and $NO_3^-$ are soluble, while most compounds of $PO_4^{3-}$ and $CO_3^-$ salts are not. (See Table 8.2 in the text.)

8.4 Solubility of Covalent Compounds
  A. "Like Dissolves Like" – Nonpolar substances such as fats and oils are normally readily soluble in nonpolar solvents such as carbon tetrachloride or benzene.

## Chapter 8 - Solutions

    B. The solubility of polar substances in a polar solvent such as water depends on its size and ability to form hydrogen bonds with water. For polar organic (carbon-based) compounds, the solubility decreases appreciably when the ratio of C to O plus N exceeds 4:1.
    C. Other terms
        1. *Hydration* – interaction of water with a solute
        2. *Hydrate* – crystalline compound with bound water
        3. *Anhydrous* – "dry" compound; hydrate minus water
        4. *Efflorescent* – compounds that become anhydrous on standing in air
        5. *Hygroscopic* – compounds that form hydrates on standing in air
        6. *Deliquescent* – compounds that are so hygroscopic that they dissolve in accumulated water on standing in air

8.5 Dynamic Equilibria
    A. A solution that contains all of the solute that it can at a given temperature is said to be *saturated*; the solution is *unsaturated* when it contains less than it can contain. At the saturation point there must be some solid *precipitate* in *dynamic equilibrium* with the solution so the rate of dissolution is just equal to the rate of precipitation.
    B. A *supersaturated* solution is not at equilibrium; it is a metastable solution that contains solute in excess of what it would contain if it were at equilibrium.

8.6 Solutions of Gases in Water
    A. Aqueous solutions of gases are fairly common, i.e., blood, "soda pop," ammonia cleansers, etc.
    B. Unlike most solids, the solubility of a gas decreases with increasing temperature but increases with increasing pressure.

8.7 Solution Concentrations
    A. *Dilute* and *concentrated* are relative terms. Chemists need more quantiative ways to describe solutions. The *concentration* of a solution expresses a ratio of the amount of *solute per amount of solution*. The amount of solute can be expressed in units of grams, mL or moles.
    B. *Molarity* (M) = moles of solute per liter of solution   M = [# moles / V (L)];  # moles = M × V(L)
    C. Percent Concentration
        1. *% by volume*     (v/v)%    =    $\frac{\text{vol. of solute}}{\text{vol. of solution}} \times 100$
        2. *% by mass*     (w/w)%    =    $\frac{\text{g of solute}}{\text{g of solution}} \times 100$
        3. *mass/volume percent*   mg %    =    $\frac{\text{mg solute}}{100 \text{ mL (or dL) solution}} \times 100$
        4. Parts per billion     (ppb)    =    $\frac{\text{mg of solute}}{\text{kg of solution}}$
           (1 cent/$10 million)
        5. Ratio concentrations
           a. Many liquids (drugs, insecticides, etc.) are purchased in concentrated form and must be diluted before use.
           b. Dilution formula: $C_s \times V_s = C_d \times V_d$;   $C_s$ = concentration of stock solution;
                                                                    $C_d$ = concentration of dilute solution.

8.8 Colligative Properties of Solutions
    A. *Colligative properties* are properties such as *boiling point elevation, freezing point depression, and osmotic pressure* that depend more on the number rather than the type of particles in solution.

### Chapter 8 - Solutions

B. Solutions have lower vapor pressures than the pure solvent. Consequently, the boiling points of solutions are elevated (0.51 °C/molal for water).
C. The freezing point of a solution is depressed (−1.86 °C/molal for water).
D. *Osmolality* = osmol/L where an *osmol* is a mole of solute particles.

8.9 Solutions and Cell Membranes: Osmosis
  A. The molecular sieve theory of *osmosis* holds that a *semipermeable membrane* has pores large enough to permit the passage of solvent water molecules but too small to permit passage of larger molecules.
  B. During osmosis, solvent will continue to flow through the semipermeable membrane from the dilute region to the more concentrated region until the *osmotic pressure* builds up on the concentrated side to halt the flow of solvent.
  C. Solutions such as 0.89% sodium chloride (0.16M) or 5.5% glucose (0.31M) are said to be *isotonic* since they exert the same osmotic pressure as cellular fluids. When a cell is placed in a solution with a lower osmotic pressure (*hypotonic*), it may rupture (*plasmolysis* or *hemolysis*); and when it is placed in a solution of higher osmotic pressure (*hypertonic*), it may shrivel (*crenation*).
  D. *Reverse osmosis* can take place by using high pressure (> osmotic pressure) to force solvent through a semipermeable membrane and out of solution. This process can be used to produce drinking water from sea water.

8.10 Colloids
  A. *Colloids* are large particles (about 0.05 to 0.25 nm in diameter) that form *colloidal dispersions*, which are between true homogeneous solutions (sugar water) and heterogeneous suspensions (sand in water).
  B. Colloids do not settle out on standing nor can they be separated by filtering through filter paper. However, colloids can be observed by the scattering of a beam of light as it passes through a colloidal dispersion, a process known as the *Tyndall effect* (after John Tyndall, 1869).
  C. The most important colloids in biological systems are *emulsions* in which either liquids or solids are dispersed in water. Substances such as soap and bile salts that stabilize emulsions are called *emulsifying agents*.

8.11 Dialysis
  A. *Dialysis* refers to the process whereby small molecules and ions pass through a semipermeable membrane (dialyzing membrane) from areas of higher concentration to areas of lower concentration.
  B. The kidneys are an example of a complex dialyzing system responsible for the removal of high concentrations of toxic waste products from the blood.

# Chapter 8 - Solutions

## DISCUSSION

Aqueous solutions are very important in chemistry and biology. The material in this chapter can be divided into three main areas. Sections 1-6 deal with the concept of solubility and the many qualitative terms that are used to describe solutions. Section 7 treats the quantitative aspects of solutions. Finally, Sections 8-11 cover the biological importance of colloids, membranes, and the processes of osmosis and dialysis.

The terms soluble/insoluble and miscible/immiscible have similar meanings and are just a few of the many qualitative terms used to describe solutions. To say something is soluble simply means that a significant amount of the solute dissolves in the solvent, whereas miscible implies that solute and solvent can be mixed in all proportions. If one wishes to be somewhat more precise in describing the amount of solute dissolved in a particular solvent, the terms unsaturated, saturated, and supersaturated can be employed. Except for saturated, these terms, too, are quite imprecise. An unsaturated solution can be further described as being dilute or concentrated, depending on how close the solution is to saturation.

If a more precise description of a solution is required, a quantitative measure of concentration must be used. The concentration of a solution always expresses a ratio of the amount of solute per a given amount of solvent or solution present. There are many terms used to express concentration, depending on the units used to express the amount of solute and solvent or solution present. The amount of solute is usually expressed in either grams or volume. Some of the most common definitions for concentration used in this text are:

**Molarity (M)** = $\dfrac{\text{moles of solute}}{\text{liter of solution}}$   where moles of solute = $\dfrac{\text{g of solute}}{\text{formula wt. solute}}$

**Volume (v/v) %** = $\dfrac{\text{volume of solute}}{\text{volume of solution}} \times 100$

**Weight (w/w) %** = $\dfrac{\text{g of solute}}{\text{g of solution}} \times 100$

Two important differences between molarity and percent concentrations should be noted. First, molarity expresses the amount of solute in moles. Therefore, the formula of the solute must be known and its formula weight used in calculating molarity. On the other hand, when using percent concentrations the nature of the solute can be ignored since it is only the gram or volume amount of solute that is needed for the calculation. A second difference between molarity and percent concentrations lies in the size of the solution sample referred to. A 1.0 molar solution refers to 1 mole per liter, whereas a 1 % solution contains 1 g per 100 g or 1 mL per 100 mL of solution. The following example illustrates the relationship of these two forms of expressing concentration.

$$1 \text{ M NaOH} = \dfrac{1 \text{ mol NaOH}}{1 \text{ L solution}} = \dfrac{40 \text{ g NaOH}}{1000 \text{ mL soln.}} = \dfrac{4 \text{ g NaOH}}{100 \text{ mL soln.}} \sim \dfrac{4 \text{ g NaOH}}{100 \text{ g soln.}}$$

⇑ Molarity = 1                                                                 ⇓ (w/w)% = 4

Note that all of the above refer to the amount of solution in the denominator. It is also possible to express concentration in terms of the amount of solvent in the denominator. In fact, most chemical handbooks express saturation limits in terms of "grams of solute"/100 g solvent.

The last part of this chapter deals with colloids, osmosis and dialysis. The subtleties that distinguish osmosis from dialysis often present difficulties. In both osmosis and dialysis there is the tendency for the concentrations on either side of the semipermeable membrane to equalize. In dialysis this involves the flow of small molecules or ions as they diffuse from a region of higher solute concentration through the membrane to a region of lower solute concentration. In osmosis, it is the flow of solvent that is involved. There is a net flow of solvent (water) from a region of higher solvent concentration (lower solute concentration and lower osmotic pressure) to a region of lower solvent concentration (higher solute concentration and higher osmotic pressure). Thus, the net flow of water is from a solution of low osmotic pressure to a solution of high osmotic pressure. These points are summarized in the following diagram:

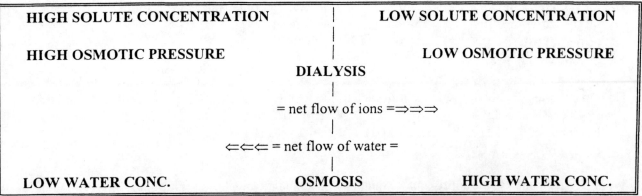

The quantitative treatment of solutions is best learned by working out problems. Remember that concentration by itself does not say anything about the amount of solute but always expresses a ratio of the amount of solute to a given amount of solvent or solution. The two most common types of quantitative problems dealing with concentrations are:
1. Calculate the concentration given the amounts of solute and solution.
2. Calculate the amount of solute needed for a given volume and concentration of the solution.

The problems at the end of Chapter 9 in the text will help improve your understanding of this material.

## Problems

1. Calculate the molarity of the solution that results if the stated amount of solute is dissolved in 1 L of solution.
   a. 29.2 g NaCl
   b. 49 g $H_2SO_4$
   c. 74.5 g $(NH_4)_3PO_4$
   d. 4.0 g NaOH
   e. 80 g of $CaBr_2$

2. Calculate the molarity (volumes given are of the solution).
   a. 5.85 g NaCl in 2 L
   b. 4.9 g of $H_2SO_4$ in 50 mL
   c. 74.5 g of $(NH_4)_3PO_4$ in 10 L
   d. 0.4 g of NaOH in 50 mL
   e. 40 g of $CaBr_2$ in 0.4 L

3. How many moles of solute are present in each of the following?
   a. 1 L of 2 M NaCl
   b. 10 mL of 0.1 M $H_2SO_4$
   c. 10 L of 0.01 M $(NH_4)_3PO_4$
   d. 50 mL of 5 M NaOH
   e. 5 mL of 0.64 M $CaBr_2$

4. How many grams of solute are present in 1 L of each of the following solutions?
   a. 2 M NaCl
   b. 0.1 M $H_2SO_4$
   c. 0.01 M $(NH_4)_3PO_4$
   d. 5 M NaOH
   e. 0.64 M $CaBr_2$

## Chapter 8 - Solutions

5. How many grams of solute are there in each of the following?
   a. 50 mL of 2 M NaCl
   b. 2 mL of 0.1 M $H_2SO_4$
   c. 10 L of 0.01 M $(NH_4)_3PO_4$
   d. 2 L of 5 M NaOH
   e. 5 mL of 0.64 M $CaBr_2$

6. How many grams of solute are present in 100 g of each of the following solutions?
   a. 10% NaCl
   b. 0.5% $H_2SO_4$
   c. 12% $(NH_4)_3PO_4$
   d. 40% NaOH
   e. 2% $CaBr_2$

7. How many grams of solute are present in each of the following solutions?
   a. 40 g of 10 % NaCl
   b. 10 mL of 0.5% $H_2SO_4$
   c. 10 g of 12% $(NH_4)_3PO_4$
   d. 1 kg of 40% NaOH
   e. 1 L of 2% $CaBr_2$

8. What are the molarity, the approximate weight percent, and the mg % concentrations of a solution in which 0.4 g of NaOH is dissolved in 100 mL of solution?

You can check your understanding of the rest of the material covered by going through the **SELF-TEST**.

## SELF-TEST

1. If a teaspoon of salt is added to a liter of water, water is the:
   a. solute     b. solvent     c. solution
2. The process of water molecules being attracted to and surrounding solute ions/particles is called:
   a. hydrogenation     b. solventation     c. salutation     d. hydrolysis     e. solvation
3. If a small amount of sugar is dissolved in a large amount of water, sugar is the:
   a. solute     b. solvent     c. solution
4. A mixture is homogeneous if:
   a. the components of the mixture are so intimately mixed that all samples of the mixture have the same composition.
   b. the mixture has a lower osmotic pressure than physiological saline.
   c. the mixture will absorb water from the air.
5. If the energy required to break the crystal lattice is greater than the energy of solvation,
   a. the solid dissolves in the solvent.          b. the solid is insoluble in the solvent.
6. Generally, if a seed crystal is added to a supersaturated solution,
   a. the crystal dissolves.
   b. the crystal causes all of the solute to precipitate from solution.
   c. solute precipitates until an unsaturated solution forms.
   d. some solute precipitates and equilibrium is established with the saturated solution.
7. An increase in temperature usually results in _____ of a solid solute.
   a. an increase in solubility          b. a decrease in solubility
   c. an increase in density             d. an increase in boiling point
8. Oil and water are:
   a. miscible     b. immiscible
9. Salts incorporating all but one of the following ions are usually soluble. Which ion is the exception?
   a. $NO_3^-$     b. $CO_3^{2-}$     c. $Na^+$     d. $NH_4^+$

## Chapter 8 - Solutions

10. Which of the following compounds would you expect to be soluble in water?
    a. $CaCO_3$  b. $PbSO_4$  c. $HgS$  d. $(NH_4)_3PO_4$
11. Salts incorporating all but one of the following ions are usually insoluble. Which ion is the exception?
    a. $CO_3^-$  b. $PO_4^{3-}$  c. $NO_3^-$  d. $S^{2-}$
12. Which term is used to describe a compound that loses water on standing in dry air?
    a. efflorescent  b. deliquescent  c. hygroscopic
13. Which is least likely to dissolve in water?
    a. ionic compound  b. polar compound  c. nonpolar compound
14. The compound $Na_2CO_3 \cdot 7 H_2O$ is a(an):
    a. hydroxide  b. hydrate  c. anhydrous compound
15. Which is generally not true? An increase in temperature increases:
    a. the rate of a reaction
    b. the solubility of a salt in water
    c. the solubility of a gas in water
16. Which compound would not be expected to dissolve in water?
    a. $CH_2CH_2CH_3$       b. $CH_3CH_2CH_2$       c. $CH_3CH_2CH_3$       d. $CH_2CH_2CHCH_2CH_2$
       $\phantom{xx}|$                                       $\phantom{xx}|$                                                                                  $\phantom{xx}|\phantom{xxx}|\phantom{xxx}|$
       OH                                       $NH_2$                                                                          OH   OH   OH
17. The solubility of sodium chloride in water is 36% at 20 °C. At the same temperature, therefore, a solution of 30 g NaCl in 1 liter of water would be classified as:
    a. dilute  b. concentrated  c. saturated  d. supersaturated
18. If a 36% solution of NaCl is saturated, and if 4 g of NaCl is used to prepare 10 mL of a solution at the same temperature, the resulting solution would be:
    a. dilute  b. concentrated  c. saturated
19. If a solution is saturated when its concentration is 5%, at the same temperature 200 g of the solution would be supersaturated if it contained _____ grams of solute.
    a. 2.5  b. 5  c. 10  d. 12
20. If the concentration of sodium hydroxide is 10%, then the amount of NaOH in 200 mL of solution is:
    a. 2 g  b. 5 g  c. 20 g  d. 50 g  e. 100 g  f. 200 g
21. In 400 mL of a 0.25 M solution of NaOH there are how many grams of NaOH?
    a. 2  b. 5  c. 10  d. 40  e. 4  f. 500
22. How many millimoles of HCl are there in 100 mL of 0.01 M HCl solution?
    a. 10  b. 1  c. $10^{-1}$  d. $10^{-2}$  e. $10^{-4}$  f. $10^{-5}$
23. What is the formula weight of a compound if 20 g of the compound in 500 mL of solution gives a 0.5 M solution?
    a. 10  b. 20  c. 40  d. 50  e. 60  f. 80
24. What is the formula weight of a compound if 20 g of the compound in 300 mL of solution gives a 0.2 M solution?
    a. 200  b. 600  c. 167  d. 333  e. 667  f. 100
25. How many grams of $(NH_4)_2SO_4$ are needed to prepare 100 mL of a 2.5M solution?
    a. 13.2  b. 26  c. 132  d. 33  e. 66  f. 100
26. You have only 30 g of $CaBr_2$ and want to prepare a 0.20M solution. How many liters of the solution can you prepare with your 30 g of $CaBr_2$.
    a. 1.0 L  b. 0.75 L  c. 0.50 L  d. 0.44 L  e. 0.20 L
27. A solution containing 100 mg of solute in 1 L of solution has a concentration of:
    a. 1 mg%  b. 10 mg%  c. 100 mg%  d. 1000 mg%
28. Which equality is correct?
    a. 1 ppm = 1000 ppb  b. 1000 ppm = 1 ppb

Chapter 8 - Solutions

29. A solution that has been diluted 1:3 is less concentrated than the solution that has been diluted:
    a. 1:2       b. 1:30       c. 1:3000
30. Compared with pure water, a solution containing one mole of sugar per kg of water:
    a. melts at and boils at a higher temperature.
    b. melts at and boils at a lower temperature.
    c. melts higher and boils lower.
    d. melts lower and boils higher.
31. The freezing point in °C of a solution that contains 2 moles of sugar per kg of water is:
    a. –2   b. 2   c. 80   d. –80   e. –3.72   f. 1.02   g. –1.02
32. Which solution has the higher osmotic pressure?
    a. 0.1 M NaCl       b. 0.5 M NaCl
33. Which aqueous solution has the highest boiling point?
    a. 0.7 M NaCl    b. 0.6 M $CaCl_2$    c. 0.4 M $Al_2(SO_4)_3$    d. 0.8 M $NH_4NO_3$
34. Which of the following aqueous solutions has the lowest freezing point?
    a. 1.0 M NaCl    b. 1.2 M $NH_4NO_3$    c. 1.0 M $CaCl_2$    d. 0.8 M $AlCl_3$
35. A solution contains particles of three sizes (., o, O ). If two of these types of particles pass through a dialyzing membrane, the two must be:
    a. . and o       b. . and O       c. o and O
36. In both dialysis and osmosis, which particles do not pass through the membrane?
    a. water       b. small molecules       c. colloids
37. The net flow of water through a semipermeable membrane as equilibrium is approached, is:
    a. from a solution of higher osmotic pressure to one of lower osmotic pressure
    b. from a solution of lower osmotic pressure to one of higher osmotic pressure
38. If two solutions with concentrations of 0.1 M sugar and 0.5 M sugar respectively are separated by a semipermeable membrane, during osmosis there is a net flow of:
    a. sugar molecules from the 0.1 M solution to the 0.5 M solution
    b. sugar molecules from the 0.5 M solution to the 0.1 M solution
    c. water molecules from the 0.1 M solution to the 0.5 M solution
    d. water molecules from the 0.5 M solution to the 0.1 M solution
39. Red blood cells undergo hemolysis if placed in a solution that has a _____ osmotic pressure than the solution inside the blood cells.
    a. higher       b. lower
40. Crenation of cells occurs when the cells are placed in a(an) _____ solution.
    a. hypotonic    b. isotonic    c. hypertonic    d. physiological saline
41. Which is not true of a colloidal dispersion?
    a. Particles may have a diameter of 10 nm.
    b. The dispersion exhibits the Tyndall effect.
    c. The dispersion does not settle out on standing.
    d. The dispersed particles can be filtered by slowly passing the dispersion through filter paper.
    e. All of the above are true of a colloidal dispersion.
42. Which type of mixture cannot exist as a colloidal dispersion?
    a. solid in solid       b. liquid in liquid       c. gas in gas
43. An emulsifying agent is used to:
    a. bring about precipitation of solute from a true solution
    b. stabilize a colloidal dispersion
    c. filter the particles of a suspension

44. In the Tyndall effect one observes:
    a. the precipitation of solute triggered by addition of a seed crystal
    b. particles slowly settling out of a suspension
    c. a beam of scattered light
45. Blood sugar levels are about 100 mg glucose ($C_6H_{12}O_6$ )/100 mL blood or 0.10%. If a person has 6.0 L of blood, how many grams of glucose does this represent? _____ How many ounces of glucose?
46. Associate the items on the right with the best term on the left.
    a.__ colligative property          1. oil and vinegar dressing
    b.__ Tyndall effect                2. deicing roadways
    c.__ osmotic presssure             3. ice cream
    d.__ dialysis                      4. kidney machine
    e.__ colloid                       5. lighthouse
    f.__ immiscible                    6. cell membranes
    g.__ emulsifying agent             7. detergent

## ANSWERS

### Problems

| | | | | |
|---|---|---|---|---|
| 1. a. 0.5 M | b. 0.5 M | c. 0.5 M | d. 0.1 M | e. 0.4 M |
| 2. a. 0.05 M | b. 1 M | c. 0.05 M | d. 0.2 M | e. 0.5 M |
| 3. a. 2 mol | b. 0.001 mol | c. 0.1 mol | d. 0.25 mol | e. 0.0032 mol |
| 4. a. 117 g | b. 9.8 g | c. 1.49 g | d. 200 g | e. 128 g |
| 5. a. 5.85 g | b. 0.0196 g | c. 14.9 g | d. 400 g. | e. 0.64 g |
| 6. a. 10 g | b. 0.5 g | c. 12 g | d. 40 g | e. 2 g |
| 7. a. 4 g | b. ~0.05 g | c. 1.2 g | d. 400 g | e. ~20 g |

8. 0.1 M, 0.4%, 400 mg%

### Self-Test

| | | | | |
|---|---|---|---|---|
| 1. b | 10. d | 19. d | 28. a | 37. b |
| 2. e | 11. c | 20. c | 29. a | 38. c |
| 3. a | 12. a | 21. e | 30. d | 39. b |
| 4. a | 13. c | 22. b | 31. e | 40. c |
| 5. b | 14. b | 23. f | 32. b | 41. d |
| 6. d | 15. c | 24. d | 33. c | 42. c |
| 7. a | 16. c | 25. d | 34. d | 43. b |
| 8. b | 17. a | 26. b | 35. a | 44. c |
| 9. b | 18. c | 27. b | 36. c | 45. 6, 0.21 |
| | | | | 46. a. 2 b. 5 |
| | | | | c. 6 d. 4 |
| | | | | e. 3 f. 1 |
| | | | | g. 7 |

# 9    Acids and Bases I

**KEY WORDS**

| | | | |
|---|---|---|---|
| *acid* | *weak acids* | *hydroxide ion* | *antacid* |
| *Arrhenius* | *ortho acids* | *salt* | *Brönsted-Lowry* |
| *hydronium ion* | *meta acid* | *neutralization* | *conjugate acid* |
| *strong acid* | *base* | *acid rain* | *conjugate base* |
| *proton donor* | *proton acceptor* | *acidic anhydride* | *alkalosis* |
| *carboxlic acid* | *amines* | *basic anhydride* | *net ionic equation* |
| *strong base* | *weak base* | *polyprotic acid* | *monoprotic acid* |
| *diprotic acid* | *triprotic acid* | | |

**SUMMARY**

9.1 Acids and Bases: Definitions and Properties
   A. Characteristic properties of *acids*.
      1. Taste sour
      2. Produce a stinging sensation
      3. Turn blue litmus red
      4. React with active metals to produce $H_2$ gas
      5. React with alkalis or bases to produce water and a salt
   B. Characteristic properties of *bases*
      1. Taste bitter
      2. Feel slippery
      3. Turn litmus from red to blue
      4. React with acids to produce water and salt
   C. *Arrhenius* (1887) proposed that acid properties are actually the properties of hydrogen ions ($H^+$) and defined acids as compounds that yield hydrogen ions in aqueous solutions. Arrhenius defined bases as compounds that yield *hydroxide ions* ($OH^-$) in aqueous solutions. *Neutralization* refers to the reaction of an acid and a base to form water ($H^+ + OH^- \rightarrow H_2O$).
   D. The *hydronium ion* ($H_3O^+$) is used to represent the hydrated proton [$H(H_2O)_4^+$ would be better but more cumbersome].
   E. *Brönsted-Lowry* defined an acid as a *proton donor* and a base as a *proton acceptor*. Thus all acids exist as *conjugate acid/conjugate base* (or c.a./c.b.) pairs.
      1. c.a. = c.b. + $H^+$
      2. The conjugate base of a strong conjugate acid (HCl) will be a weak base ($Cl^-$). Similarly, the c.b. of a weak c.a. ($HPO_4^{2-}$) will be a strong base ($PO_4^{3-}$).
      3. $H_2PO_4^- + OH^- \rightarrow HPO_4^{2-} + H_2O$
          c.a.      c.b.       c.b.       c.a.
          stronger pair      weaker pair

9.2 Strong and Weak Acids
   A. *Strong acids* are those that ionize completely, or nearly so, in aqueous solutions.
        $HCl + H_2O \rightarrow H_3O^+ + Cl^-$

# Chapter 9 - Acids and Bases I

B. **Weak acids** ionize only slightly in aqueous solutions. (A 1.0 M acetic acid solution is only 0.42% dissociated.)

$$CH_3COOH + H_2O \rightleftharpoons CH_3COO^- + H_3O^+$$

C. Some common acids
1. *Monoprotic*: HCl, HNO$_3$, CH$_3$COOH (acetic acid)
2. *Diprotic*: H$_2$SO$_4$, H$_2$CO$_3$
3. *Triprotic*: H$_3$PO$_4$, C$_3$H$_4$OH(COOH)$_3$ (citric acid)
4. *Polyprotic*: Acids with more than one ionizable proton.

D. Formulas of inorganic acids traditionally have the hydrogen first (HCl, HNO$_3$, H$_2$SO$_4$) whereas for organic acids the hydrogen is traditionally written last as part of the *carboxylic acid* (–COOH) group. However, you should keep in mind that structurally the hydrogen is bonded to the oxygen in both HNO$_3$ (as NO$_2$OH) and HCOOH.

## 9.3 Names of Some Common Acids

A. Naming:

| Acid | Anion | Acid (Example) | | Anion (Example) | |
|---|---|---|---|---|---|
| 1. hydro___ic acid | ___ide | HCl | hydro<u>chlor</u>ic acid | Cl$^-$ | <u>chlor</u>ide |
| 2. hypo___ous acid | hypo___ite | HClO | hypo<u>chlor</u>ous acid | ClO$^-$ | hypo<u>chlor</u>ite |
| 3. ___ous acid | ___ite | HClO$_2$ | <u>chlor</u>ous acid | ClO$_2^-$ | <u>chlor</u>ite |
| 4. ___ic acid | ___ate | HClO$_3$ | <u>chlor</u>ic acid | ClO$_3^-$ | <u>chlor</u>ate |
| 5. per___ic acid | per___ate | HClO$_4$ | per<u>chlor</u>ic acid | ClO$_4^-$ | per<u>chlor</u>ate |

B. **Strong acids** give up their protons readily.
1. HCl – hydrochloric acid
2. HBr – hydrobromic acid
3. HI – hydroiodic acid
4. HNO$_3$ – nitric acid
5. H$_2$SO$_4$ – sulfuric acid

C. **Weak acids** do not give up their protons readily.
1. H$_3$PO$_4$ – phosphoric acid (moderately strong)
2. HF – hydrofluoric acid
3. CH$_3$COOH – acetic acid
4. H$_2$CO$_3$ – carbonic acid
5. NH$_4^+$ – ammonium ion
6. HCN – hydrogen cyanide or hydrocyanic acid
7. H$_3$BO$_3$ – boric acid

D. **Ortho and meta acids** – Any acid with 3 or more replaceable hydrogens can have water (H$_2$O) removed to form the meta acid.
1. H$_3$PO$_4$ – phosphoric acid (ortho form)      (PO$_4^{3-}$ – phosphate ion)
2. (HPO$_3$)$_3$ – meta phosphoric acid

## 9.4 Some Common Bases

A. Arrhenius defined bases as compounds that yield *hydroxide ions* (OH$^-$) in aqueous solutions.

B. Some compounds produce OH$^-$ ions by reacting with water.

$$HOH + :NH_3 \rightarrow NH_4^+ + OH^-$$
$$H_2O + CO_3^{2-} \rightarrow HCO_3^- + OH^-$$

C. Organic *amines* (CH$_3$–NH$_2$) are related to ammonia and are common weak bases in living cells.

D. Common bases

Chapter 9 - Acids and Bases I

    1. *Strong bases*: NaOH (lye), KOH, Ca(OH)$_2$ (lime)
    2. *Weak bases*: NH$_3$, CH$_3$COO$^-$ ion, CN$^-$ ion, etc.

9.5 Acidic and Basic Anhydrides
    A. Nonmetal oxides are *acidic anhydrides*.
$$SO_3 + H_2O \rightarrow H_2SO_4$$
    B. Metal oxides are *basic anhydrides*
$$CaO + H_2O \rightarrow Ba(OH)_2$$

9.6 Neutralization Reactions
    A. An acid reacts with an equivalent amount of base in a *neutralization* reaction to produce water and a *salt*.
$$HCl\ (aq) + NaOH\ (aq) \rightarrow H_2O + NaCl\ (aq)$$
    B. Net ionic equation for neutralization
$$H_3O^+ + Cl^- + Na^+ + OH^- \rightarrow 2\ H_2O + Na^+ + Cl^-$$
$$\text{or}\quad H_3O^+ + OH^- \rightarrow 2\ H_2O$$

9.7 Reactions of Acids with Carbonates and Bicarbonates
    A. Carbonates (CO$_3^{2-}$) and bicarbonates (HCO$_3^-$) are salts of the weak and unstable carbonic acid (H$_2$CO$_3$).
        1. $H_2CO_3 + OH^- \rightarrow HCO_3^- + H_2O$
        2. $H_2CO_3 + 2\ OH^- \rightarrow CO_3^{2-} + 2\ H_2O$
    B. Carbonic acid is not very stable.
$$H_2CO_3\ (aq) \rightarrow H_2O + CO_2\ (g)$$
    C. *Acid rain* and marble statues
        1. Sulfur – containing coal produces SO$_2$.
        2. SO$_2$ reacts with O$_2$ to produce SO$_3$.
        3. SO$_3$ combines with water to form H$_2$SO$_4$ (acid rain).
        4. Limestone and marble contain CaCO$_3$.
$$CaCO_3 + H_2SO_4\ (aq) \rightarrow Ca^{2+}(aq) + CO_2\ (g) + H_2O + SO_4^{2-}(aq)$$

9.8 Acids, Bases, and Human Health
    A. Strong acids and bases are corrosive poisons that can cause chemical "burns."
    B. Most acids and bases produced in this country are used by industry, but some can be found around the house.
        1. H$_2$SO$_4$ – automobile batteries, some drain cleaners
        2. HCl (muriatic acid) – toilet bowl cleaner
        3. CaO (lime) – mortar and cement
        4. NaOH (lye) – drain and oven cleaner
    C. *Acid Rain* – Burning sulfur-containing fuels generates sulfur oxides which can react with moisture in the air to form sulfuric acid, which can dissolve calcium carbonate in marble and limestone.
$$SO_3 + H_2O \rightarrow H_2SO_4\ ;\quad CaCO_3 + H_2SO_4 \rightarrow CaSO_4 + CO_2\ (g)$$
    D. *Antacids* – The stomach secretes HCl to aid in digestion of food. Overindulgence and stress can lead to hyperacidity and ulcers. Ulcers are holes in the mucosal lining of the stomach caused by excess acidity or certain bacteria. All antacids are bases, and overuse can make the blood too alkaline, causing *alkalosis*.
        1. Alka-Seltzer – sodium bicarbonate + citric acid + aspirin
        2. Tums – flavored calcium carbonate

# Chapter 9 - Acids and Bases I

    3. Phillips "Milk of Magnesia" – suspension of magnesium hydroxide
    4. Maalox – combination of aluminum and magnesium hydroxides
    5. Rolaids – aluminum sodium dihydroxy carbonate
  E. Proteins (eyes, skin, lungs) are readily denatured by exposure to acids or bases. One should exercise care and wear protective clothing (including goggles!) when working with acids and bases.

## DISCUSSION

We will begin by reviewing the definitions of acids. Acids were defined in the chapter in three ways: as compounds that yield hydrogen ions; as compounds that yield hydronium ions in aqueous solutions; and as compounds that act as proton donors. It is now generally accepted that a hydrogen ion does not exist in solution as an independent unit. Thus the second and third definitions are attempts to be a bit more accurate in describing the action of an acid.

A hydrogen ion and a proton are identical species. The Bohr picture of a hydrogen atom is:

A hydrogen ion is the hydrogen atom minus its outer shell electron. For hydrogen, with only one electron, that means that the ion is simply the hydrogen nucleus. The hydrogen nucleus contains a single proton. Therefore, $H^+ = p$. The preferred term for this species is proton, but the symbol most commonly used is $H^+$. The proton or hydrogen ion does not exist as an independent species in solution. In aqueous solutions, protons from acids are transported by water molecules. This is where the hydronium ion comes in. It's like transferring food from your plate to your mouth. It's the food that's transferred, but a fork carries it from one place to another. In reactions involving acids in aqueous solutions, the proton is being transferred, but a water molecule carries the proton from one place to another.

The many reactions discussed in this chapter can be classified under just a few headings. First, there are what we could call "the defining equations." These are the equations that say "this compound is an acid," or "this compound is a base." Since acids and bases were defined in a number of ways in the chapter, the defining equation can be written in a number of ways.

Equation for an Arrhenius acid:           $HCl \rightarrow H^+ + Cl^-$

Equation for a Brönsted-Lowry acid:      $HCl + H_2O \rightarrow H_3O^+ + Cl^-$

Both equations indicate that HCl is an acid, but the second one emphasizes that the proton is transferred and not simply released. The extent to which this reaction proceeds to the right determines whether the acid is called strong or weak. Table 9.1 in the text lists several important strong acids and is worth committing to memory.

Equations for an Arrhenius base:        $NaOH \rightarrow Na^+ + OH^-$

Equations for a Brönsted–Lowry base:   $OH^- + H_3O^+ \rightarrow H_2O + H_2O$
                                             $NH_3 + H_2O \rightarrow NH_4^+ + OH^-$

Notice that the equation for ammonia qualified under both definitions: It shows the release of hydroxide ion (Arrhenius) and it also shows ammonia picking up a proton (Brönsted-Lowry). For sodium hydroxide,

# Chapter 9 - Acids and Bases I

we use two different equations. To satisfy the Arrhenius definition, we simply show sodium hydroxide dissociating to produce the hydroxide ion in solution. To satisfy the Brønsted–Lowry definition, we concentrate on the hydroxide ion from sodium hydroxide and show it picking up a proton.

The remaining reactions can be categorized into three groups.

1. Neutralization: Acid plus base yields salt plus water.
   $$HCl + NaOH \rightarrow NaCl + H_2O$$
2. Reaction of active metal with acid yields salt plus hydrogen gas.
   $$Mg + H_2SO_4 \rightarrow MgSO_4 + H_2$$
3. Reaction of carbonate (or bicarbonate) with acid yields salt plus carbon dioxide plus water.
   $$2\,HCl + Na_2CO_3 \rightarrow 2\,NaCl + CO_2 + H_2O$$
   $$HCl + NaHCO_3 \rightarrow NaCl + CO_2 + H_2O$$

We note again that the very important reaction of carbonic acid:

$$H_2CO_3 \rightarrow CO_2 + H_2O$$

is not a general reaction of acids or even a general reaction of weak acids. It is a chemical property peculiar to carbonic acid. The equation indicates the special instability of carbonic acid. If this compound is produced in a reaction, e.g.,

$$HCl + NaHCO_3 \rightarrow NaCl + H_2CO_3$$
$$\Downarrow \text{ It (immediately) decomposes.}$$
$$H_2O + CO_2$$

Notice that when another weak acid, such as acetic acid, is produced in a reaction:

$$HCl + CH_3COONa \rightarrow NaCl + CH_3COOH$$

the acid does not decompose. We call your attention to this property of carbonic acid because the equilibrium between carbon dioxide, carbonic acid, and bicarbonate ion play an extremely important role in controlling the acidity of the blood. This subject will be discussed in Chapter 11.

## SELF-TEST

1. According to the theory of Arrhenius, the properties of acids are the properties of:
   a. $H^+$      b. $OH^-$      c. $H_2O$      d. $NH_3$
2. According to the theory of Arrhenius, the properties of bases are the properties of:
   a. $H^+$      b. $OH^-$      c. $H_2O$      d. $NH_3$
3. According to the Brønsted-Lowry theory, an acid is
   a. a proton donor          b. a proton acceptor
   c. a hydronium ion donor   d. a hydronium ion acceptor
4. According to the Brønsted-Lowry theory, a base is
   a. a proton donor          b. a proton acceptor
   c. a hydronium ion donor   d. a hydronium ion acceptor
5. If litmus paper changes color from blue to red when placed in an aqueous solution, the solution is:
   a. acidic      b. basic      c. neutral
6. Which is a monoprotic acid?
   a. $H_2CO_3$      b. $H_2SO_4$      c. $CH_3COOH$      d. $H_3PO_4$

# Chapter 9 - Acids and Bases I

7. Which is the weakest acid?
   a. HCl   b. $HNO_3$   c. $H_2SO_4$   d. $H_2CO_3$
8. Which is a strong acid?
   a. $H_2CO_3$   b. $CH_3COOH$   c. $H_3PO_4$   d. $HNO_3$   e. HCN
9. According to the equation $C_6H_5OH + H_2O \rightleftharpoons C_6H_5O^- + H_3O^+$, the compound $C_6H_5OH$ is a:
   a. strong acid   b. strong base   c. weak acid   d. weak base
10. According to the equation shown in question 9, which set constitutes a conjugate acid-base pair?
    a. $C_6H_5OH$ and $H_2O$   b. $C_6H_5OH$ and $C_6H_5O^-$   c. $C_6H_5OH$ and $H_3O^+$
11. According to the equation $C_6H_5OH + OH^- \rightarrow C_6H_5O^- + H_2O$, the compound $C_6H_5OH$ is the:
    a. acid   b. base   c. salt   d. buffer
12. The conjugate partner of a strong acid is a:
    a. strong acid   b. weak acid   c. strong base   d. weak base
13. According to the Brønsted-Lowry theory, a weak acid:
    a. holds on to its protons relatively tightly
    b. has a very weak hold on its protons
14. The formula for phosphoric acid is:
    a. $H_2PO_3$   b. $H_3PO_3$   c. $H_2PO_4$   d. $H_3PO_4$
15. The name of the acid HBr is:
    a. hydrobromic acid   b. bromic acid   c. bromous acid   d. bromate acid
16. If $ClO_3^-$ is the chlorate ion, $HClO_3$ is:
    a. hydrochloric acid   b. chloric acid   c. chlorous acid   d. chlorate acid
17. If $ClO_2^-$ is the chlorite ion, chlorous acid would be:
    a. $H_2ClO_3$   b. $HClO_3$   c. $HClO_2$   d. $H_3ClO_4$
18. Identify the correct formula for sodium hypochlorite.
    a. $Na_2ClO_3$   b. $NaClO_3$   c. $NaClO_2$   d. $NaClO$
19. Identify the correct formula for potassium perchlorate.
    a. $K_2ClO_3$   b. $KClO_3$   c. $KClO_4$   d. $KClO$
20. Identify the correct formula for sodium chlorate.
    a. $Na_2ClO_3$   b. $NaClO_3$   c. $NaClO_2$   d. $NaClO$
21. Which oxide when added to water would produce an acidic solution?
    a. $K_2O$   b. $BaO$   c. $SO_2$   d. $MgO$   e. $Na_2O$
22. The metaborate ion is $BO_2^-$. Orthoboric acid is:
    a. $BO_2^{3-}$   b. $BO_3^{3-}$   c. $HBO_2$   d. $H_3BO_3$
23. Which is regarded as a strong base in aqueous solution?
    a. KOH   b. $Mg(OH)_2$   c. $NH_3$
24. Which base exists primarily in un-ionized form in aqueous solution?
    a. KOH   b. $Mg(OH)_2$   c. $NH_3$
25. Which is not highly ionized in aqueous solution?
    a. NaOH   b. $Ca(OH)_2$   c. $H_2SO_4$   d. $CH_3COOH$   e. $(NH_4)_3PO_4$
26. Muriatic acid is:
    a. $H_3PO_4$   b. $H_2SO_4$   c. $H_2CO_3$   d. HCl
27. Which set of reactants does not produce a gas when mixed?
    a. $NaHCO_3$ and HCl
    b. NaOH and HCl
    c. Mg and HCl
    d. All of the reactions yield a gaseous product.

## Chapter 9 - Acids and Bases I

28. Which is not a product of the reaction: $Na_2CO_3 + HNO_3 \rightarrow$ ?
    a. $H_2$   b. $Na^+$   c. $CO_2$   d. $H_2O$   e. $NO_3^-$
29. The equation, $H_2CO_3 \rightarrow CO_2 + H_2O$, indicates that carbonic acid is:
    a. a weak acid   b. a strong acid   c. an unstable acid
30. Lye is the common name for:
    a. NaOH   b. $NH_3$   c. HCl   d. $H_2CO_3$
31. Which equation describes a neutralization reaction?
    a. $Cu + 2\,Ag^+ \rightarrow 2\,Ag + Cu^{2+}$
    b. $2\,HNO_3 + Zn \rightarrow Zn(NO_3)_2 + H_2$
    c. $CH_3COOH + NaOH \rightarrow CH_3COONa + H_2O$
    d. All of the equations are for neutralization reactions.
32. The net ionic equation that describes a neutralization reaction is:
    a. $H^+ + OH^- \rightarrow H_2O$
    b. $HCO_3^- + H^+ \rightarrow H_2CO_3$
    c. $H_2CO_3 \rightarrow CO_2 + H_2O$
    d. $M + 2\,H^+ \rightarrow M_2^+ + H_2$
33. Which of the following has the highest concentration of H+ ion in aqueous solution?
    a. 0.30 M $H_2CO_3$   b. 0.20 M HCl   c. 0.50 M HCOOH   d. 1.0 M $NH_3$
34. The conjugate acid of $HPO_4^{-2}$ is:
    a. $H_3PO_4$   b. $H_2PO_4^-$   c. $HPO_3^{-2}$   d. $H_2PO_4$   e. $HPO_3^{-2}$
35. The conjugate base of $HPO_4^{-2}$ is:
    a. $H_3PO_4$   b. $H_2PO_4^-$   c. $HPO_3^{-2}$   d. $H_2PO_4$   e. $PO_4^{-3}$
36. Which process is described by the following equation?
    $H_2SO_4\,(aq) + CaCO_3\,(s) \rightarrow CaSO_4(aq) + H_2O + CO_2(g)$
    a. erosion of limestone by acidic pollutants
    b. neutralization of stomach acid by an antacid
    c. the oxidation of an active metal by an acid
37. Which is not found in antacid preparations?
    a. $CaCO_3$   b. $Mg(OH)_2$   c. NaOH   d. $NaHCO_3$
38. Hydrochloric acid is found:
    a. in the stomach   b. in toilet bowl cleaners   c. in both a and b
39. Which is the bicarbonate ion?
    a. $H_2CO_3$   b. $H_2CO_4^-$   c. $HCO_3^{-2}$   d. $HCO_3^-$   e. $CO_3^{-2}$
40. Which solution does not cause severe chemical burns on contact with skin?
    a. concentrated sulfuric acid
    b. concentrated hydrochloric acid
    c. concentrated sodium hydroxide solution
    d. concentrated sodium chloride solution
    e. All cause severe chemical burns.

## ANSWERS

| | | | | |
|---|---|---|---|---|
| 1. a | 9. c | 17. c | 25. d | 33. b |
| 2. b | 10. b | 18. d | 26. d | 34. b |
| 3. a | 11. a | 19. c | 27. b | 35. e |
| 4. b | 12. d | 20. b | 28. a | 36. a |
| 5. a | 13. a | 21. c | 29. c | 37. c |
| 6. c | 14. d | 22. d | 30. a | 38. c |
| 7. d | 15. a | 23. a | 31. c | 39. d |
| 8. d | 16. b | 24. c | 32. a | 40. d |

# 10 Acids and Bases II

## KEY WORDS

equivalent weight   titration         ion product of water      acidosis
equivalence point   end point         hydrolysis                alkalosis
normality (N)       pH, pOH           buffer solution           indicator dye
neutralization      standard          dilution                  reversible reaction
equilibrium constant  $K_{eq}, K_a, K_b, K_w$   Henderson-Hasselbalch   equilibrium
neutral             acidic            common ion effect

## SUMMARY

10.1 Concentrations of Acids and Bases

   A. **Dilution** problems (Recall: Concentration × Volume = Amount of Solute)

      1. It is frequently necessary to prepare dilute solutions from more concentrated stock solutions.

      2. $V_{con} \times M_{con} = V_{dil} \times M_{dil}$, or **amount** in concentrated solution = **amount** in diluted solution.

   B. Equivalents and normality (optional)

      1. Not all acids and bases are *equivalent* in the number of protons donated or accepted. $H_2SO_4$ donates 2 protons per molecule, while HCl can donate only 1. $H_2SO_4$ is said to have 2 equivalents per mole.

| Common Acid or Base | Equivalents per Mole (*n*) |
|---|---|
| HCl, $HNO_3$, NaOH, $NH_3$ | 1 |
| $H_2SO_4$, $H_2CO_3$, $Ca(OH)_2$, $Mg(OH)_2$ | 2 |
| $H_3PO_4$ | 3 |

      2. **Equivalent weight** = Formula weight/*n*

      3. **Normality (N)**   $N$ = equivalents/L of solution;   $N = n \times$ Molarity

10.2 Acid-Base Titrations

   A. For acid-base reactions, the *equivalence point* occurs when the same number of equivalents of acid and base have been added so that *neutralization* has occurred.

   B. **Titration** is a process for determining the amount of an unknown substance by adding an equivalent amount of a reactive substance of known concentration. Titrations can be used with many types of systems providing there is some *indicator dye* that produces a detectable change (*end point*) to signal that the *equivalence point* has been reached. Phenolphthalein is a common indicator that is colorless in its conjugate acid form (acidic solutions) but turns pink in its conjugate base form (basic solutions).

   C. Typical steps in titrating an unknown acid with a standard base:

      1. A carefully measured aliquot of the acid sample is placed in a flask.

      2. A few drops of an indicator dye are added to the acid sample.

      3. A *standard* base solution is then added slowly from a burette until the dye changes color. This is the end point of the titration and should also represent the equivalence point.

      4. Calculations: At the equivalence point: # equivalents of acid = # equivalents of base,

        or $N_a \times V_a = N_b \times V_b$, so that $N_a = N_b \times \dfrac{V_b}{V_a}$

# Chapter 10 - Acids and Bases II

10.3 The pH Scale
   A. Water can act as both an acid and a base.
   1. $H_2O + H_2O \rightleftharpoons H_3O^+ + OH^-$
      55.5 M          0.0000001 M    0.0000001 M
   2. Pure water at 25 °C is a neutral solution having
      $[H_3O^+] = [OH^-] = 0.0000001\ M = 1.0 \times 10^{-7}\ M$
   B. *Ion product of water*: $K_w$
   1. $K_w = [H_3O^+] \times [OH^-] = 1.0 \times 10^{-14}$
   2. This equation, defining the ion product of water, applies to any aqueous solution at 25 °C. Thus a solution 0.1 M in $H_3O^+$ must also be $10^{-13}$ M in $OH^-$.
   C. The pH scale
   1. $pH = -\log [H_3O^+]$     $pOH = -\log [OH^-]$
   2. pH + pOH = 14
   3. **pH < 7, acidic;  pH = 7, neutral;  pH > 7, basic**
   4.

   | $[H_3O^+]$ | $[OH^-]$ | pH | pOH | Solution |
   |---|---|---|---|---|
   | $10^{-7}$ | $10^{-7}$ | 7 | 7 | neutral |
   | $10^{-4}$ | $10^{-10}$ | 4 | 10 | acidic |
   | $10^{-9}$ | $10^{-5}$ | 9 | 5 | basic |
   | 1.0 | $10^{-14}$ | 0 | 14 | acidic |

   5. Students of chemistry should be familiar with pH and be able to fill in a table like the one above when given any one of the first four values. When the $[H_3O^+]$ is not a simple power of ten, it is necessary to use a "log table" or a calculator with a "log" function to determine the pH.
      e.g. if $[H_3O^+] = 6.3 \times 10^{-8} = 10^{0.8} \times 10^{-8} = 10^{-7.2}$, thus pH = 7.2

   | $[H_3O^+] =$ | $6.3 \times 10^{-7}$ | $6.3 \times 10^{-8}$ | $6.3 \times 10^{-9}$ |
   |---|---|---|---|
   | pH = | 6.2 | 7.2 | 8.2 |

   Note that a change of one pH unit always implies a tenfold change in $[H^+]$ concentration.
   D. Measuring pH
   1. Many organic indicator dyes are different colors in their conjugate acid and conjugate base forms and can be used to estimate ±1 pH unit.
   2. pH meters measure $[H_3O^+]$ concentration electronically to about 0.01 pH units.

10.4 Equilibrium Calculations
   A. Equilibrium constant ($K_{eq}$) – Consider a general *reversible reaction*: $aA + bB \Leftrightarrow cC + dD$
   $$K_{eq} = \frac{[C]^c [D]^d}{[A]^a [B]^b}$$  where [ ] = molar concentration at **equilibrium**
   B. The *equilibrium constant* for a reaction is a constant at the given temperature.
   C. Ionization of weak acids: $K_a$
   1. Consider the dissociation of a general weak acid HA.
      $HA + H_2O \Leftrightarrow H_3O^+ + A^-$
   $$K_{eq} = \frac{[H_3O^+][A^-]}{[HA][H_2O]}$$
   2. The $[H_2O]$ in dilute solutions is always ~ 55.5 M. Define $K_a$ as
   $$K_a = K_{eq} \times [H_2O] = \frac{[H_3O^+][A^-]}{[HA]}$$
   3. The larger the value of $K_a$, the stronger the acid.

4. Calculate [$H^+$] from $K_a$ for a 1 M solution of HA.

$$HA \Leftrightarrow H^+ + A^-$$

initial [ ]  1.0  0  0
equilibrium [ ]  1–x  x  x

$$K_a = \frac{[H^+][A^-]}{[HA]} = \frac{(x)(x)}{(1.0-x)} \sim \frac{(x)^2}{1.0} \quad \text{since for weak acids } x <<< 1.0$$

$x = \sqrt{K_a}$ ; The [$H^+$] concentration of a 1.0 M weak acid solution is equal to the square root of its $K_a$.

D. Equilibria involving weak bases: $K_b$
  1. Consider the equilibrium expression for a general conjugate base.
  $$A^- + H_2O \Leftrightarrow HA + OH^-$$
  therefore, $K_{eq} = \dfrac{[HA][OH^-]}{[A^-][H_2O]}$

  2. Define $K_b$ as
  $$K_b = K_{eq} \times [H_2O] = \frac{[HA][OH^-]}{[A^-]}$$

  3. The larger the value of $K_b$, the stronger the base. Note: $K_a \times K_b = K_w = [H^+][OH^-] \sim 10^{-14}$
  Note: Defining $pK_a = -\log K_a$ and $pK_b = -\log K_b$, then $pK_a + pK_b = 14$
  These relationships are illustrated in the following Table.

| Ionization Constants for Some Weak Acids in Water at 25 °C | | | | | | |
|---|---|---|---|---|---|---|
| Name | c.a. | c.b. | $K_a$ | $K_b$ | $pK_a$ | $pK_b$ |
| Phosphoric acid | $H_3PO_4$ | $H_2PO_4^-$ | $7.5 \times 10^{-3}$ | $1.3 \times 10^{-12}$ | 2.1 | 11.9 |
| Hydrofluoric acid | HF | $F^-$ | $6.6 \times 10^{-4}$ | $1.5 \times 10^{-11}$ | 3.2 | 10.8 |
| Acetic acid | $CH_3COOH$ | $CH_3COO^-$ | $1.8 \times 10^{-5}$ | $5.6 \times 10^{-10}$ | 4.7 | 9.3 |
| Carbonic acid | $H_2CO_3$ | $HCO_3^-$ | $4.2 \times 10^{-7}$ | $2.4 \times 10^{-8}$ | 6.4 | 7.6 |
| Dihydrogen Phosphate | $H_2PO_4^-$ | $HPO_4^{-2}$ | $6.2 \times 10^{-8}$ | $1.6 \times 10^{-7}$ | 7.2 | 6.8 |
| Hydrocyanic acid | HCN | $CN^-$ | $6.2 \times 10^{-10}$ | $1.6 \times 10^{-5}$ | 9.2 | 4.8 |
| Ammonia | $NH_4^+$ | $NH_3$ | $5.6 \times 10^{-10}$ | $1.8 \times 10^{-5}$ | 9.3 | 4.7 |
| Monohydrogen Phos. | $HPO_4^{-2}$ | $PO_4^{-3}$ | $3.6 \times 10^{-13}$ | $2.8 \times 10^{-2}$ | 12.4 | 1.6 |

10.5 Salts in Water: Acidic, Basic, or Neutral?
  A. Many salts dissolve in water to produce acidic or basic solutions as the weak conjugate acid (c.a.) or conjugate base (c.b.) reacts with water (*hydrolysis*) to produce new $H_3O^+$ or $OH^-$ ions.
   1. Ammonium chloride in solution (acidic)
   $$NH_4^+ + H_2O \rightleftharpoons H_3O^+ + NH_3$$
   2. Sodium cyanide in solution (basic)
   $$CN^- + H_2O \rightleftharpoons HCN + OH^-$$
  B. Rules
   1. Salt of a strong acid + strong base → neutral salt solution
      HCl                NaOH              (NaCl)
   2. Salt of a strong acid + weak base → acidic salt solution
      HCl                $NH_3$              ($NH_4Cl$)
   3. Salt of a weak acid + strong base → basic salt solution

## Chapter 10 - Acids and Bases II

         HCN            NaOH        (NaCN)

4. Salt of a weak acid + weak base → ? (depends on their relative strengths)

### 10.6 Buffers: Control of pH

A. A *buffer solution* is one that resists changes in pH on addition of acid or base.

B. Chemically, a buffer is a solution of a weak conjugate acid–conjugate base pair. The conjugate acid (c.a.) neutralizes incoming base, and the conjugate base (c.b.) neutralizes incoming acid.

C. Calculations involving buffers

  1. Buffer solutions resist changes in pH upon addition of either acid or base. Buffer solutions are made by using a mixture of a weak acid and its conjugate base taking advantage of the *common ion effect*. The buffer effect of the buffer solution works best when the $[H^+]$ is approximately equal to the $K_a$ value of the conjugate acid, or pH ~ $pK_a$ of the acid.

  2. From the expression for $K_a$, we note that the $[H^+] = K_a$ when the [c.b] = [c.a.].

$$K_a = [H]^+ \times \frac{[A^-]}{[HA]}$$

  3. Henderson-Hasselbalch equation

    a. From $K_a$: $\quad [H^+] = K_a \times \frac{[HA]}{[A^-]}$

    b. Taking the negative of the log of both sides:

$$-\log[H^+] = -\log K_a + \log \frac{[A^-]}{[HA]}$$

$$\text{or} \quad pH = pK_a + \log \frac{[A^-]}{[HA]} \quad \text{(the \textbf{Henderson-Hasselbalch} equation)}$$

  4. The *Henderson-Hasselbalch* equation states that the pH of a c.b./c.a. pair will be numerically equal to the $pK_a$ of that c.b./c.a. pair when [c.b.] = [c.a.]. Thus the pH optimum of a buffer is equal to the $pK_a$ of the c.a. in the buffer.

| [c.b.]/[c.a.] = | 0.1 | 1.0 | 10.0 |
|---|---|---|---|
| pH = | $pK_a - 1.0$ | $pK_a$ | $pK_a + 1.0$ |

### Titration Curve of Weak Acid

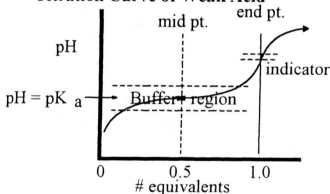

### 10.7 Buffers in Blood

A. Blood plasma pH is ~ 7.4. If it falls below 6.8 (*acidosis*) or rises above 7.8 (*alkalosis*), it can lead to death.

B. Our bodies are acid factories. The stomach produces HCl, our muscles produce lactic acid, and respiration produces $H_2CO_3$. To maintain blood pH, buffers are needed. Three of the most important are:
1. Bicarbonate/carbonic acid system

$$HCO_3^- + H^+ \Leftrightarrow H_2CO_3 \Leftrightarrow H_2O + CO_2$$
(kidneys)                                                         (lungs)

The ratio of $HCO_3^-/H_2CO_3$ must be maintained at about 20:1 for this system to buffer at pH 7.4. This ratio is maintained in part because excess $HCO_3^-$ can be removed by the kidneys and excess $CO_2$ removed via the lungs.
2. $HPO_4^{2-}/H_2PO_4^-$ (monohydrogen phosphate/dihydrogen phosphate)
3. The c.b. ($-COO^-$) and c.a. ($-NH_3^+$) pairs of protein molecules

10.8 Acidosis and Alkalosis
    A. *Acidosis*: blood pH < 7.35
        1. Prolonged, strenuous exercise
        2. Starvation, diabetes mellitus
    B. *Alkalosis*: blood pH > 7.45

## DISCUSSION

The use of equivalents and normality may seem to introduce unnecessary confusion now that you are just getting comfortable with molarity. However, equivalents are a more convenient unit for expressing the relative amounts of reacting substances. You will find that expressing concentrations in terms of equivalents/L, or normality, will make acid-base titration calculations easier to handle. We offer below a comparison of calculations of moles and molarity and equivalents and normality. Sulfuric acid is used as the example, and we propose to calculate the concentration of a solution containing 9.8 g of $H_2SO_4$ in a total of 2 L of solution.

| Moles and Molarity | Equivalents and Normality |
|---|---|
| Formula wt. = 98 | Equivalent wt. = (Formula wt./2) = 49 |
| Gram form. wt. (GFW) = 98 g/mole | Gram equiv. wt. (GEW) = 49 g/equiv. |
| Number of moles: weight/GFW | Number of equivalents: weight/GEW |
| Example: $\dfrac{9.8 \text{ g}}{98 \text{ g/mol}}$ = 0.1 mol | Example: $\dfrac{9.8 \text{ g}}{49 \text{ g/equivalent}}$ = 0.2 equiv. |
| Molarity: $\dfrac{\text{moles of solute}}{\text{liters of solution}}$ | Normality: $\dfrac{\text{equivalents of solute}}{\text{liters of solution}}$ |
| Example: $\dfrac{0.1 \text{ mole}}{2 \text{ L}}$ = **0.05 M** | Example: $\dfrac{0.2 \text{ equivalent}}{2 \text{ L}}$ = **0.1 N** |
| M × 2 = N for $H_2SO_4$ (since there are 2 equiv./mol) ||

We could have calculated the molarity and then used the relationship of N = 2 × M to determine the normality. The 2 in that relationship and the 2 in the denominator of the calculation of equivalent weight are both derived from the fact that sulfuric acid is a diprotic acid, i.e., it supplies two protons.

# Chapter 10 - Acids and Bases II

The other major concept involving calculations that was introduced in this chapter is the pH scale. We could introduce more mathematics by asking you to work with more difficult logarithms in dealing with pH, such as examples 10.9 and 10.10 of the text, but what we really want you to do is to become comfortable with the concept of pH. For that reason, we'll stick to calculations involving simple powers of ten (i.e., 0.001 or $10^{-3}$) so you can clearly see the pattern for conversion from concentration units to pH or pOH units. You should be able to interconvert among pH, $[H^+]$, pOH, and $[OH^-]$ by using the definition of pH (or pOH) and the ion product constant of water. Thus, if you know $[H^+]$ equals $10^{-6}$ M, you also know that the pH is 6, the pOH is 8 (because pH + pOH = 14), and $[OH^-]$ equals $10^{-8}$ M.

Similarly:   If $[OH^-] = 10^{-3}$ M,   pOH = 3,   pH = 11,   $[H^+] = 10^{-11}$ M.
             If pH = 9,   $[H^+] = 10^{-9}$ M,   pOH = 5,   $[OH^-] = 10^{-5}$ M.
             If pOH = 1,   $[OH^-] = 10^{-1}$ M,   pH = 13,   $[H^+] = 10^{-13}$ M.

Although pOH is as easy to calculate as pH, the latter value is almost always quoted. Therefore, you will find it most useful to gear your thinking to the pH scale. Remember that low pH means that there are lots of protons available and high pH means that there are few protons available. Low pH means acidic: high pH means basic. The variations in pH, pOH, $[H^+]$, and $[OH^-]$ can be summarized as follows.

|        | Acidic | Neutral | Basic |
|--------|--------|---------|-------|
| pH     | 1      | 7       | 13    |
| $[H^+]$ | $10^{-1}$ | $10^{-7}$ | $10^{-13}$ |
| $[OH^-]$ | $10^{-13}$ | $10^{-7}$ | $10^{-1}$ |
| pOH    | 13     | 7       | 1     |

The treatment of dynamic equilibrium was introduced in Chapter 9 and here in Chapter 10 we are introduced to many quantitative uses of $K_{eq}$. We can use Le Chatelier's principle to qualitatively predict the shifts in equilibrium distributions when a stress is placed on the system. In order to treat dynamic equilibria in a quantitative manner, it is necessary to be able to write equilibrium constant expressions for any given balanced, reversible chemical reaction. We will first demonstrate how to write equilibrium expressions with a few simple rules, and then we will look at the subject of equilibrium calculations. The latter requires us to understand the meanings of various types of equilibrium expressions and to be able to solve simple algebraic problems with them.

The formula for the equilibrium constant ($K_{eq}$) of a generalized balanced chemical equation
$$aA + bB \Leftrightarrow cC + dD$$ is given by the **Law of Mass Action** as:

$$K_{eq} = \frac{[C]^c [D]^d}{[A]^a [B]^b}$$   where [ ] refers to the molar concentration at equilibrium of that species.

Expressed in words, the equilibrium constant is defined as the product of the equilibrium concentrations of all products of the reaction divided by the product of the equilibrium concentrations of all reactants, each raised to the power corresponding to its stoichiometric coefficient in the balanced chemical equation.

The value of $K_{eq}$ is a constant for a given reaction at the given temperature. The numerical value of the equilibrium constant reveals a lot about the tendency for the reaction to form products as written.

# Chapter 10 - Acids and Bases II

$K_{eq} > 10^4$ — Reaction tends to go to completion as written.
$K_{eq} > 1$ — Reaction favors the formation of products.
$K_{eq} = 1$ — Reaction favors the formation of neither products nor reactants.
$K_{eq} < 1$ — Reaction favors the retention of the reactants.
$K_{eq} < 10^{-4}$ — Forward reaction tends not to occur to any large extent.

For equilibria in aqueous solution, the concentration of water is not included in the expression even though it may be involved in the reaction, that is,

$$HCN + H_2O \Leftrightarrow H_3O^+ + CN^-$$

is treated as

$$HCN \Leftrightarrow H^+ + CN^-$$

The $[H_2O]$ is not included because for most aqueous solutions, the concentration of water is essentially constant at 55.5 M. The same argument applies to equilibria involving solids. A solid is not really part of the solution, and its concentration in the solid is a constant, so solids are also left out of equilibrium constant expressions.

The following examples illustrate these rules.

---

**Gases**

$$N_2 + 3H_2 \Leftrightarrow 2NH_3 \qquad\qquad H_2 + Cl_2 \Leftrightarrow 2HCl$$

$$K = \frac{[NH_3]^2}{[N_2][H_2]^3} \qquad\qquad K = \frac{[HCl]^2}{[H_2][Cl_2]}$$

**Solutions**

$$Ag^+ + 2NH_3 \Leftrightarrow Ag(NH_3)_2^+ \qquad\qquad HCN \Leftrightarrow H^+ + CN^-$$

$$K = \frac{[Ag(NH_3)_2^+]}{[Ag^+][NH_3]^2} \qquad\qquad K = \frac{[H^+][CN^-]}{[HCN]}$$

**Solids**

$$BaSO_4 \Leftrightarrow Ba^{2+} + SO_4^{2-} \qquad\qquad Ag^+ + Cl^- \Leftrightarrow AgCl$$

$$K = [Ba^{2+}][SO_4^{2-}] \qquad\qquad K = \frac{[1]}{[Ag^+][Cl^-]}$$

The constants for different types of equilibrium processes are given special names:
$K_a$ = acid dissociation or ionization constant
$K_b$ = base ionization constant
$K_{sp}$ = solubility product constant

---

We will study the solubility product constant later. The remainder of this special topic deals with equilibrium calculations involving solutions of weak acids, weak bases, or both (buffers).

## Chapter 10 - Acids and Bases II

Equilibrium calculations involving weak acids or weak bases are generally of two types: Calculate $K_a$ or $K_b$ given the concentrations, or calculate $[H^+]$ or $[OH^-]$ given the initial concentration and $K_a$. Buffers are solutions of both a weak acid and its conjugate base. The conjugate base is frequently supplied as the salt of the weak acid. Thus, buffer calculations illustrate the common ion effect. Buffer problems can be expressed in terms of the resulting $[H^+]$ or pH. Frequently, buffer problems refer to the pH of the solution and not its $[H^+]$. The Henderson-Hasselbalch equation is a logarithmic rearrangement of the acid ionization constant expression to state explicitly how the pH of a buffer system varies with the ratio of [c.b.]/[c.a.].

---

Henderson-Hasselbalch equation $\qquad pH = pK_a + \log \dfrac{[c.b.]}{[c.a.]}$

---

Note that when the [c.b.] = [c.a.], pH = $pK_a$, which is the same as saying $[H^+] = K_a$.
Both of these kinds of buffer calculations are illustrated in the text and in Examples 3 and 4 that follow.

Examples
1. Calculate $K_a$ given the concentrations of each species.

   Consider the reaction: $\qquad HCN \Leftrightarrow H^+ + CN^-$
   Calculate the $K_a$ for hydrocyanic acid, given that the equilibrium concentrations are:

   $[H^+] = [CN^-] = 7.9 \times 10^{-6}$ M; $\qquad$ [HCN] = 0.10 M

   a. Write the correct equilibrium expression:
   $$K_a = \dfrac{[H^+][CN^-]}{[HCN]}$$

   b. Insert the appropriate concentrations and solve for $K_a$:

   $$K_a = \dfrac{(7.9 \times 10^{-6})(7.9 \times 10^{-6})}{(0.10)} = 6.2 \times 10^{-10}$$

2. Calculate the $[H^+]$ or $[OH^-]$ given the value of $K_a$ and the initial total concentration of the weak acid or base. This type of problem can be considerably more difficult to work than problems like the one in Example 1 unless we can assume that the acid or base is so weak that the equilibrium concentration of the un-ionized acid or base is not significantly different from the initial total concentration of the acid or base. Again, consider the reaction of hydrocyanic acid, but now we want to calculate the $[H^+]$ of a 1.0 M HCN solution.

   a. Write out the reaction, and express both the initial and equilibrium concentrations.

   |  | HCN | $\Leftrightarrow$ | $H^+$ | + | $CN^-$ |
   |---|---|---|---|---|---|
   | initial [ ] | 1.0 | | 0 | | 0 |
   | at equilibrium [ ] | 1.0 – x | | x | | x |

   b. Write the equilibrium constant expression as given in 1.a above.
   c. Insert the given data and solve for $[H^+]$ = x.

$$K_a = 6.2 \times 10^{-10} = \frac{(x)(x)}{(1.0-x)} \sim \frac{(x)(x)}{(1.0)} \sim \frac{(x)^2}{(1.0)}$$

or $x^2 = 6.2 \times 10^{-10}$ or $x = 2.5 \times 10^{-5} = [H^+]$

Note: You can always check to determine if the assumption was correct. Is $1.0 - x \sim 1.0$?
$1.000000 - .000025 = .999975 \sim 1.0$. Yes, the assumption was justified in this instance.
(It is also possible to solve such equations without making such assumptions by using the quadratic equation, but you will not be required to do this in this text.)

3. Calculate the $[H^+]$ of a solution that is 0.1 M in HCN and 0.1 M in KCN.
   a. Write out the ionization constant expression.
   
   $$K_a = \frac{[H^+][CN^-]}{[HCN]}$$
   
   b. Insert the appropriate concentrations where
   
   $[H^+]$ = x
   $[CN^-]$ = concentration of the soluble salt, [KCN] = 0.1 M
   $[HCN]$ = concentration of the initial acid, [HCN] = 0.1 M
   
   $$K_a = \frac{(x)(0.1)}{(0.1)} \text{ or } K_a = x = [H^+] = 6.2 \times 10^{-10}$$
   
   Note: The hydrogen ion concentration will always be numerically equal to the value of $K_a$ when the buffer consists of equal amounts of a conjugate acid and its conjugate base.

4. Calculate the pH of a solution that is 0.1 M in HCN and 0.01 M in KCN. The $pK_a$ of hydrocyanic acid is 9.2.
   a.
   
   $$pH = pK_a + \log \frac{[CN^-]}{[HCN]}$$
   
   b. $pK_a$ = 9.2    $[CN^-]$ = 0.01 M    $[HCN]$ = 0.1 M
   
   c. $pH = 9.2 + \log \frac{(0.01)}{(0.1)} = 9.2 + \log(0.1) = 9.2 - 1.0 = 8.2$

Your understanding of this material can be reviewed by working the questions at the end of the chapter. The **SELF-TEST** will give you additional practice on equilibrium calculations.

**Problems** (The following problems supplement those at the end of the chapter in the textbook.)

1. Oxalic acid ($H_2C_2O_4$) is a diprotic acid.
   a. Calculate the formula weight and the equivalent weight of oxalic acid.
   b. Calculate the number of equivalents in 9.0 g of oxalic acid; in 450g; in 18.0 g.
   c. Calculate the normality of the following aqueous solutions of oxalic acid: 9.0 g in 1 L of solution; 450 g in 5 L; 18 g in 500 mL.
   d. What are the molarities of the solutions in part c?

## Chapter 10 - Acids and Bases II

2. In many instances, particularly when the common mineral acids like $H_2SO_4$, $H_3PO_4$, $HNO_3$ are involved, solutions of the acids are prepared by diluting the concentrated acids. Review Examples 10.1 and 10.2 in the text, then try the following.
    a. Concentrated $H_2SO_4$ is 36 N. What volume of this acid should be used in preparing 500 mL of 12 N $H_2SO_4$? 2 L of 3.6 N $H_2SO_4$? 100 mL of 9 M $H_2SO_4$?
    b. What concentration of solution results when concentrated $H_2SO_4$ is diluted as indicated? 10 mL diluted to 100 mL; 0.5 L diluted to 2.5 L; 100 mL diluted to 0.2 L?
3. Determine the unknown concentration from the given titration data.
    a. If 20 mL of 0.5 N HCl is required to titrate 50 mL of NaOH, what is the normality of the NaOH?
    b. If 15 mL of 0.4 N $H_2SO_4$ is required to titrate 45 mL of NaOH, what is the normality of the NaOH?
    c. If 30 mL of 0.4 M $H_2SO_4$ is required to titrate 60 mL of NaOH, what is the molarity of the NaOH?
    d. If 100 mL of 0.1 N $H_2SO_4$ is required to titrate 200 mL of $Ba(OH)_2$, what is the concentration of the barium hydroxide solution in normality? In molarity?

4. Fill in the blanks in the table.

|   | pH | pOH | [H$^+$] | [OH$^-$] | Acidic or Basic? |
|---|----|-----|---------|----------|------------------|
| a. | 2 | --- | --- | --- | --- |
| b. | --- | 7.6 | --- | --- | --- |
| c. | --- | --- | $10^{-10}$ M | --- | --- |
| d. | --- | --- | --- | 0.01 M | --- |

## SELF-TEST

1. If 4 equivalents of a substance are dissolved in 200 mL of solution, the concentration is:
    a. 1.25 N   b. 2 N   c. 4 N   d. 8 N   e. 20 N
2. If 2 g of an acid in 0.5 L of solution gives a 0.1 N solution, what is the equivalent weight of the acid?
    a. 1   b. 2   c. 4   d. 10   e. 20   f. 40   g. 100   h. 200   i. 400
3. If a sample contains 1 milliequivalent of $H_2SO_4$, how many grams of $H_2SO_4$ does it contain?
    a. 0.0049   b. 0.0098   c. 0.0196   d. 0.049   e. 0.098   f. 0.196   g. 49   h. 98   i. 196
4. Twenty milliequivalents of a compound per liter of solution gives a solution with a concentration of:
    a. 20 N   b. 2 N   c. 0.2 N   d. 0.02 N   e. 0.002 N
5. If 72 g of $H_3X$ in 1 L of solution gives a 3 N solution, the equivalent weight of $H_3X$ is:
    a. 3   b. 6   c. 8   d. 12   e. 16   f. 24   g. 48   h. 72
6. When used with sulfuric, hydrochloric or nitric acids, the term dilute refers to a concentration of:
    a. 1 M   b. 1 N   c. 6 M   d. 6 N   e. 12 M   f. 36 N
7. How many mL of a stock 6.0 N HCl solution are required to prepare 50.0 mL of a 0.10 N HCl solution?
    a. 0.833   b. 6.0   c. 8.33   d. 1.2   e. 1.6   f. 2.4
8. How many mL of a stock 8.0 N KCN solution are required to prepare 10.0 mL of a 0.010 N KCN soln?
    a. 0.10   b. 0.125   c. 0.40   d. 0.08   e. 0.0125   f. 0.008
9. A standard base is:
    a. sodium hydroxide
    b. a solution of base of known concentration
    c. a basic solution with pH 7

## Chapter 10 - Acids and Bases II

10. A solution is considered strongly acidic if its pH is:
    a. 4   b. 6   c. 7   d. 8   e. 13   f. 0
11. What is the pH of a 0.0001 M solution of HCl?
    a. $10^4$   b. $10^{-4}$   c. $10^{10}$   d. $10^{-10}$   e. 4   f. –4   g. 10
12. What is the pH of a 0.001 M solution of NaOH?
    a. $10^3$   b. –11   c. $10^{11}$   d. $10^{-11}$   e. 3   f. –3   g. 11
13. What is the [OH⁻] concentration in a solution with pH 4?
    a. 10 M   b. 4 M   c. $10^{-10}$ M   d. $10^{-4}$ M
14. A solution is classified as a basic solution if:
    a. [OH⁻] = $10^{-8}$ M   b. [H⁺] = $10^{-4}$ M   c. pH = 7   d. pH = 9
15. Which of the following is a typical indicator?
    a. a mixture of a weak acid and its salt   b. litmus   c. a burette
16. Consider these solutions:
    (1) 0.001 M NaOH   (2) a solution with pH 9   (3) a solution with pOH 6   (4) $10^{-5}$ M HCl
    If the solutions are arranged left to right in the order of increasing acidity, the order is:
    a. 1 2 3 4
    b. 4 3 2 1
    c. 1 4 3 2
    d. 4 3 1 2
    e. 2 1 3 4
17. In a titration an indicator is used to:
    a. buffer the solution
    b. make the end point of the titration visible
    c. neutralize the acid or base present
18. Which of the following salts, when dissolved, results in a solution with a pH greater than 7?
    a. $NH_4Cl$   b. $(NH_4)_2SO_4$   c. KBr   d. KCN   e. $KNO_3$
19. Which of the following salts, when dissolved, results in a solution with a pH less than 7?
    a. $NH_4Cl$   b. $Na_2SO_4$   c. KCl   d. $K_3PO_4$   e. NaCN
20. If the pH of a solution of a salt is 6, the salt must be one that was formed from:
    a. a strong acid and a strong base
    b. a strong acid and a weak base
    c. a weak acid and a strong base
    d. a weak acid and a weak base
21. The pH of a solution of a salt obtained from the titration of acetic acid with potassium hydroxide is :
    a. somewhat acidic   b. somewhat basic   c. neutral
22. An aqueous solution of $NH_4NO_3$ is slightly acidic because of the following equilibrium:
    a. $H_2O \rightarrow H^+ + OH^-$
    b. $NO_3^- + H_2O \rightleftharpoons HNO_3 + OH^-$
    c. $NH_4^+ + H_2O \rightleftharpoons NH_3 + H_3O^+$
23. A student titrated 15.0 mL of an unknown acid solution with 25.0 mL of a 0.100 M KOH solution. What is the normality of the unknown acid?
    a. 0.115   b. 0.150   c. 0.230   d. 0.167   e. 0.126
24. A student titrated 20.0 mL of an unknown acid solution with 32.4 mL of a 0.115 M KOH solution. What is the normality of the unknown acid?
    a. 0.153   b. 0.084   c. 0.224   d. 0.347   e. 0.186

## Chapter 10 - Acids and Bases II

25. A buffer is:
    a. used to establish the end point of a titration
    b. used to maintain the pH of a solution
    c. a combination of a strong acid and a strong base

26. If a solution is buffered at pH 6, addition of a small amount of base will result in a pH of approximately:
    a. 4    b. 6    c. 8

27. Addition of a small amount of acid to a solution buffered with $HPO_4^{-2}/H_2PO_4^-$ results in the following shift in equilibrium:
    a. $H^+ + HPO_4^{2-} \rightarrow H_2PO_4^-$
    b. $H_2PO_4^- + H^+ \rightarrow H_3PO_4$
    c. $H_2PO_4 \rightarrow H^+ + HPO_4^{2-}$

28. Which of these buffers does not play a major role in stabilizing the pH of the blood?
    a. $CH_3COOH/CH_3COO^-$    b. $H_2CO_3/HCO_3^-$    c. $H_2PO_4^-/HPO_4^{2-}$

29. A titration experiment revealed that a $H_3PO_4$ solution was 0.12 N. What is the molarity of this acid solution?
    a. 0.12 M    b. 0.24 M    c. 0.04 M    d. 0.36 M    e. 0.06 M

30. What is the pH of a solution whose pOH = 4.7?
    a. 4.7    b. 7.0    c. 9.3    d. 11.7    e. 9.7

31. Write the equilibrium constant expression for each of the following reactions.
    a. $N_2 + O_2 \Leftrightarrow 2\,NO$ (all gases)
    b. $2\,P(s) + 3\,Cl_2(g) \Leftrightarrow 2\,PCl_3(g)$
    c. $H_3PO_4 \Leftrightarrow H^+ + H_2PO_4^-$

32. The $K_a$ for hydrofluoric acid is $6.6 \times 10^{-4}$, and that for hydrocyanic acid is $6.2 \times 10^{-10}$. Which is the weaker acid?
    a. hydrofluoric acid    b. hydrocyanic acid

33. Calculate the $[H^+]$ of a 0.0001 M solution of HCN ($K_a = 6.2 \times 10^{-10}$).
    a. $2.0 \times 10^{-5}$    b. $2.0 \times 10^{-7}$    c. $2.5 \times 10^{-7}$    d. $2.5 \times 10^{-6}$    e. $2.5 \times 10^{-8}$

34. Determine the $K_a$ for an acid HX if the $[H^+]$ of a 0.200 M solution of HX is 0.00003 M.
    a. $9.0 \times 10^{-10}$    b. $3.0 \times 10^{-9}$    c. $3.0 \times 10^{-7}$    d. $2.5 \times 10^{-7}$    e. $4.5 \times 10^{-9}$

35. Determine the $K_a$ for an acid HX if the pH of a 0.100 M solution of HX is 4.5.
    a. $1.0 \times 10^{-8}$    b. $9.0 \times 10^{-10}$    c. $3.0 \times 10^{-7}$    d. $2.5 \times 10^{-8}$    e. $4.5 \times 10^{-9}$

36. What is the $[H^+]$ of a solution that is 0.40 M in HF and 0.20 M in NaF? ($K_a = 6.6 \times 10^{-4}$).
    a. $2.5 \times 10^{-7}$    b. $1.3 \times 10^{-3}$    c. $6.0 \times 10^{-8}$    d. $2.5 \times 10^{-7}$    e. $2.5 \times 10^{-7}$

37. What is the pH of a solution that is 0.20 M in HF and 0.10 M in NaF? ($pK_a = 3.2$).
    a. 3.2    b. 3.0    c. 3.5    d. 2.2    e. 2.9

(Problems 38 – 45 refer to acetic acid; $K_a = 1.8 \times 10^{-5}$; $pK_a = 4.7$)

38. The $[H^+]$ of a 1.0 M acetic acid solution is:
    a. 1.0 M    b. 0.3 M    c. 0.0042 M    d. $1.8 \times 10^{-5}$ M

39. The hydrogen ion concentration of a 0.30 M acetic acid solution is:
    a. 0.10 M    b. 0.023 M    c. $1.8 \times 10^{-6}$ M    d. 0.0023 M

40. The $[H^+]$ of a 0.20 M acetic acid solution that is also 0.20 M in sodium acetate is:
    a. 0.10 M    b. $1.8 \times 10^{-5}$ M    c. $1.8 \times 10^{-6}$ M    d. 0.0013 M

41. The hydrogen ion concentration of a 0.30 M acetic acid solution that is also 0.60 M in potassium acetate is:
    a. $1.8 \times 10^{-5}$ M    b. 0.10 M    c. $9 \times 10^{-6}$    d. $3.6 \times 10^{-5}$

## Chapter 10 - Acids and Bases II

42. The pH of a solution that is 0.30 M in acetic acid that is also 0.20 M in sodium acetate is:
    a. 0.2  b. 4.7  c. 4.9  d. 5.0  e. 4.5
43. The pH of a 0.10 M acetic acid solution that is also 1.0 M in sodium acetate is:
    a. 4.7  b. 5.7  c. 3.7  d. 1.0
44. The pH of a 1.0 M acetic acid solution that is also 0.01 M in sodium acetate is:
    a. 2.7  b. 3.7  c. 4.7  d. 5.7  e. 6.7  f. 1.0
45. What ratio of acetate/acetic acid is needed to prepare a buffer at pH 5.7?
    a. 100:1  b. 10:1  c. 1:1  d. 1:10  e. 1:100
46. What is the [$OH^-$] of a 0.00167 M HCl solution?
    a. 0.00167  b. $6.0 \times 10^{-12}$  c. $3.0 \times 10^{-7}$  d. $6.0 \times 10^{-11}$  e. $6.0 \times 10^{-10}$
47. What is the pH of a 0.00167 M HCl solution?
    a. 1.67  b. 1.12  c. 3.0  d. 0.60  e. 2.78
48. What is the pH of a 0.04 M KOH solution?
    a. 1.40  b. 9.6  c. 3.9  d. 12.6  e. 10.4
49. What is the [$H^+$] of a solution at pH 3.5?
    a. 0.00167  b. $8.84 \times 10^{-3}$  c. $5.3 \times 10^{-5}$  d. $3.16 \times 10^{-4}$  e. $6.0 \times 10^{-4}$
50. A condition called acidosis is diagnosed when the blood pH:
    a. falls below 7.35
    b. rises above 7.45
    c. is anywhere outside the range of 7.35–7.45

## Chapter 10 - Acids and Bases II

## ANSWERS

### Problems
1. a. formula weight = 90; equivalent weight = 45
   b. 0.2; 10; 0.4
   c. 0.2 N; 2 N; 0.8 N
   d. 0.1 M; 1 M; 0.4 M
2. a. 167 mL; 0.2 L; 50 mL
   b. 3.6 N or 1.8 M; 7.2 N or 3.6 M; 18 N or 9 M
3. a. 0.2 N;  b. 0.13 N:  c. 0.4 M;  d. 0.05 N, 0.025 M

4. pH Table

| | pH | pOH | [H$^+$] | [OH$^-$] | Acidic or Basic |
|---|---|---|---|---|---|
| a. | 2 | 12 | $10^{-2}$ M | $10^{-12}$ M | acidic |
| b. | 6.4 | 7.6 | $4 \times 10^{-7}$ M | $2.5 \times 10^{-8}$ M | acidic |
| c. | 10 | 4 | $10^{-10}$ M | $10^{-4}$ M | basic |
| d. | 12 | 2 | $10^{-12}$ M | 0.01 M | basic |

### Self–Test

1. e
2. f
3. d
4. d
5. f
6. d
7. a
8. e
9. b
10. f

11. e
12. g
13. c
14. d
15. b
16. a
17. b
18. d
19. a
20. b

21. b
22. c
23. d
24. e
25. b
26. b
27. a
28. a
29. c
30. c

31. see below
32. b
33. c
34. e
35. a
36. b
37. e
38. c
39. d
40. b

41. c
42. e
43. b
44. a
45. b
46. b
47. e
48. d
49. d
50. a

31. a. $K = \dfrac{[NO]^2}{[N_2][O_2]}$   b. $K = \dfrac{[PCl_3]^2}{[Cl_2]^3}$   c. $K = \dfrac{[H^+][H_2PO_4^-]}{[H_3PO_4]}$

# 11 Electrolytes

**KEY WORDS**

| | | | | |
|---|---|---|---|---|
| *electricity* | *metal* | *ionization* | *battery* | *electrolysis* |
| *anode* | *conductor* | *electrolyte* | *electroplating* | *precipitation* |
| *cathode* | *nonmetal* | *nonelectrolyte* | *activity series* | *solubility product* |
| *anion* | *nonconductor* | *volt* | *corrosion* | *fuel cell* |
| *cation* | *dissociation* | *edema* | *hypertension* | *colligative properties* |
| | | | | *electrochemical cell* |

**SUMMARY**

11.1 Early Electrochemistry
  A. In 1752 Benjamin Franklin demonstrated that lightning was a form of electricity.
  B. In 1800 Volta invented a battery that produced electric current. Soon many chemists were using this process of *electrolysis* to split compounds.
  C. Terms
    1. *Electricity* – flow of electrons
    2. *Electrolysis* – splitting of compounds by means of electricity
    3. *Anode* – positively charged electrode; site of oxidation
    4. *Cathode* – negatively charged electrode; site of reduction
    5. *Anions* – negative ions attracted to the anode
    6. *Cations* – positive ions attracted to the cathode

11.2 Electrical Conductivity
  A. Most *metals* are good *conductors* of electricity because they have mobile electrons in conduction bands that permit outer electrons to flow while the nucleus and inner electrons remain fixed.
  B. Most *nonmetals* are *nonconductors* in their solid state, but may conduct electricity as melts or when in solutions.
  C. Electrolytes are salts that form ions in solution that can carry an electric current.
    1. Strong *electrolytes* – good conductors in aqueous solution (strong acids and bases, soluble salts)
    2. Weak electrolytes – conduct electricity but do so poorly (weak acids and bases, slightly soluble salts)
    3. *Nonelectrolytes* – *nonconductors* (substances that do not produce any ions in solution)
  D. Most molecular compounds (glucose) are nonelectrolytes or weak electrolytes.

11.3 The Theory of Electrolytes: Ionization and Dissociation
  A. Arrhenius (1887) and modern theory of electrolytes
    1. *Electrolytes* form ions in solution by *dissociation* or *ionization*.
    2. The algebraic sum of all positive and negative charges is zero, so the electrolyte solution as a whole is neutral.
    3. Each particle (ion or undissociated molecule) in solution makes the same contribution to colligative property effects (boiling-point elevation, osmotic pressure, etc.).
    4. Weak electrolytes produce few ions in solution.
    5. *Nonelectrolytes* exist as molecular rather than ionic forms in solution.
  B. Boiling-point elevation, freezing-point depression, and osmotic pressure are *colligative properties* that depend on the number (not kind) of particles in a given amount of solvent. Since electrolytes

# Chapter 11 - Electrolytes

    break into ions in solution, on a mole-for-mole basis electrolytes give rise to larger effects than do nonelectrolytes.
  C. Freezing-point depression of 1 kg of water is −1.86 °C per mole of dissolved particles. Boiling point elevation of 1 kg of water is +0.52 °C per mole of dissolved particles.
  D. An ionic solid dissociates in water, while a polar molecule interacts with water, causing it to ionize.

11.4 Electrolysis: Chemical Change Caused by Electricity
  A. *Electrolysis* is the process of using electricity to bring about chemical change. It is used for the preparation of many metals, *electroplating*, and in the medical removal of hair and warts.
  B. Electrolysis of molten NaCl
    1. Anode – oxidation:     $2\ Cl^- \rightarrow Cl_2 + 2\ e^-$
    2. Cathode – reduction:   $2\ e^- + 2\ Na^+ \rightarrow 2\ Na$

11.5 Electrochemical Cells: Batteries
  A. Electricity can cause chemical change, and chemical change can produce electricity.
    1. When a reactive metal is placed in contact with the ions of a less reactive metal (see the activity series below), it will give up its electrons (oxidation) while it reduces the less reactive metal ion. The *volt* is a measure of electrical potential.
    2. The redox half-reactions can be placed in separate compartments, so the electrons exchanged must flow through an external circuit. This flow of electrons is an electric current produced by the *electrochemical cell*. A *battery* is technically a series of such cells.
  B. Some batteries use irreversible reactions and when the reactions are used up the battery is "dead." Other batteries, like your car battery, can be "recharged" by connecting the battery to an external electric energy source and forcing the electrons to flow in the opposite direction.
  C. Note: Ionic processes in and around cells also give rise to electrical potentials.
    1. ECG (electrocardiograph) – monitors the electrical changes associated with the beating of the heart.
    2. EEG (electroencephalograph) – monitors electrical activity of the brain.
  D. Fuel Cells
    1. A *fuel cell* is a device in which chemical reactions are used to produce electricity directly.
    2. Spacecraft use fuel cells based on the reaction of hydrogen with oxygen to produce water.
        $2\ H_2 + O_2 \rightarrow 2\ H_2O$

11.6 The Activity Series
  A. Some very reactive metals react with water to produce hydrogen; others react with steam or acids; and some do not react with water or acids to produce hydrogen (see Table 11.2).

    Most active:   Li, Na, K, Rb, Cs, Ca   – react with water, producing hydrogen
         ↓          Mg, Al, Zn, Cr, Fe     – react with steam, producing hydrogen
         ↓          Ni, Sn, Pb             – react with acids, producing hydrogen
    Least active:  Cu, Ag, Hg, Pt, Au      – do not react with acids to form hydrogen

  B. It is possible to arrange these metals according to their relative activity in an "*activity series,*" with reactive metals like K at the top and unreactive metals like Au at the bottom.
    1. The position of a metal in such a table reflects its tendency to give up electrons.
    2. A metal can transfer electrons to the ions of any metal below it.
        $Fe + Cu^{2+} \rightarrow Fe^{2+} + Cu$     (Copper metal is deposited on the iron.)

C. **Corrosion** is the chemical attack of a metal by substances in the environment.
1. The rusting of iron is an electrochemical process requiring iron, water, $O_2$, and an electrolyte to complete the circuit.
$$2\,Fe + O_2 + 2\,H_2O \rightarrow 2\,Fe(OH)_2$$
$$4\,Fe(OH)_2 + O_2 + 2\,H_2O \rightarrow 4\,Fe(OH)_3$$
2. Silver tarnish is $Ag_2S$ formed when silver is oxidized to $Ag^+$, which reacts with $H_2S$ in the air.
$$2\,Ag^+ + H_2S \rightarrow Ag_2S + 2\,H^+$$
Aluminum, being more reactive than silver, can be used to reduce the silver tarnish back to free silver metal.

## 11.7 **Precipitation**: The **Solubility Product** Relationship; $K_{sp}$

A. A salt will start precipitating from solution when its solubility has been exceeded.

1. $BaSO_4(s) \rightleftharpoons Ba^{2+}(aq) + SO_4^{2-}(aq)$
                            0.00001 M     0.00001 M    at 18 °C

$K_{sp} = [Ba^{2+}][SO_4^{2-}] = 1 \times 10^{-10}$ at 18 °C or solubility product constant, $K_{sp} = 1 \times 10^{-10}$

NOTE: $K_{sp}$ is a special type of $K_{eq}$. The concentration of a solid is a constant and not included in equilibrium constant expressions (refer to Selected Topic A).

2. It is important to remember that $BaSO_4$ will precipitate whenever the product of the two ions reaches $1 \times 10^{-10}$, not when both ion concentrations reach $10^{-5}$ M.

B. Solubility products differ greatly in magnitude. In the list below silver acetate is the most soluble and mercury sulfide is the least soluble, but all of these are relatively insoluble materials.

| Compound | Formula | $K_{sp}$ |
|---|---|---|
| Silver acetate | $AgC_2H_3O_2$ | $2.0 \times 10^{-3}$ |
| Calcium carbonate | $CaCO_3$ | $4.8 \times 10^{-9}$ |
| Silver chloride | $AgCl$ | $1.2 \times 10^{-10}$ |
| Silver chromate | $Ag_2CrO_4$ | $2.4 \times 10^{-12}$ |
| Mercury sulfide | $HgS$ | $3.0 \times 10^{-53}$ |

C. Many very insoluble precipitates can be made to go into solution by adding substances that tie up or remove one of the precipitating ions.

Zinc sulfide:    $K_{sp} = 1.1 \times 10^{-21}$     $ZnS \rightleftharpoons Zn^{2+} + S^{2-}$

ZnS will dissolve in acid solutions because the equilibrium can be shifted to the right by the removal of sulfide.
$$S^{2-} + 2\,H_3O^+ \rightarrow H_2S(g) + 2\,H_2O$$

D. Our teeth and bones are largely calcium phosphate salts. The $K_{sp}$ for calcium phosphate is $4 \times 10^{-27}$. However, the concentrations of $Ca^{2+}$ and $PO_4^{3-}$ ions in the body are also quite low. Osteoporosis, brittle bone disease caused by loss of calcium, is common in adults who consume few dairy products or who absorb calcium from foods less efficiently. Teeth can dissolve under acidic conditions that reduce the phosphate ion concentration, such as:
1. Chronic acidosis
2. Loss of salivary glands
3. Bulimia – binge eating followed by induced vomiting
4. Plaque

## Chapter 11 - Electrolytes

11.8 The Salts of Life: Minerals
  A. A variety of inorganic compounds are necessary for proper growth and repair of body tissues.
  B. Salt intake in our diets is a potential concern. NaCl is essential to life, but high salt levels increase water retention that cause swelling (*edema*) and high blood pressure (*hypertension*).
  C. A partial listing of elements (other than H, C, N, and O) essential for living systems follows.
    Ca – bones, teeth, blood clotting, milk
    P – bones, teeth, ATP, nucleic acids, membranes
    K – intracellular cation, muscle contraction
    S – amino acids methionine and cysteine
    Na – extracellular cation, osmotic pressure
    Cl – anion, HCl in gastric juice
    Mg – enzyme cofactor, nerve impulses
    Fe – hemoglobin for $O_2$ transport, cytochromes
    F – teeth enamel, bones
    I – thyroxin in thyroid (goiter)
    Co – vitamin $B_{12}$
    Zn, Mn, Mo, Cn, Cr, Se, V, As – enzyme cofactors

## DISCUSSION

Much of the terminology of this chapter (see question 1 at the end of the chapter) was introduced originally in Chapter 2. The phenomena of freezing-point depression, boiling-point elevation, and osmotic pressure were originally discussed in Chapter 8 and oxidation/reduction in Chapter5. Thus, this chapter ties together many ideas that may have seemed unrelated. Some review of relevant material in the earlier chapters may prove helpful in understanding the material in the present chapter. For anyone new to the subject, electrolytes seem to cause more confusion when they are weak than when they are strong. The confusing culprit is usually the solubility product relationship, a concept introduced with weak electrolytes.

If, while covering the material on solubility products, you find that you were distracted by the arithmetic involved in multiplying powers of 10, read through Appendix I in the text. The mathematical manipulation of exponential numbers is quite straightforward. If it's causing problems, you probably just need to refresh your memory on the rules of exponential arithmetic.

It is also possible that your first encounter with the formula for solubility products has left you unnecessarily confused. We'll attempt to clarify the concept by using examples.

The product of the ion concentrations of barium sulfate ($BaSO_4$) is written: $[Ba^{2+}][SO_4^{2-}]$
The product of the ion concentrations of barium phosphate ($Ba_3(PO_4)_2$) is: $[Ba^{2+}]^3[PO_4^{3-}]^2$

The exponents introduced in the second product are a reflection of the structure of the compound. Each unit of barium sulfate, $BaSO_4$, produces one barium ion ($Ba^{2+}$) and one sulfate ion ($SO_4^{2-}$) in solution. Each unit of barium phosphate, $Ba_3(PO_4)_2$, produces three barium ions and two phosphate ions ($PO_4^{3-}$) in solution. The solubility product is defined in such a way that this difference between the two compounds is taken into account. The simplest way to state this is to say that the exponents in the solubility product formula are equal to the subscripts, which indicate the combining ratio of the ions in the compound formula.

For a given salt (at a given temperature, although we are ignoring temperature for the moment), the product of the ion concentrations cannot exceed the value of the solubility product constant. Such

# Chapter 11 - Electrolytes

constants are determined experimentally and are listed in chemical handbooks and in Table 11.3. The symbol for solubility product constant is $K_{sp}$. Frequently, you would not calculate a solubility product constant or $K_{sp}$, since its value has to be determined experimentally or supplied to you. What we have been doing in this chapter is calculating the product of ion concentrations for various solutions. We then compare this calculated value with the known $K_{sp}$ for a given salt. If the calculated value exceeds the $K_{sp}$, we conclude that precipitation of the salt occurs. If the calculated value is equal to or smaller than the $K_{sp}$, no precipitation occurs, and in the latter case more of the salt can be dissolved in the solution.

Evaluating when one exponential number exceeds or is less than another can sometimes present difficulties. In problems involving $K_{sp}$ we routinely compare numbers with negative exponents. Remember that the value of a number with a large negative exponent is smaller than the value of a number with a small negative exponent.

$10^{-12}$ is smaller than $10^{-6}$
$1.7 \times 10^{-21}$ is larger than $1.2 \times 10^{-22}$
$9.7 \times 10^{-9}$ is smaller than $2.5 \times 10^{-5}$

Finally, let's take a quick look at what may be one last source of confusion. The solubility product relationship has been defined in such a way that it incorporates exponents for those salts whose formulas indicate a combining ratio involving more than one cation or anion, e.g., $Ca_3(PO_4)_2$.

For calcium phosphate, the solubility product is:

$$K_{sp} = [Ca^{2+}]_3 \cdot [PO_4^{3-}]_2$$

If the concentration of calcium ion in a solution is 0.01 M and the concentration of phosphate ion is adjusted to 0.0001 M, the product of the ion concentrations for calcium phosphate is:

$$(0.01)^3 (0.0001)^2 = (0.000001)(0.00000001) = 0.00000000000001$$

But most scientists would immediately convert those concentrations to exponential form:

$$0.01 = 10^{-2} \text{ and } 0.0001 = 10^{-4}.$$

The product of the ion concentrations may then be written:

$$(10^{-2})^3 \cdot (10^{-4})^2 = (10^{-6}) \times (10^{-8}) = 10^{-14}$$

If you don't use exponential numbers regularly, you may find the appearance of exponents within exponents confusing. Scientists would simply argue that it is easier to raise $10^{-2}$ to the third power (multiply the exponents to get $10^{-6}$). If you raise 0.01 to the third power by longhand multiplication, you would have to do something like this:

```
                    .01    } — total of 4 decimal places
                  × .01    
                  .0001       4 decimal places
 total of 6 decimal places — {
                  × .01
  6 decimal places  .000001
```

*You get the same answer (0.000001 = $10^{-6}$), but it probably took longer.*

## Chapter 11 - Electrolytes

**Example:** $K_{sp}$ **problems.** The $K_{sp}$ for calcium phosphate is $4 \times 10^{-27}$ (at 37 °C). Will calcium phosphate precipitate at 37 °C from a solution in which calcium ion concentration is 0.01 M and phosphate ion concentration is 0.0001 M?
Steps:
1. What is the formula for calcium phosphate? $Ca_3(PO_4)_2$
2. What is the solubility product formula for $Ca_3(PO_4)_2$? $[Ca^{2+}]^3 \cdot [PO_4^{3-}]^2$
3. What is the product for the given ion concentrations?
   $(0.01)^3(0.0001)^2$ or $(10^{-2})^3(10^{-4})^2 = (10^{-6})(10^{-8}) = 10^{-14}$
4. Does this exceed the $K_{sp}$ for calcium phosphate?
   Yes, $10^{-14}$ is larger than $4 \times 10^{-27}$.
   **Calcium phosphate will precipitate from this solution.**

Sometimes, instead of just stating the ion concentrations as we did here, these concentrations are given indirectly. In these instances the concentration of the soluble salt that was dissolved to produce the ion in solution is given. We would say 0.01 mole of calcium chloride (which yields 0.01 mole of calcium ion) and 0.0001 mole of sodium phosphate (which yields 0.0001 mole of phosphate ion) were dissolved in 1 L of solution. The results would be the same as given in the example. We are not so much interested in your becoming expert at working solubility product problems as we are in your understanding the chemical explanations of such processes as bone growth and tooth decay. The following problems are presented for those of you who want to double-check your understanding of the solubility product relationship.

## Problems

1. Write the solubility product formulas for the following compounds. Note: These problems do not involve arithmetic evaluations, just which ions are involved and what the proper exponents are.
   a. $Na_2SO_4$   b. $BaCO_3$   c. $Ag_2S$.   d. $Ba(OH)_2$   e. $AlF_3$   f. $Mg_3(PO_4)_2$   g. $MgNH_4PO_4$
   h. copper(I) hydroxide   i. copper(II) hydroxide   j. iron(III) phosphate
   k. iron(II) carbonate   l. calcium phosphate   m. calcium hydrogen phosphate
2. The $K_{sp}$ of barium carbonate ($BaCO_3$) is $5.1 \times 10^{-9}$. Will precipitation of this salt occur from a solution in which both the barium ion and carbonate ion concentrations were originally 0.001 M?
3. The $K_{sp}$ of aluminum sulfide ($Al_2S_3$) is $2 \times 10^{-7}$. Will precipitation occur if the concentration of aluminum ion is 0.1 M and sulfide ion is 0.01 M?
4. A solution is prepared by dissolving 0.1 mole of sodium nitrate ($NaNO_3$) and 0.01 mole of barium chloride ($BaCl_2$) in one liter of the solution. If the $K_{sp}$ of barium nitrate, $Ba(NO_3)_2$, is $4.5 \times 10^{-3}$, will this salt precipitate from solution?

## SELF-TEST

1. Carbon tetrachloride ($CCl_4$) is a:
   a. strong electrolyte   b. weak electrolyte   c. nonelectrolyte
2. Identify the missing product in the following reaction.
   $Mg(s) + H_2O \rightarrow Mg(OH)_2 + ?$
   a. Mg   b. $O_2$   c. $H_2$   d. $Mg(OH)_2$   e. MgO
3. Which is <u>not</u> a strong electrolyte?
   a. NaCl   b. $NaHCO_3$   c. HCl   d. NaOH   e. $NH_3$
4. Identify the <u>most</u> reactive metal in the following list.
   a. Cu   b. Na   c. Fe   d. Mg   e. Li

## Chapter 11 - Electrolytes

5. Identify the least reactive metal in the following list.
    a. Cu    b. Na    c. Fe    d. Mg    e. Li
6. Which of the following metals will not react with acid to release hydrogen?
    a. Ca    b. Zn    c. Fe    d. Ni    e. Ag
7. Which will cause the electric light of the conductivity apparatus to glow only weakly?
    a. sodium acetate    b. acetic acid    c. sodium hydroxide
8. Which will not conduct an electric current efficiently?
    a. ionic solid    b. metallic solid    c. ionic melt    d. solution of ions
9. Which conducts a current most efficiently in solution?
    a. strong acid    b. weak base    c. insoluble salt    d. sugar
10. If one mole of each of the following salts was dissolved in 0.50 L of water, which of the solutions would show the greatest freezing point depression?
    a. NaCl    b. $NH_4NO_3$    c. $MgCl_2$    d. $K_3PO_4$
11. Estimate the freezing point of a 6.0 M NaCl solution.
    a. −6 °C    b. −3 °C    c. −11 °C    d. −8 °C    e. −15 °C
12. Estimate the boiling point of a 6.0 M NaCl solution.
    a. 106 °C    b. 103 °C    c. 111 °C    d. 108 °C    e. 115 °C
13. Which is not true of nonelectrolytes?
    a. Nonelectrolytes fail to yield ions in solution.
    b. When dissolved, nonelectrolytes have no effect on freezing point, boiling point, or osmotic pressure of the solution.
    c. Nonelectrolytes do not conduct an electric current in solution.
14. The osmotic pressure of a solution is greatest if it contains
    a. one mole of particles whose formula weight is 50
    b. one mole of particles whose formula weight is 400
    c. two moles of particles whose formula weight is 20
    d. All of these solutions would have the same osmotic pressure.
15. A solution of hydrogen chloride in water conducts electricity because, in water, hydrogen chloride undergoes a process called:
    a. dissociation    b. ionization    c. melt formation
16. Soluble salts conduct electricity in solution because they undergo a process called:
    a. dissociation    b. ionization    c. melt formation
17. When sodium sulfide dissolves in water, it yields two sodium ions and one sulfide ion. Therefore, a solution of sodium sulfide:
    a. is not electrically neutral as a whole
    b. is electrically neutral as a whole but still conducts an electric current
    c. is electrically neutral as a whole and does not conduct an electric current
18. In electrolysis, oxidation occurs at the:
    a. anode    b. cathode
19. In the reaction, $MgCl_2 \rightarrow Mg + Cl_2$, which element is formed at the anode?
    a. Mg    b. $Cl_2$    c. $MgCl_2$
20. In electroplating, the metal is plated:
    a. on the anode    b. on the cathode
    c. on neither the anode nor cathode    d. on both the anode and cathode
21. Electrical batteries generate a current by taking advantage of:
    a. a difference in the tendency of chemical species to give up electrons
    b. the ability of a solution to build up an excess of positive charge
    c. the ability of a solution to build up an excess of negative charge

Chapter 11 - Electrolytes

22. A car battery is often a lead-acid storage battery that uses lead (Pb) and lead oxide ($PbO_2$) that form lead sulfate ($PbSO_4$). Which is the cathode in a lead-acid storage battery?
    a. Pb     b. $PbO_2$     c. $PbSO_4$
23. The efficiency of converting fossil fuels to heat to generate electricity is about __?__ %.
    a. 10%     b. 20%     c. 35%     d. 60%     e. 90%
24. The efficiency of an electrochemical cell to generate electricity is about __?__ %.
    a. 20%     b. 30%     c. 40%     d. 60%     e. 90%
25. The electrical activity of the <u>brain</u> is measured by:
    a. an electrolysis apparatus            b. a battery
    c. an electrocardiograph                d. an electroencephalograph
26. A salt will precipitate from solution if the product of the ion concentrations for the salt:
    a. exceeds the solubility product constant for the salt
    b. is less than the solubility product constant for the salt
    c. is equal to the solubility product constant for the salt
27. If the solubility product constant for a salt is $3 \times 10^{-9}$, precipitation of the salt will occur if the product of the ion concentrations in the solution is:
    a. $9.4 \times 10^{-10}$     b. $2.0 \times 10^{-9}$     c. $4.6 \times 10^{-8}$
28. The concentration of calcium ion in a solution is $10^{-2}$ M. The concentration of fluoride ion in the solution is $10^{-1}$ M. What is the ion product for $CaF_2$ in this solution?
    a. $10^{-1}$     b. $10^{-2}$     c. $10^{-3}$     d. $10^{-4}$     e. $10^{-5}$     f. $10^{-6}$     g. $10^{-7}$
29. The solubility product constant for $CaF_2$ is $2.7 \times 10^{-11}$. Will calcium fluoride precipitate from the solution described in Question 28?
    a. yes     b. no
30. If bone formation occurs as calcium phosphate precipitates from solution, removal of phosphate ion from the solution will result in:
    a. more rapid bone formation            b. decrease in the rate of bone formation
    c. no change in the precipitation of calcium phosphate
31. The formation of dental caries is favored by:
    a. a decrease in the pH at the surface of the tooth
    b. an increase in the pH at the surface of the tooth
    c. an increase in the phosphate ion concentration at the surface of the tooth
32. Iodide salts are important to the:
    a. oxygen transport system of the body     b. the functioning of the thyroid gland
    c. proper development of bones and teeth
33. The ions of which element are incorporated in the hemoglobin molecule?
    a. iron     b. iodine     c. calcium     d. phosphorus     e. sodium
34. Classify each of the following as a strong (s), weak (w), or nonelectrolyte (n).
    a. $NaNO_3$     b. $CH_3COOH$     c. $(NH_4)_2SO_4$     d. glucose     e. $NH_3$

# Chapter 11 - Electrolytes

## ANSWERS

### Problems

1. a. $[Na^+]^2[SO_4^{2-}]$    b. $[Ba^{2+}][CO_3^{2-}]$    c. $[Ag^+]^2[S^{2-}]$    d. $[Ba^{2+}][OH^-]^2$
   e. $[Al^{3+}][F^-]^3$    f. $[Mg^{2+}]^3[PO_4^{3-}]^2$    g. $[Mg^{2+}][NH_4^+][PO_4^{3-}]$
   h. $CuOH$, $[Cu^+][OH^-]$    i. $Cu(OH)_2$, $[Cu^{2+}][OH^-]^2$    j. $FePO_4$, $[Fe^{3+}][PO_4^{3-}]$
   k. $FeCO_3$, $[Fe^{2+}][CO_3^{2-}]$    l. $Ca_3(PO_4)_2$, $[Ca^{2+}]^3[PO_4^{3-}]^2$    m. $CaHPO_4$, $[Ca^{2+}][HPO_4^{2-}]$

2. Ion product = $(10^{-3})(10^{-3}) = 10^{-6}$    The ion product is greater than the solubility product constant; the salt will precipitate.

3. Ion product = $(10^{-1})^2(10^{-2})^3 = 10^{-8}$    The ion product is smaller than the solubility product constant; the salt will not precipitate.

4. Ion product = $(10^{-2})(10^{-1})^2 = 10^{-4}$    The ion product is smaller than the solubility product constant; the salt will not precipitate.

### Self-Test

1. c
2. c
3. e
4. e
5. a
6. e
7. b
8. a
9. a
10. d
11. c
12. a
13. b
14. c
15. b
16. a
17. b
18. a
19. b
20. b
21. a
22. b
23. c
24. e
25. d
26. a
27. c
28. d
29. a
30. b
31. a
32. b
33. a
34. s, w, s, n, w

# A    Inorganic Chemistry

**KEY WORDS**

| | | | |
|---|---|---|---|
| noble gas | allotropes | nonmetals | acid rain |
| alkali metal | photochemical smog | groups | halogen |
| alkaline earth | metals | ozone | synergistic |
| organicchemistry | basic oxide | industrial smog | transition metal |
| inorganic chemistry | greenhouse effect | metalloid | hard water |
| s-, p-, d-, f-blocks | main group | lanthanide | actinide |
| | | global warming | |

**SUMMARY**

*Organic chemistry* is based on the compounds of carbon.
*Inorganic chemistry* is the chemistry of all the other elements.

A.1 Using the Periodic Table to Write Electronic Configurations
  A. Elements are arranged in the periodic table into *groups* of elements with similar properties. The periodic table can also be viewed as elements arranged in *"blocks"* according to the type of orbital being filled (*s, p, d* or *f*) see below and refer also to Chapter 2.

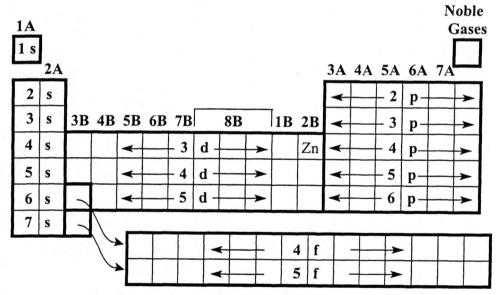

**Electronic configurations and the periodic table**

Use the order-of-filling given below to write the electron configuration.

$1s^2 \quad 2s^2 2p^6 \quad 3s^2 3p^6 \quad 4s^2 3d^{10} 4p^6 \quad 5s^2 4d^{10} 5p^6 \quad 6s^2 4f^{14} 5d^{10} 6p^6$

  B. The periodic table as four *"blocks"* of elements.
   1. *s–block*: ns subshell, left *main group* elements.
   2. *p–block*: np subshell, right main group elements.
   3. *d–block*: (n–1)p subshell, *transition* elements.
   4. *f–block*: (n–2)f subshells; 4f = *lanthanides*; 5f = *actinides*

**Selected Topic A - Inorganic Chemistry**

A.2 The *s*–Block Elements
   A. Group 1A: the *alkali* metals ($ns^1$)
      1. Group 1A elements of lithium (Li), sodium (Na), potassium (K), rubidium (Rb), cesium (Cs), and francium (Fr) comprise the alkali family. The alkalis are soft *metals* with low melting points. They are highly reactive and form basic hydroxides.
      2. Lithium salts are used in the treatment of manic-depressive psychoses. Sodium is found in common table salt (NaCl). Sodium and potassium ions help maintain osmotic pressure and the electrical potential of cells.
   B. Group 2A: the *alkaline earth* Elements ($ns^2$)
      1. The alkaline earth elements are beryllium (Be), magnesium (Mg), calcium (Ca), strontium (Sr), barium (Ba), and radium (Ra). These *metals* are fairly reactive, form 2+ ions, and have *basic oxides* and hydroxides.
      2. Be – very poisonous; forms covalent rather than ionic bonds
         Mg – chlorophyll (photosynthesis), enzymes, nerves
         Ca – bones and teeth; $CaCO_3$ in limestone; *hard water*

A.3 The *p*–Block Elements
   A. Group 3A: boron and aluminum ($ns^2 np^1$)
      1. Group 3A consists of a *metalloid* boron (B) and metals aluminum (Al), gallium (Ga), indium (In), and thallium (Tl).
      2. Aluminum
         a. Aluminum is light and strong, and is often used in place of steel.
         b. Aluminum is more reactive than iron and "rusts" more readily, but $Al_2O_3$ forms a hard, protective surface layer on aluminum, while iron oxide is porous and flaky.
         c. Al is the most abundant metal in Earth's crust. It occurs primarily as bauxite ($Al_2O_3 \cdot xH_2O$). Pure aluminum can be obtained by the Hall process, the electrolysis of melted aluminum oxide. This requires a lot of electricity; it is far cheaper to recycle Al. Hall discovered this procedure in 1886 while a college student.
   B. Group 4A: some compounds of carbon ($ns^2 np^2$)
      1. Group 4A consists of carbon (C), silicon (Si), germanium (Ge), tin (Sn) and lead (Pb).
      2. Carbon forms millions of different compounds with hydrogen, nitrogen, and oxygen. The study of hydrocarbons and their derivatives is called organic chemistry.
      3. Inorganic carbon
         a. Carbon exists in multiple *allotropic* forms: graphite, diamond, and the recently discovered Fullerenes or "Buckyballs" (Nobel Prize, 1996). *Allotropes* are modifications of an element that can exist in more than one form in the same physical state.
         b. When burned, carbon compounds form CO and/or $CO_2$. CO is an invisible, odorless, tasteless, but deadly gas. It combines with hemoglobin in the blood and prevents oxygen transport.
         c. $CO_2$ is the product of complete combustion and respiration. It is not poisonous, but $CO_2$ levels in the atmosphere are increasing and contribute to the "*greenhouse effect*," which could cause a *global warming* trend.
         d. Many minerals, such as limestone, are composed of a carbonate salt (limestone = $CaCO_3$).
   C. Group 5A: some nitrogen compounds ($ns^2 np^3$)
      1. Group 5A consists of nitrogen (N), phosphorus (P), arsenic (As), antimony (Sb), and bismuth (Bi).

**Selected Topic A - Inorganic Chemistry**

2. Nitrogen makes up 78% of the atmosphere, but must be "fixed" before it is a useful source of nitrogen for living systems.
   a. Some nitrogen is fixed by lightning.
   $$N_2 + O_2 \rightarrow 2\,NO \text{ ( and in subsequent steps to ) } \rightarrow NO_2 \rightarrow HNO_3$$

   b. Fritz Haber, a German, discovered how to fix nitrogen industrially on the eve of World War I.
   $$3\,H_2 + N_2 \rightarrow 2\,NH_3$$

   This $NH_3$ was used to make ammonium nitrate, an explosive. However, today the Haber process is used to make nitrogen fertilizer to increase food production.

3. Air pollution
   a. Automobiles also fix nitrogen, leading to the production of NO and $NO_2$. Sunlight can decompose $NO_2$, leading to ozone production. The *photochemical smog* in cities such as Los Angeles consists of unburned hydrocarbons and nitrogen oxides from automobiles and usually occurs in dry sunny climates.

   Automobile engine:　$N_2 + O_2 \rightarrow 2\,NO$
   　　　　　　　　　$2\,NO + O_2 \rightarrow 2\,NO_2$
   Warm air:　　　　$NO_2 + \text{sunlight} \rightarrow NO + O$
   　　　　　　　　　$O + O_2 \rightarrow O_3 \text{ (ozone)}$

   b. $NO_2$ is an irritant to the eyes and respiratory system. *Ozone* ($O_3$) is a toxic pollutant.
   c. These nitrogen oxides also contribute to *acid rain*, but most of the NO and $NO_2$ in the atmosphere is produced by natural processes.

D. Group 6A: compounds of oxygen and sulfur ($ns^2np^4$)
   1. Group 6A consists of *nonmetals* oxygen (O), sulfur (S), selenium (Se), tellurium (Te), and polonium (Po).
   2. Oxygen is very common. It is found as free $O_2$ in the atmosphere and in combined forms in the Earth's crust, water, carbohydrates, proteins, etc. Ozone ($O_3$) is an allotropic form of oxygen. Oxygen reacts with many elements to form oxides.
      a. The oxides of metals are generally basic oxides of nonmetals are acid oxides.
      b. Ozone near the surface of Earth is a toxic and harmful pollutant.
   3. Sulfur occurs in nature in both combined and elemental forms. Sulfur compounds are also important components of proteins and drugs. Elemental sulfur occurs as $S_8$, a ring of eight S atoms.
      a. Sulfur reacts with many metals and hydrogen to form sulfides ($S^{2-}$). $H_2S$ is a toxic gas with the characteristic odor of rotten eggs.
      b. Sulfur can also have oxidation numbers of +4 as in $SO_2$ or +6 as in $SO_3$.
   4. *Industrial (London) smog* is a mixture of smoke, fog, soot, sulfur oxides, and sulfuric acid.
   5. *Acid rain* is caused when sulfur oxides combine with moisture to produce the corresponding acids. Burning sulfur-containing coal releases large amounts of sulfur oxides into the air. Sulfur dioxide and particulate matter have a combined *synergistic* effect; that is, the combined effect is greater than the sum of the parts taken separately. Acid rain can corrode metals, decompose buildings and statues, and make lakes so acidic that all life is destroyed.

E. Group 7A: the halogens ($ns^2np^5$)
   1. All *halogens* ("salt formers") have seven valence shell electrons. They exist as diatomic molecules and usually form –1 anions. This family of elements react with metals to form salts; NaCl, NaF, KF, etc.

2. The members of this family are:
   a. Fluorine ($F_2$) is a gas; small amounts of fluoride are incorporated in tooth enamel to prevent cavities; most fluoride salts are poisons; NaF is a common rat poison; HF is an acid used to etch glass.
   b. Chlorine ($Cl_2$) is a gas; $Cl^-$ is the major anion in living cells; $Cl_2$ is used in water treatments to kill bacteria; NaOCl is a common bleaching ingredient; HCl is used for digestion in the stomach.
   c. Bromine ($Br_2$) is a liquid; AgBr is used in photographic film; HBr is a strong acid.
   d. Iodine ($I_2$) is a solid; iodized salt is used to prevent goiter; it is needed for proper thyroid function.

F. Group 8A: the noble gases ($ns^2np^6$)
   1. This relatively new group of elements is exceptionally resistant to chemical reactions and exists in nature as monatomic gases.
   2. The *noble gases* are unreactive due to their filled outer valence electronic structures ($1s^2$ or $ns^2np^6$).
   3. Members of the *noble gas* family are:
      a. Helium (He) – formed by alpha decay of heavy radioactive elements; nonflammable; good lifting power; also used in breathing mixtures for deep-sea divers
      b. Neon (Ne) – used in lighted signs for advertising
      c. Argon (Ar) – most abundant noble gas in the atmosphere; used in lightbulbs
      d. Krypton (Kr) and Xenon (Xe) – rare and expensive; several compounds of xenon and krypton have been prepared; $XeF_4$.
      e. Radon (Rn) – radioactive

A.4 The *d*–Block Elements
   A. All B–Group elements are metals. There are 10 *transition metals* in each of the 4th through 6th periods of the periodic table, which correspond to filling the 3rd through 5th *d* orbitals, respectively. Most transition metals exhibit more than one oxidation state, and many form colored compounds ($MnO_4^-$ : deep purple color).
   B. A number of transition metals are essential to life: Fe, Cu, Zn, Co, Mn, V, Cr, Mo. Iron is found in hemoglobin and the cytochromes; cobalt is a component of vitamin $B_{12}$; and many of the others are cofactors for enzymes.

# DISCUSSION

There are no major new chemical concepts introduced in this Selected Topic. We have paused in our theoretical development of the subject to undertake a brief descriptive survey of several families of elements. We had previously been looking below the surface of matter--developing a picture of its atomic and molecular structure – evaluating phenomena such as evaporation or precipitation with models such as the kinetic molecular theory. Now we are looking at various elements as we might actually encounter them in our environment. We are making the point that structural theories were developed, for the most part, to account for observed properties of compounds. For example, the observed chemical inertness of the noble gases led to the development of the octet rule. The unifying theme of this chapter is that chemistry is not merely a classroom subject, but a collection of observable phenomena. Your understanding of the theory of chemistry enhances your appreciation of what you observe around you.

Several groups of compounds were identified by family names in this chapter. The periodic table shown below identifies the most common groups by their names.

## Selected Topic A - Inorganic Chemistry

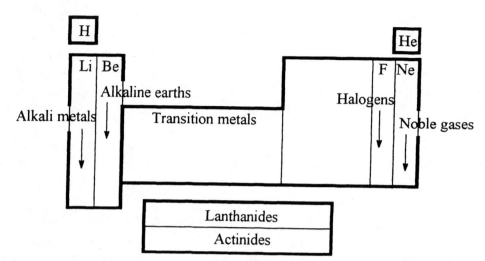

The questions at the end of the chapter in the text will lead you through a detailed review of the descriptive material in the chapter. After you've covered those questions, try the **SELF-TEST**.

## SELF-TEST

1. Potassium is considered to be a(an):
    a. alkali metal   b. alkaline earth metal   c. transition metal   d. noble gas   e. halogen
2. Which is a halogen?
    a. I   b. Ca   c. Cs   d. C   e. Co
3. Which is a transition element?
    a. Ca   b. Co   c. Cs   d. Cl   e. Ar
4. Which is an alkaline earth metal?
    a. C   b. Cs   c. Ca   d. Cl   e. Co
5. Which element is characterized by its extreme nonreactivity?
    a. Al   b. Ne   c. Na   d. $F_2$
6. One of the noble gases is formed in the alpha decay of radioactive elements. Which one?
    a. He   b. Ne   c. Ar   d. Kr   e. Xe   f. Rn
7. The compounds of which element are generally excluded from studies of inorganic chemistry?
    a. B   b. C   c. K   d. Ar   e. U
8. The noble gases have been employed for a number of purposes. Which of the following does not apply to one or more of the noble gases?
    a. used to provide the lift for lighter-than-air ships
    b. used to dilute oxygen in breathing mixtures
    c. used to provide unreactive atmospheres for various purposes
    d. used to absorb sunlight in the upper atmosphere
9. Which hydrohalic acid is involved in the digestive process?
    a. HF   b. HCl   c. HBr   d. HI   e. $H_2SO_4$
10. A mixture of nitrogen oxides, unburned hydrocarbons, and higher-than-normal levels of ozone are the characteristics of:
    a. industrial smog
    b. photosynthesis
    c. photochemical smog

**Selected Topic A - Inorganic Chemistry**

11. The term "inert gas" is no longer used for the noble gases because:
    a. helium can cause the bends.
    b. the noble gases have formed some compounds such as $XeF_4$
    c. the noble gas radon–226 is radioactive.
    d. neon signs glow.
12. Chlorine and fluorine (or their compounds) are added to drinking water in some areas. Which of the following is not an objective of this water treatment?
    a. the destruction of bacteria
    b. strengthening of tooth enamel
    c. reduction of the incidence of goiter
13. Which family of elements forms oxides that are classified as acidic?
    a. alkali metals      b. alkaline earth metals      c. nonmetals
14. Which condition is associated with industrial rather than photochemical smog?
    a. sunny, dry weather
    b. combustion of high-sulfur coal
    c. operation of automobiles
15. Which of these compounds is one of the primary pollutants associated with industrial-type smog?
    a. NO       b. $O_3$       c. $SO_2$       d. $N_2$
16. The greenhouse effect results from an increase in the concentration of one of the following substances in the atmosphere. Which one?
    a. $CO_2$       b. $NO_2$       c. $O_3$       d. $SO_2$       e. particulates
17. Because of the greenhouse effect, the temperature of Earth is expected to:
    a. increase      b. decrease
18. In which location is ozone considered beneficial to life on Earth?
    a. in the upper atmosphere
    b. at ground level
    c. Ozone is extremely toxic and is never considered beneficial to life.
19. Which pollutant does not contribute significantly to the formation of acid rain?
    a. $SO_3$       b. $NO_2$       c. CO
20. Which is a combination that exhibits a harmful, synergistic effect?
    a. sulfur dioxide and particulate matter as pollutants
    b. oxygen and nitrogen as a breathing mixture
    c. ozone and aerosol propellants in the upper atmosphere
21. Nitrogen fixation is accomplished by:
    a. some bacteria       b. lightning       c. automobile engines
    d. the Haber process   e. all of these    f. none of these
22. Nitrates are used as:
    a. fertilizers       b. explosives       c. both a and b       d. neither a nor b
23. Which is not an allotropic form of carbon?
    a. Fullerenes       b. carbon dioxide       c. diamond       d. graphite
    e. All of the above are allotropic forms of carbon.
24. Partial but incomplete combustion of carbon yields:
    a. CO       b. $CO_2$       c. $CH_4$
25. Of the following toxic gases, which is colorless, odorless, and tasteless?
    a. $NO_2$       b. $H_2S$       c. $NH_3$       d. CO
26. Which pollutant could be responsible for the observed lowering of the Earth's average temperature?
    a. $O_3$       b. $CO_2$       c. particulates
27. Which is an allotrope of oxygen?
    a. S       b. $O_3$       c. CO       d. $CO_2$       e. SO       f. O

## Selected Topic A - Inorganic Chemistry

28. Alkali metal ions:
    a. contribute to the osmotic pressure of fluids in living systems
    b. maintain electrical potentials in cells
    c. both a and b
29. The presence of certain ions in water causes the water to be classified as hard. Which ion is **not** included in this group?
    a. $Mg^{2+}$     b. $Fe^{2+}$     c. $Fe^{3+}$     d. $K^+$     e. $Ca^{2+}$
30. Which alkaline earth metal ion is found in chlorophyll?
    a. Be     b. Mg     c. Ca     d. Sr     e. Ba     f. Ra
31. In transition elements:
    a. the outer electron shell is a perfect octet
    b. inner electron shells are not completely filled
    c. both of the above are true
32. Which transition metal is not recognized as necessary for good health?
    a. Fe     b. Co     c. Mn     d. Hg
33. Identify the family with the characteristic, outer-shell electronic structure $ns^2$
    a. alkali metal     b. alkaline earth     c. transition metal     d. halogen     e. noble gas
34. Identify the family with the characteristic, outer-shell electronic structure $ns^2np^5$
    a. alkali metal     b. alkaline earth     c. transition metal     d. halogen     e. noble gas
35. Identify the family with the characteristic, outer-shell electronic structure $ns^2np^6$
    a. alkali metal     b. alkaline earth     c. transition metal     d. halogen     e. noble gas
36. Identify the chemical family for the element zinc.
    a. alkali metal     b. alkaline earth     c. transition metal     d. halogen     e. noble gas
37. Identify the chemical family for the element iodine.
    a. alkali metal     b. alkaline earth     c. transition metal     d. halogen     e. noble gas
38. Identify the chemical family for the element argon.
    a. alkali metal     b. alkaline earth     c. transition metal     d. halogen     e. noble gas
39. Identify the chemical family for the element samarium.
    a. alkali metal     b. alkaline earth     c. transition metal     d. lanthanide     e. noble gas
40. Identify the chemical family for the element silver.
    a. alkali metal     b. alkaline earth     c. transition metal     d. halogen     e. noble gas

## ANSWERS

| | | | | |
|---|---|---|---|---|
| 1. a | 9. b | 17. a | 25. d | 33. b |
| 2. a | 10. c | 18. a | 26. c | 34. d |
| 3. b | 11. b | 19. c | 27. b | 35. e |
| 4. c | 12. c | 20. a | 28. c | 36. c |
| 5. b | 13. c | 21. e | 29. d | 37. d |
| 6. a | 14. b | 22. c | 30. b | 38. e |
| 7. b | 15. c | 23. b | 31. b | 39. d |
| 8. d | 16. a | 24. a | 32. d | 40. c |

# 12 The Atomic Nucleus

## KEY WORDS

| | | | | | |
|---|---|---|---|---|---|
| *alpha decay* | *nucleus* | *fission* | *radioisotopes* | *curie* | *Geiger counter* |
| *radioactivity* | *proton* | *beta decay* | *gamma decay* | *roentgen* | *scintillation* |
| *chemical rxns.* | *neutron* | *fusion* | *ionizing radiation* | *rad* | *medical imaging* |
| *nuclear rxns.* | *electron* | *half-life* | *transmutation* | *rem* | $LD_{50}$/30 days |
| *chain reaction* | *penetrating* | *X-ray* | *CAT, PET, MRI* | *plasma* | *ultrasonograph* |
| *critical mass* | *power* | | *radioactive decay* | *positron* | *electron capture* |

## SUMMARY

### 12.1 Discovery of Radioactivity
A. Wilhelm Roentgen (1895) – discovery of ***X rays***.
B. Henri Becquerel – discovery of ***radioactivity***
C. Marie and Pierre Curie – studies on radioactivity, discovery of radium.

### 12.2 Types of Radioactivity
A. Chemical reactions involve changes in the outer electrons of an atom, while nuclear reactions involve changes within the nucleus of an atom. Some nuclei are unstable and undergo spontaneous reactions called radioactive decay.
B. Rutherford classified three types of radioactivity.
  1. ***Alpha decay***  α particles  $^4$He  4 amu  2+ charge
  2. ***Beta decay***   β particles  $_{-1}e$  1/1837 amu  1–
  3. ***Gamma decay***  γ rays  0 amu  0
C. Two other types of radioactivity
  1. ***Positron***  $β^+$ particle  $_{+1}e$  1/1837 amu  1+
  2. ***Electron capture***  E.C.
D. Examples of the major types of ***radioactive decay***:
  1. **Alpha (α)** decay is characterized by the giving off of particles consisting of 2 protons and 2 neutrons.
  $$^{226}_{88}Ra \rightarrow {}^{222}_{86}Rn + {}^{4}_{2}He$$
  2. **Beta (β)** decay is characterized by the emission of a *nuclear electron* as a neutron is converted to a proton.
  $$^{14}_{6}C \rightarrow {}^{14}_{7}N + {}^{0}_{-1}e$$
  3. **Gamma (γ)** decay is characterized by the emission of high-energy photons ("*nuclear X-rays*") and occurs when an excited nucleus drops to a lower energy state, emitting the energy difference as "light.".

### 12.3 Penetrating Power of Radiation
A. The intensity of all point radiation sources decreases with the square of the distance from the source, i.e. the intensity is only 1/4 the strength at twice the distance. However, the penetrating power of different types of radiation varies widely. This is due in part to their masses but also to

## Chapter 12 - The Atomic Nucleus

their charges. A charged particle will undergo more interactions with matter and will lose energy faster than an uncharged particle.   ***penetrating power*** :   $\alpha < \beta < \gamma$

B. External α–particles are stopped easily, but internal α-emitters are very dangerous because of the large number internal ions that will be produced even over the short distances that α-particles can travel in flesh.

12.4 Radiation Measurement
  A. Curie, roentgen, rad, rem
   1. ***Curies*** (Ci) – 1 Ci = $3.7 \times 10^{10}$ disintegrations per second (~ activity of 1 g Ra).
   2. ***Roentgen*** (R) – a measure of exposure to ***ionizing radiation***; 1R = amount of gamma or X-rays required to produce ions carrying 2.1 billion units of electrical charge in 1 cc of dry air at 0 °C and 1 atm.
   3. Radiation absorbed dose (***rad***) – 1 rad = 100 ergs ($2.4 \times 10^{-6}$ cal) of radiant energy absorbed per gram of tissue; 500 rads would be lethal to most people; average yearly dose absorbed from medical and dental X-rays is about 1 rad. Our exposure to cosmic rays and natural radioisotopes amounts to about 100–600 millirads/year.
   4. Roentgen equivalent in humans (***rem***) – a measure of the biological damage produced by a particular dose of radiation. A radiation worker should not be exposed to more than 500 mrem/year.
  B. Radiation Detectors
   1. ***Geiger counter*** – radiation ionizes gas, causing current signal.
   2. ***Scintillation*** counter – radiation produces a photon, which is amplified by a photomultiplier and counted as an electronic signal.
   3. Film badge – radiation "blackens" film.

12.5 Half-life
  A. Radioactivity is dependent on the ***isotope*** involved but generally independent of any outside influences such as temperature or pressure.
  B. ***Radioactivity*** is a random process but, for large populations, it has a predictable half-life characteristic of that isotope. The ***half-life*** is that period of time during which one-half of the radioactive atoms undergo decay. The half-life of radioisotopes varies from less than a second to millions of years.
  C. The fraction of the original radioactive sample remaining after "n" half-lives is given by $(1/2)^n$ or ½, 1/4, 1/8, 1/16, and so forth.

12.6 Radioisotopic Dating
  A. The half-lives of some isotopes can be used to estimate the age of rocks and archaeological artifacts.
   1. Rocks: Uranium–238 decays with a half-life of 4.5 billion years, eventually decaying to lead–206. The ratio of Pb–206 to U–238 remaining can therefore be used to date very old rocks back to the origin of earth.
   2. Artifacts: Carbon–14 is formed in the upper atmosphere and decays to nitrogen–14 with a half-life of 5730 years. The amount of C–14 remaining can be used to estimate the age of organic objects as old as 50,000 years. C–14 dating has been useful to detect forgeries of ancient artifacts.
   3. Tritium: Hydrogen–3 has a half-life of 12.3 years and is useful in dating materials less than 100 years old.

# Chapter 12 - The Atomic Nucleus

12.7 Artificial Transmutation and Induced Radioactivity
   A. Nuclear changes can also be brought about by the bombardment of stable nuclei with subatomic particles.
   $$^{9}_{4}Be + ^{4}_{2}He \rightarrow ^{12}_{6}C + ^{1}_{0}n$$
   B. **Transmutations** have led to the discovery of fundamental particles such as the neutron.
   C. A common application of artificial transmutations is to produce unstable radioactive isotopes by the bombardment of other isotopes with subatomic particles.
   $$^{27}Al + \alpha \text{ particle} \rightarrow ^{30}P + \text{a neutron}$$
   D. Many isotopes used in medicine undergo decay by electron capture (E.C.).
   $$^{125}I + e^- \rightarrow ^{125}Te + X\text{-rays}$$

12.8 Fission and Fusion
   A. Albert Einstein (1905) derived a relationship between matter and energy as part of his theory of relativity.
   $$E = mc^2 \quad \text{where} \quad E = \text{energy}$$
   $$m = \text{mass}$$
   $$c = \text{speed of light}$$
   B. Nuclear energy changes are many orders of magnitude greater than chemical energy changes. One gram of matter is equivalent to the burning of approximately 21 billion kilocalories, or $9 \times 10^{13}$ J.
   C. Nuclear *fission* refers to the process of splitting heavy atomic nuclei into major fragments. Enormous amounts of energy are released during this process.
   D. Neutron bombardment of uranium–235 produces Xe–143, Sr–90 plus three free neutrons which can be used to split other U–235 atoms to create a **chain reaction**. The **critical mass** is the minimum amount of fissionable material needed to sustain a chain reaction.
   E. Binding energy is the energy that the nucleus no longer has; it represents the energy given up when the nucleus formed. It is related to the mass defect (isotopic mass – mass number) by $E = mc^2$. The most stable nuclei are in the vicinity of iron ($Z = 26$), implying that large amounts of energy are also available from combining very light nuclei to produce heavier nuclei, that is, nuclear *fusion*. Hydrogen is the most abundant element in the universe. In stars like our sun, very high temperatures permit light nuclei to fuse as hydrogen forms helium, for example. Temperatures of 50,000,000 °C must be maintained to promote fusion reactions. Hydrogen bombs utilize a small fission bomb to create the high temperatures needed for the fusion process. In fusion reactors, magnetic fields are used to control the *plasma* of nuclei and free electrons.
   F. Many radioactive daughter isotopes may be produced as nuclear fallout from nuclear fission reactions. Several of these daughter isotopes are dangerous when they are incorporated into the food chain. Strontium–90 can substitute for calcium and is incorporated into the bone; Cesium–137 mimics potassium; Iodine–131 is concentrated in the thyroid.

12.9 Nuclear Medicine
   A. Radiation and living things
      1. Cells that are constantly and rapidly being replaced are more susceptible to radiation damage than are other cells that are replaced less frequently.
      2. The $LD_{50}/30$ days for whole-body exposure in humans is about 400 rems. *$LD_{50}/30$* is the lethal dose required to kill 50% of the populations within 30 days.
      3. Acute radiation syndrome is characterized by a short latent period followed by nausea, vomiting, a drop in white blood cell count, fever, diarrhea, hair loss, and death.

## Chapter 12 - The Atomic Nucleus

    B. Nuclear Tools in Medical Diagnosis
        1. Iodine–131 – size, shape, and activity of the thyroid gland
        2. Cobalt–57 – vitamin $B_{12}$ uptake
        3. Iron–59 – formation and lifetime of red blood cells (RBCs)
        4. Gadolinium–153 – bone mineralization, osteoporosis
        5. Technetium–$99^m$ – good γ-source for imaging of brain, lungs, cardiovascular system
        6. Cobalt–60 is a powerful gamma emitter used in radiation therapy to destroy localized malignancies.
        7. See Table 12.4 for other examples.
    C. The radiation used in cancer treatment can cause radiation sickness, nausea and vomiting, and can lead to birth defects and to some forms of leukemia.

12.10 Medical Imaging
    A. Modern computer-aided *medical imaging* methods provide a means of looking at internal organs without resorting to surgery.
    B. There are many such techniques, each utilizing a different energy source.
        1. *CAT* scan – computer-aided tomography of X-ray and γ scans
        2. *PET* – positron emission tomography
        3. *Ultrasonography* – echo analysis of high-frequency sound waves
        4. *MRI* – nonionizing, nonradioactive, low-energy nuclear magnetic resonance imaging

12.11 Other Applications
    A. Radioactive isotopes can substitute chemically for their nonradioactive counterparts but have the advantage of being easily detected.
    B. The radioisotopes of $^{14}C$ and $^3H$ have helped unravel complex metabolic pathways.
    C. Radioisotopes and animals and vegetables
        1. Irradiation of foodstuffs for preservation
        2. Purposeful mutations of plants by irradiation has led to improved strains of many crops.

12.12 The Nuclear Age Revisited
    A. The atomic bomb
        1. Much of the early work on nuclear fission was carried out by German scientists.
        2. During WW II the United States initiated the Manhattan Project to study atomic energy.
            a. August 6, 1945, a uranium bomb was dropped on Hiroshima.
            b. August 9, 1945, a plutonium bomb was dropped on Nagasaki. Pu–239 is fissionable and can be made from the more abundant isotope, U–238. August 14, 1945, Japan surrendered.
    B. More lives have been saved through nuclear medicine than lost by nuclear bombs which have not been used in warfare since 1945.
    C. Nuclear power continues to represent a plentiful source of energy.
        1. Nuclear power plants use the same fission reactions as nuclear bombs. However, boron and cadmium control rods absorb neutrons to control the reactions. The heat released generates steam, which is used to produce electricity without air pollution. The United States has about 100 nuclear power plants.
        2. The advantage is that there is no soot, ash, or sulfur oxides.
        3. The disadvantages are high construction costs, limited and expensive fuel, radioactive waste products, waste heat that causes thermal pollution, radiation leakage and accidents, and the possibility of a major accident like Three Mile Island (1979) and Chernobyl (1986).

# Chapter 12 - The Atomic Nucleus

## DISCUSSION

Many new words were introduced in this chapter. Review those listed under the **KEY WORDS** section to test your recall of the correct meaning of each term. Two types of quantitative material were introduced: balancing of nuclear equations and problems dealing with half-lives. The key to balancing nuclear equations is to remember that the sum of atomic numbers (Z) and atomic mass numbers (A) must be equal on each side of the equation. The proper chemical symbol can then be deduced from the correct atomic number once it has been determined. It is also helpful to memorize the characteristics of α and β (β⁻ or e⁻) decay with respect to the effect that they have on the atomic and mass numbers (Z and A) for the newly produced daughter nucleus.

| Decay Type | Particle | Parent | Daughter |
|---|---|---|---|
| alpha α | $^{4}_{2}He$ | $^{A}_{Z}X \longrightarrow$ | $^{(A-4)}_{(Z-2)}X$ |
| beta β | $^{0}_{1-}e$ | $^{A}_{Z}X \longrightarrow$ | $^{(A)}_{(Z+1)}X$ |

## Problems

Balance the following equations by supplying the missing component.

1. $^{14}_{7}N + ? \rightarrow\ ^{18}_{9}F$

2. $^{90}_{38}Sr \rightarrow ? +\ ^{0}_{1-}e$

3. $^{238}_{92}U \rightarrow\ ^{4}_{2}He + ?$

4. $^{60}_{27}Co \rightarrow ? + \beta^{-}$

5. $^{56}_{26}Fe +\ ^{2}_{1}H \rightarrow\ ^{54}_{25}Mn + ?$

6. $^{239}_{94}Pu +\ ^{1}_{0}n \rightarrow ? +\ ^{0}_{1-}e$

7. $? +\ ^{1}_{1}H \rightarrow\ ^{4}_{2}He +\ ^{4}_{2}He$

8. $^{3}_{1}H +\ ^{2}_{1}H \rightarrow\ ^{4}_{2}He + ?$

9. $^{241}_{95}Am + ? \rightarrow\ ^{243}_{97}Bk + 2\ ^{1}_{0}n$

10. $^{235}_{92}U +\ ^{1}_{0}n \rightarrow\ ^{103}_{42}Mo + ? + 2\ ^{1}_{0}n$

11. $? +\ ^{1}_{0}n \rightarrow\ ^{141}_{56}Ba +\ ^{91}_{36}Kr + 3\ ^{1}_{0}n$

12. Write an equation for the alpha decay of thorium–228.
13. Protactinium–234 has a half-life of one minute. After 5 minutes, how many micrograms of this isotope remain in a sample that originally contained 80 μg?

## Chapter 12 - The Atomic Nucleus

14. If the half-life of polonium–218 is 3 minutes and a sample originally contains 64 mg of this isotope, how much of the isotope remains after 9 minutes?
15. Radioactive $^{154}$Tm has a half-life of 5 seconds. After 20 seconds, how many milligrams of this isotope remain in a sample that originally contained 160 mg?
16. Radioactive nitrogen–13 has a half-life of 10 minutes. After an hour, how much of this isotope would remain in a sample that originally contained 96 mg?

**SELF-TEST** (Refer to a periodic table.)

1. Isotopes of the same element have the same:
    a. number of neutrons    b. atomic number    c. atomic weight
2. Which form of nuclear radiation most closely resembles X-rays?
    a. alpha particles    b. beta rays    c. gamma rays
3. Which is the least penetrating radiation?
    a. alpha particles    b. beta rays    c. gamma rays
4. Which is the most penetrating radiation?
    a. alpha particles    b. beta rays    c. gamma rays
5. In general, which type of radiation is most useful when diagnostic scanning of an internal organ is desired?
    a. alpha particles    b. beta rays    c. gamma rays    d. cosmic rays
6. Which of the radiation-detecting devices signals when the radiation ionizes gas molecules and causes an electric current to flow?
    a. Geiger counter    b. scintillation counter    c. film badge
7. Which type of radiation is not a stream of charged particles?
    a. alpha particles    b. beta rays    c. gamma rays    d. cosmic rays
8. Which is considered an ionizing radiation?
    a. α particles    b. β rays    c. γ rays    d. cosmic rays    e. all of these
9. Which type of radiation can be stopped by a sheet of paper?
    a. alpha particles    b. beta rays    c. gamma rays
10. If the intensity of radiation is 32 units at a distance of 1 m from the source, then the intensity at 4 m from the source is:
    a. 8 units    b. 2 units    c. 4 units    d. 20 units    e. 16 units
11. Which describes the activity of the radioactive source?
    a. curie    b. roentgen    c. rad    d. rem    e. LD$_{50}$/30 days
12. Which is used to measure the dose of radiation absorbed by tissue?
    a. curie    b. roentgen    c. rad
13. A minimum lethal whole-body dose of radiation for most human beings would be in the range:
    a. 5–10 rads    b. 50–100 rads    c. 500–1000 rads    d. 5000–10,000 rads
14. You are exposed to ionizing radiation:
    a. because of atmospheric testing of nuclear weapons
    b. when your teeth are X-rayed by a dentist
    c. simply because you live on Earth
    d. for all of these reasons
    e. in none of these instances
15. Which hydrogen isotope contains two neutrons?
    a. protium ($^1$H)    b. deuterium ($^2$H)    c. tritium ($^3$H)
16. Which is an isotope of $^{16}$O?
    a. $^{16}$C    b. $^{18}$O    c. $^{16}$N    d. $^{16}$O

# Chapter 12 - The Atomic Nucleus

17. For which of the following would Sr be an improper chemical symbol?
    a. $^{90}_{38}X$      b. $^{88}_{38}X$      c. $^{90}_{39}X$

18. If $^{173}Yb$ emits a β particle, the isotope produced is: Cl, Ar, Sc, Ti, V, Mn, Mo, Tc, Ru, Lu, Rh, Pd

19. Which isotope is particularly useful for both diagnostic and therapeutic work with the thyroid gland?
    a. cobalt–60      b. iodine–131      c. technetium–99m

20. The isotope with ideal properties for a large number of diagnostic scanning uses, including the brain, is:
    a. I–131      b. Tc–99m      c. U–235      d. U–238      e. Co–60

21. Which process does this equation illustrate? $^2H + {}^2H \rightarrow {}^4He$
    a. fission      b. fusion      c. radioactivity

22. Which of the following properties makes technetium–99m a good isotope for diagnostic scanning procedures?
    a. It does not emit alpha or beta particles.
    b. It emits extremely high-energy gamma rays.
    c. It has a very long half-life.
    d. All of these reasons.

23. Which of the following nuclear events would not be described as an example of transmutation?
    a. emission of an alpha particle
    b. emission of a beta particle
    c. emission of a neutron

24. Which medical imaging technique involves the use of ionizing radiation?
    a. PET      b. MRI      c. ultrasonography

25. What isotope is produced if a single neutron is ejected when $^{242}Cm$ is bombarded by an α-particle?
    a. $^{242}Am$      b. $^{246}Cf$      c. $^{246}Cm$      d. $^{245}Cf$      e. $^{243}Bk$

# ANSWERS

## Problems

1. $^4_2He$
2. $^{90}_{39}Y$
3. $^{234}_{90}Po$
4. $^{60}_{28}Ni$
5. $^4_2He$
6. $^{240}_{95}Am$
7. $^7_3Li$
8. $^1_0n$
9. $^4_2He$
10. $^{131}_{50}Sn$
11. $^{234}_{92}U$
12. $^{228}_{90}Th \rightarrow {}^4_2He + {}^{224}_{88}Ra$
13. 2.5 μg
14. 8 mg
15. 10 mg
16. 1.5 mg

## Self-Test

| | | | | |
|---|---|---|---|---|
| 1. b | 7. c | 13. c | 19. b | 25. d |
| 2. c | 8. e | 14. d | 20. b | |
| 3. a | 9. a | 15. c | 21. b | |
| 4. c | 10. b | 16. b | 22. a | |
| 5. c | 11. a | 17. c | 23. c | |
| 6. a | 12. c | 18. Lu | 24. a | |

# 13 Hydrocarbons

**KEY WORDS**

| | | | | |
|---|---|---|---|---|
| IUPAC | methyl | combustion | hydrogenation | polymer |
| organic | ethyl | substitution | halogenation | plastics |
| inorganic | propyl | alkene | hydration | polyethylenes |
| saturated | isopropyl | unsaturated | polymerization | polyesters |
| alkane | butyl | monomer | alkyne | polyamides |
| isomers | s-butyl | addition | aromatic | plasticizers |
| alkyl group | t-butyl | halides | chloroform | homologous |
| ortho- | meta- | para- | carcinogen | CFCs, HFCs |
| freons | refrigerants | perhalo | polyvinylchlorides | resonance |
| stem | polyurethanes | benzene | chloroflurocarbon | isobutyl |
| aryl group | aliphatic | natural gas | octane-rating | petroleum |
| condensed structural formula | | | | |

**SUMMARY**

*Organic* versus *Inorganic* – Compounds obtained from plants and animals were called *organic* because they were isolated from organized (living) systems. Today, thousands of organic compounds have been synthesized in the laboratory, and organic chemistry is defined simply as the study of carbon compounds.

The physical and chemical properties of organic and inorganic compounds are different. Organic compounds tend to be insoluble in water, have low melting and boiling points, are flammable, have low densities, and utilize covalent bonds. Many *inorganic* compounds, such as NaCl, are water soluble, have high melting and boiling points, are not flammable, have high densities, and utilize ionic bonds.

13.1 Alkanes: Structures and Names
   A. A hydrocarbon is a molecule composed of only C and H. *Alkanes* are *saturated* hydrocarbons, with each carbon bonded to four other atoms; there are no double or triple bonds.
   B. Alkanes form a *homologous* series with the formula $C_nH_{2n+2}$.
      1. Methane       $CH_4$
      2. Ethane        $C_2H_6$ or $CH_3CH_3$
      3. Propane       $C_3H_8$ or $CH_3CH_2CH_3$
   C. Alkanes with 4 or more carbons have isomers. *Isomers* are different compounds with the same molecular formula. $C_4H_{10}$ = $CH_3$–$CH_2$–$CH_2$–$CH_3$    and    $CH_3$–$CH$–$CH_3$
                                                                    Butane                        $CH_3$ Isobutane
   D. Naming alkanes – All alkanes have the "-ane" ending with a standard "*stem*" used to indicate the number of carbon atoms. These "**stems**" are as follows:
         meth – 1    eth – 2    prop – 3    but – 4    pent – 5
         hex – 6     hept – 7   oct – 8     non – 9    dec – 10
   E. Condensed Structural Formulas
      1. Structural formulas show the bonding of all atoms explicitly by using bond lines as shown here for ethane and pentane.

              H H                              H H H H H
     Ethane H–C–C–H         Pentane  H–C–C–C–C–C–H
              H H                              H H H H H

2. **Condensed structural formulas** are more convenient and easier to write because all (or almost all) bond lines are omitted. Hydrogens and other groups are written next to the carbon atom to which they are attached.
   Ethane   $CH_3CH_3$        Pentane    $CH_3CH_2CH_2CH_2CH_3$

F. Alkyl groups
   1. An *alkyl* group results when a hydrogen atom is removed from an alkane ($CH_4 \rightarrow CH_3-$).
   2. Alkyl groups and are named by replacing the "*-ane*" ending with "*-yl.*"
      *methyl*   $CH_3-$         *ethyl*   $CH_3-CH_2-$         *propyl*   $CH_3-CH_2-CH_2-$
      *isopropyl*   $CH_3-CH-CH_3$      *butyl*   $CH_3-CH_2-CH_2-CH_2-$
      *s-butyl*   $CH_3-CH-CH_2-CH_3$

      *isobutyl*   $CH_3-CH-CH_2-$       *t-butyl*   $CH_3-\underset{CH_3}{\overset{CH_3}{C}}-$
                   $\ \ \ \ \ \ \ \ \ CH_3$

## 13.2 IUPAC Nomenclature
A. The International Union of Pure and Applied Chemistry (*IUPAC*) has established formal rules for naming compounds.

|  *Substituents --- Stem --- Ending*  |

B. IUPAC rules for alkanes
   1. Branched carbon chain molecules are named as substituents of the "parent" compound (the longest continuous carbon chain in the molecule).
   2. All names end in "*-ane*" for saturated hydrocarbons with "stems" as given in 13.1.D above
   3. Hydrocarbon substituents or *alkyl* groups are named by replacing the "*-ane*" with "*-yl.*"
   4. The lowest possible Arabic numerals are used to indicate the positions of substituents attached to the parent chain.
   5. If the same alkyl group appears more than once, the numbers of all the carbons to which it is attached are expressed.
   6. The additional prefixes *di-*, *tri-*, and *tetra-* are used if 2, 3, or 4 substituents of the same type are attached to the parent chain.
   7. Common non-alkyl substituents are:
      nitro = $-NO_2$,   hydroxy = $-OH$,   fluoro = $-F$,   chloro = $-Cl$,   bromo = $-Br$
   8. Substituents should be listed in alphabetical order.
      $CH_3-CH-CH_2-CH-CH_2-CH_3$
      $\ \ \ \ \ \ \ Cl\ \ \ \ \ \ \ \ \ \ CH_3$        2–Chloro–4–methylhexane

## 13.3 Properties of Alkanes
A. Physical properties of alkanes
   1. 1–4 C (gases) ; 5–16 C (liquids) ; > 17 C (solids)
   2. All alkanes are insoluble in and less dense than water.
B. Physiological properties of alkanes
   1. Methane – physiologically inert
   2. 2–6 C alkanes – anesthetics
   3. Liquid alkanes – dermatitis from contact with skin, chemical pneumonia
   4. Heavier alkanes – emollients (skin softeners), petroleum jelly

## 13.4 Chemical Properties: Reactions of Alkanes
A. Alkanes are not reactive toward strong acids, strong bases, or most oxidizing or reducing agents.
B. Paraffin ("little affinity") wax is a mixture of solid alkanes.

## Chapter 13 - Hydrocarbons

C. Reactions:
1. Combustion: $CH_4 + 2\,O_2 \rightarrow CO_2 + 2\,H_2O + $ heat
   Note: Combustion with insufficient oxygen can produce toxic carbon monoxide, CO.
   $2\,CH_4 + 3\,O_2 \rightarrow 2\,CO + 4\,H_2O + $ heat
2. Substitution: $H_3C-H + Cl-Cl \rightarrow H_3C-Cl + H-Cl$
   or simply $CH_4 + Cl_2 \rightarrow CH_3Cl + HCl$

### 13.5 Halogenated Hydrocarbons
A. Halogenated hydrocarbons, or *alkyl halides*, are compounds in which one or more hydrogens have been replaced by halogen atoms (F, Cl, Br, or I).
   1. Common names: Alkyl group followed by the appropriate halide: ethyl bromide = $CH_3-CH_2-Br$.
   2. IUPAC names treat the halogen as a substituent on the parent compound: bromoethane.
B. Halogenated hydrocarbons rarely occur in nature, but synthetic alkyl halides have a multitude of uses, from pesticides to *plastics* in our modern society.
C. Polyhalogenated hydrocarbons: It is also possible to replace all hydrogens by halogens to form *perhalo* compounds, $CF_3CF_3$.
   1. Methane series
      | | | |
      |---|---|---|
      | Methyl chloride | ($CH_3Cl$) | – refrigerant |
      | Methylene chloride | ($CH_2Cl_2$) | – solvent |
      | *Chloroform* | ($CHCl_3$) | – early anesthetic; solvent; *carcinogen* (cancer producer) |
      | Carbon tetrachloride | ($CCl_4$) | – dry-cleaning solvent; possible carcinogen (at high temperatures, $CCl_4$ can react with water to form the deadly gas phosgene ($COCl_2$). |
   
   2. Ethane series – Isomers
      Ethyl chloride ($CH_3CH_2Cl$) – only one form
      Dichloroethane – two isomers:      $ClCH_2CH_2Cl$      $Cl_2CHCH_3$
                                             1,2–Dichloroethane    1,1–Dichloroethane
      Trichloroethane – two isomers:     $CCl_3CH_3$         $CHCl_2CH_2Cl$
                                             1,1,1–Trichloroethane   1,1,2–Trichloroethane

D. *Chlorofluorocarbons* (*CFCs*) are frequently used as *refrigerants* (*freons*) and in aerosol cans.
   1. $CF_2Cl_2$ (a freon) and related compounds are very stable, but can be broken down by UV radiation to produce halide radicals (Cl·), which, in turn, break down the ozone layer.
   2. CFCs have been banned in the United States for use in aerosols but are still used extensively as refrigerants.
   3. Fluorinated hydrocarbons (HFCs) break down in the lower atmosphere instead of the stratosphere and are being used as replacements for CFCs.

### 13.6 Cycloalkanes
A. It is possible to form ring or cyclic structures with compounds of 3 or more carbons. These cycloalkanes have properties similar to their noncyclic counterparts.

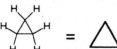

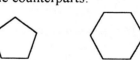

        Cyclopropane       Cyclobutane    Cyclopentane    Cyclohexane

B. Cyclopropane ($C_3H_6$) is an anesthetic.

# Chapter 13 - Hydrocarbons

13.7 Alkenes: Structures and Names
  A. *Alkenes* are *unsaturated* hydrocarbons characterized by a carbon–carbon double bond, $\text{C=C}$. The double bond atoms are planar, and there is no free rotation about the double bond.
  B. Nomenclature
    1. All names end in "*-ene*" with "*stem*"s as given in 13.1.D.
    2. The parent compound is the longest chain of atoms that contains the double bond.
    3. The position of the double bond is indicated by the first carbon involved in the double bond.
    4. Substituents are identified as discussed for alkanes above (13.4 and 13.5).
        $CH_3-C=C-CH-CH_3$
        H  H  Br   4-Bromo-2-pentene      Note: The double bond gets the lowest number.
  C. Common alkenes
    1. Ethene: $CH_2=CH_2$ (also called ethylene). Over 25 billion pounds of ethene are produced in the U.S. It is used in the production of plastics (polyethylene) and antifreeze (ethylene glycol).
    2. Propene, $CH_3CH=CH_2$ (also called propylene), is used extensively for plastics and isopropyl alcohol.
  D. The double bond of alkenes, like the ring structures of cycloalkanes, imposes geometric restrictions that can be lead to geometric or *cis–trans* isomers (See also - Chapter 18). The requirements for cis–trans isomers are:
    1. Restricted rotation within the molecule (double bonds or ring formation)
    2. Two nonidentical groups on each of the doubly bonded carbons

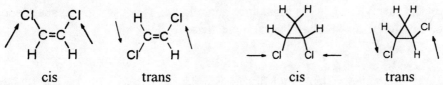

      cis          trans          cis          trans

13.8 Properties of Alkenes
  A. The physiological properties of alkenes are similar to those of alkanes.
  B. 1. Ethylene – anesthetic, used for ripening of fruit
     2. Butadiene, $CH_2=CH-CH=CH_2$ – found in coffee
     3. Carotene – vitamin A and vision

13.9 Chemical Properties: Reactions of Alkenes
  A. *Combustion*:  $C_2H_4 + 3\,O_2 \rightarrow 2\,CO_2 + 2\,H_2O + \text{heat}$
  B. *Addition* reactions:
    1. *Hydrogenation*    $CH_2=CH_2 + H_2 \rightarrow CH_3-CH_3$
    2. *Halogenation*     $CH_2=CH_2 + Br_2 \rightarrow Br-CH_2-CH_2-Br$
                          (brown-red)        (colorless)
    3. *Hydration*        $CH_2=CH_2 + H-OH \rightarrow CH_3-CH_2-OH$   (alcohol)
    4. *Polymerization*   $n\,\text{C=C} + n\,\text{C=C} \rightarrow -(-\text{C}-\text{C}-\text{C}-\text{C}-)_n-$

13.10 Polymerization
  A. The word *polymer* comes from the Greek term for "many parts." The building blocks of polymers are called *monomers* ("one part").
  B. There are two general types of polymerization reactions:
    1. Addition polymerization – uses an addition reaction in such a way that the polymeric product contains all the atoms of the starting monomeric units. Alkenes react by addition reactions and

## Chapter 13 - Hydrocarbons

are common starting materials for most addition polymers. Some addition polymers are *polyethylenes*, polypropylene, polystyrene, "Teflon," and the "PVCs" (see Table 13.11)

2. Condensation polymerization – uses a condensation reaction that results in the formation of a nonpolymeric "by-product." The "by-product" is often water, which is produced when ester and amide linkages are formed. Condensation polymers include *polyamides* (nylon), *polyesters* (Dacron), and **polyurethanes** (foam rubber). Natural materials such as proteins, cellulose, and starch are condensation polymers.

3. Synthetic polymers represent over half of the compounds produced by the chemical industry.

C. *Plasticizers:* Hard and brittle polymers can be made more flexible by the addition of small molecule internal lubricants called *plasticizers*.

1. *Polyvinyl chloride* (PVC) is normally hard and brittle but can be used for garden hoses, auto seat covers, etc. by the addition of plasticizers.
2. Plasticizers are usually lost by evaporation, and with age the plastic becomes brittle and breaks.

### 13.11 Alkynes

A. *Alkynes* are unsaturated hydrocarbons characterized by a carbon-carbon triple bond, $-C\equiv C-$.
B. The rules for naming alkynes are similar to those for naming alkenes, except that the "-yne" ending is used.
C. The physical and chemical properties of alkynes are very similar to those of the corresponding alkenes.
D. Ethyne, $HC\equiv CH$ (acetylene), is the most common alkyne. It is used in oxyacetylene torches and as the starting material for vinyl and acrylic plastics.

### 13.12 Benzene

A. *Benzene* was first isolated in 1825 by Michael Faraday who determined it had the empirical formula of $CH_2$. Later it was determined to have the molecular formula $C_6H_6$.
B. In 1865 Kekule proposed a cyclic structure for benzene. However, it has been determined that all of the bonds in benzene are identical and that there are no real double bonds in benzene. Benzene doesn't react by addition reactions, but rather by *substitution* reactions like alkanes. The structure of benzene cannot be described by a single "Lewis" structure and is said to have *resonance* forms. Resonance compounds *are not* oscillating between their resonance forms; there is really just one hybrid form. However, you will see benzene represented by each of the structures shown below.

### 13.13 Structure and Nomenclature of Aromatic Hydrocarbons

A. Compounds containing the benzene ring-type structure are called *aromatic* compounds. Non aromatic compounds are *aliphatic* compounds. Many have pleasant aromas; others stink.
B. Many aromatic compounds go by their common names.

    Toluene      o-Xylene      Naphthalene

The *ortho-* (*o-*), *meta-* (*m-*), and *para-* (*p-*) prefixes refer to the (1,2-), (1,3-), and (1,4-) disubstituted benzenes, respectively.

C. Aromatic substituents (*aryl groups*) are also possible:   phenyl         benzyl

D. Benzpyrene is a polycyclic aromatic carcinogen found in coal tar and charcoal-broiled steaks.

Benzpyrene

13.14 Uses of Benzene and Benzene Derivatives
   A. All aromatic hydrocarbons are insoluble in water. Benzene, toluene, and the xylenes are common organic solvents. Aromatic compounds react by substitution (not addition) on the benzene ring.
   B. Nitrobenzene is used to manufacture aniline, which, in turn, is used to make many dyes and drugs.
   C. The aromatic ring of benzene is also present in the amino acids (Phe, Tyr, and Trp) and vitamin related cofactors riboflavin, vitamin K, and folic acid.
   D. Physiological properties
      1. Most aromatic hydrocarbons present toxic hazards. Prolonged exposure to benzene may cause leukemia.
      2. Toluene is somewhat less toxic than benzene.

13. Essay: Petroleum and Natural Gas
   A. *Petroleum* is a complex mixture of hydrocarbons produced by the decomposition of animal and vegetable matter entrapped in the earth's crust. One barrel of crude oil equals 42 gallons. It consists largely of a mixture of alkanes consisting of 1 to 40 carbons that are in turn used to produce most *petrochemicals*. Crude oil is the liquid and natural gas is the gaseous parts of petroleum. These are separated by fractional *distillation* into various groups by size:

   | # C s | Fraction |
   |---|---|
   | 1–4 | *Natural gas* (80% $CH_4$, 10% $C_2H_6$) |
   | 5–12 | *Gasoline* |
   | 12–16 | *Kerosene* |
   | 15–18 | Heating oil |
   | 17–up | Lubricating oil |
   | 20–up | *Paraffins*, etc. |

   B. Fractions containing 12 or more carbons can be converted into more valuable gasoline by a process called *cracking*, which involves heating in the absence of air. The unsaturated hydrocarbons are the starting materials for making plastics and detergents.
   C. The "*octane-rating*" was established in 1927 when it was observed that isooctane was best at minimizing "*engine knock.*" It was assigned an octane rating of "*100 octane.*" Heptane, a straight, 7-carbon alkane caused a very bad "knock" and was assigned "0 octane." A gasoline rated "90 octane" performed the same as a mixture of 90% isooctane and 10% heptane.
   D. *Tetraethyllead* is an "octane booster." When added at as little as 1 part per 1000 parts of gasoline, it can increase the octane reading from 55 to 90 or more. However, lead is toxic and tetraethyllead is being replaced by other "octane boosters" or "oxygenates" such as methyl *t*-butyl ether, methanol, ethanol, and *t*-butyl alcohol.

# Chapter 13 - Hydrocarbons

## DISCUSSION

Much of Chapter 13 has been devoted to the subject of nomenclature. In many ways, nomenclature is a game like Monopoly or poker or baseball. There are rules to be learned, and in the beginning it sometimes seems that you'll never remember all of them. However, by the time you finish working through the questions at the end of the chapter, you should have a pretty good grasp of the rules. We strongly recommend that you do take the time to work these problems. You'll find that an investment of time now will be paid back in understanding of later chapters. The basic rules for naming hydrocarbons provide a foundation for naming all other families of organic compounds. The following summary restates briefly the more important rules for naming hyrocarbons.

**IUPAC Nomenclature** involves three parts.

---

*Substituents --- Stem --- Ending*

**Stems** – length of the longest C chain.

| | | | | |
|---|---|---|---|---|
| meth – 1 | eth – 2 | prop – 3 | but – 4 | pent – 5 |
| hex – 6 | hept – 7 | oct – 8 | non – 9 | dec – 10 |

**Ending** – identifies the functional class. ( Note: The last four will be covered in future chapters.)

| alkane | alkene | alkyne | alcohol | aldehyde | ketone | carboxylic acid |
|---|---|---|---|---|---|---|
| "-ane" | "-ene" | "-yne" | "-ol" | "-al" | "-one" | "-oic" |

**Substituents:** Alkyl groups: replace the "-ane" ending with "-yl."

*methyl* $CH_3-$  *ethyl* $CH_3-CH_2-$  *propyl* $CH_3-CH_2-CH_2-$

*isopropyl* $CH_3-CH-CH_3$  *butyl* $CH_3-CH_2-CH_2-CH_2-$

*s-butyl* $CH_3-CH-CH_2-CH_3$

*isobutyl* $CH_3-CH-CH_2-$ / $CH_3$  *t-butyl* $CH_3-\underset{CH_3}{\overset{CH_3}{C}}-$

Non-alkyl: nitro = $-NO_2$, hydroxy = $-OH$, fluoro = $-F$, chloro = $-Cl$, bromo = $-Br$

---

**Halocarbon nomenclature:** The halogens are named as substituents of a parent hydrocarbon molecule.
  Examples: chloromethane, 1-bromo-2-methylpropane, fluoroethene, 1,3,5-trichlorobenzene
**Common names:** Alkyl halide--compounds are named as derivatives of the halides.
  Examples: methyl chloride, isobutyl bromide, vinyl fluoride, cyclopentyl iodide

---

Addition polymers are made by reacting alkenes. Condensation polymerization involves chemistry we will encounter in our study of alcohols, amines, and carboxylic acids. Polyester formation differs from simple esterification only in the number of functional groups per molecule that react. In simple esterification, an alcohol with one hydroxyl group reacts with an acid with one carboxyl group:

$$R-\underset{}{\overset{O}{\overset{\|}{C}}}-OH + HO-R' \rightarrow R-\underset{}{\overset{O}{\overset{\|}{C}}}-O-R' + H_2O \quad \textbf{Esterification}$$

In **polymerization** the only difference is that each molecule must have at least 2 functional groups. In polyester formation, for example, one molecule may have 2 hydroxyl groups and the other 2 carboxyl groups on every molecule may have one hydroxyl and one carboxyl group.

HOH₂C—⌬—CH₂OH  +  HO—C(=O)—⌬—C(=O)—OH   or simply   HO—C(=O)—⌬—CH₂OH

# Chapter 13 - Hydrocarbons

The reaction of both functional groups on each on each molecule ties hundreds or thousands of these monomers together to form the polymer.

## Problems:

1. Draw and name the nine isomeric alkanes containing a total of seven carbon atoms.
2. a. Draw and name the thirteen isomeric alkanes (not including geometric isomers) containing a total of six carbon atoms.
    b. Which of the compounds in part (a) can exist as cis–trans isomers.
3. Draw and name the seven isomeric alkynes containing a total of six carbon atoms.
4. Draw and name the eight compounds containing a benzene ring and 3 additional saturated carbon atoms.
5. Draw and name each of the six compounds containing a total of five saturated carbon atoms and incorporating a ring.

The rest of the chapter material is reviewed in the **SELF-TEST**.

## SELF-TEST

1. The simplest hydrocarbon is:
    a. $CH_2$   b. $C_2H_4$   c. $C_2H_6$   d. $CH_3$   e. $CH_4$
2. The one element necessarily present in every organic compound is:
    a. hydrogen   b. oxygen   c. carbon   d. nitrogen   e. sulfur
3. A series of carbon compounds in which each member differs by $-CH_2-$ from the preceding member of the series is known as a(n):
    a. aromatic series   b. homologous series   c. hydrocarbon series   d. paraffin series
4. Compounds containing only carbon and hydrogen are known as:
    a. methane   b. hydrocarbons   c. carbohydrates   d. isomers   e. aromatic
5. Which term would be associated with the term paraffins?
    a. alkanes   b. alkenes   c. alkynes   d. alcohols
6. Compounds comprised of the same number and kinds of atoms but different in their atomic arrangement are known as:
    a. isotopes   b. isomers   c. homologs   d. allotropes
7. In ethylene the two carbons are joined by a(n):
    a. ionic bond   b. single bond   c. double bond   d. triple bond
8. Restricted rotation about double bonds results in:
    a. geometric isomerism   b. fused ring compounds   c. aromatic compounds
9. Which compound does <u>not</u> contain a double bond?
    a. acetylene   b. butene   c. cyclohexene   d. propylene
10. In the name cyclohexane, the prefix "*cyclo-*" means that:
    a. the carbon atoms are joined in a ring
    b. the compound is explosive
    c. the compound is a derivative of benzene
    d. each carbon is attached to every other carbon atom
    e. the carbons have a valence of three
11. Which is an unsaturated hydrocarbon?

   a.
   ```
   H  H
   |  |
   H-C-C-H
   |  |
   Cl Cl
   ```
   b. $H_3C-\overset{H}{C}=O$
   c. $H-\overset{H}{C}=\overset{H}{C}-H$
   d.
   ```
   H H
   | |
   H-C-C-O-H
   | |
   H H
   ```

## Chapter 13 - Hydrocarbons

12. Benzene and its derivatives are commonly known as:
    a. alkenes     b. aromatics     c. cycloparaffins     d. alkanes
13. Which is <u>not</u> an acceptable structure for benzene?
    a.                 b.                 c.

14. How many compounds having the formula $C_3H_8$ are possible?
    a. 1    b. 2    c. 3    d. 4    e. 5    f. 6
15. Which is a paraffin?

    a. $CH_2=CHCH_2CH_2CH_2CH_3$          b. (benzene ring)
    c. $CH_3CH_2CH_2CH_2CH_2\ CH_2CH_2CH_2CH_3$      d. $HC{\equiv}CCH_2CH_2CH_2CH_3$

16. In each part of this question, a group of structures (a through d) is presented. All except one of the structures in each group represent the same compound. Pick out the one structure that is actually a different compound. For simplicity, we are drawing only the carbon skeletons.

    A.
           C    C–C          C            C    C         C    C
    a. C–C–C–C–C    b. C–C–C–C–C–C    c. C–C–C–C–C–C    d. C–C–C–C–C
                                     C                                                   C

    B.
                                  C                     C                    C
    a. C=C–C–C–C    b. C=C–C–C    c. C–C–C–C=C    d. C–C–C–C
              C                              C                                                 C

    C.
           C    C          C    C        C    C    C          C    C
    a. C–C–C–C=C    b. C=C–C–C–C    c. C=C–C–C    d. C–C–C–C=C

17. Which is the correct name for   $CH_3-\underset{CH_3}{\overset{}{C}}-CH_3$ ?
    a. butyl     b. isobutyl    c. sec-butyl    d. t-butyl

18. Which alkene exists as a pair of cis–trans isomers?
    a. $CH_3CH_2CH_2\underset{CH_3}{C}=CHCH_3$    b. $\underset{H_3C}{CH_3}C=\underset{CH_2CH_3}{CCH_3}$    c. $CH_3\underset{H_3C}{CH}-\underset{CH_3}{CH}CHCH_3$    d. $CH_3C=CHCH_2CH_3$ $\underset{CH_3}{}$

19. Which is <u>not</u> a reasonable structure for a dimethylbenzene?
    a.              b.              c.              d.

20. The IUPAC name for   $CH_3CH_2CH_2CH_2\underset{CH_2CH_2CH_3}{CHCH_3}$ is:
    a. 2-propylhexane     b. 5-methylheptane     c. 4-methyloctane     d. 5-methyloctane

21. The IUPAC name for $CH_3CH_2\underset{H_2C-CH_3}{C}=CH_2$ is:
    a. 2-ethyl-1-butene     b. 2-ethyl-2-butene
    c. 3-ethyl-3-butene     d. 3-methyl-3-pentene

## Chapter 13 - Hydrocarbons

22. Propylene is:

    a. △   b. △(larger)   c. ⬠   d. CH$_2$=CH–CH$_3$   e. HC≡C–CH$_3$

23. Which compound is a cycloalkane?

    a. CH$_3$CH$_2$CH$_2$CH$_2$CH$_3$   b. CH$_2$=CH$_2$   c. HC≡CCH$_2$CH$_3$   d. ⬠   e. ⬡

24. Acetylene is a(n):
    a. alkane   b. alkene   c. alkyne   d. aromatic compound   e. paraffin

25. Which is not a gas?
    a. methane   b. ethene   c. acetylene   d. octane

26. If hexane and water are mixed, the result is:
    a. a clear solution of hexane dissolved in water
    b. a layer of hexane sitting on top of a layer of water
    c. a layer of water sitting on top of a layer of hexane

27. If just two positions on a benzene ring are substituted, how many isomers are possible?
    a. 1   b. 2   c. 3   d. 4   e. 5   f. 6   g. depends on the substituents

28. Two adjacent substituents on a benzene ring are said to be:
    a. ortho to one another     b. meta to one another
    c. para to one another      d. 1,1-disubstituted

29. Addition reactions are characteristic of:
    a. alkanes   b. alkenes   c. aromatic compounds

30. Substitution reactions are characteristic of:
    a. alkanes   b. alkenes   c. alkynes   d. aromatics

31. The reaction of bromine, Br$_2$, with benzene results in bromobenzene plus:
    a. benzobrome   b. H$_2$   c. Br·   d. HBr

32. The raw material from which most hydrocarbons are obtained is:
    a. gasoline   b. petroleum   c. animal fats   d. vegetable oils

33. Methane gas can cause death through:
    a. chemical pneumonia   b. carcinogenesis   c. asphyxiation

34. Which act as emollients?
    a. gaseous alkanes
    b. low boiling liquid alkanes
    c. high boiling liquid alkanes

35. Which hydrocarbon is used as an effective anesthetic?
    a. methane   b. benzene   c. cyclopropane

36. Hydrogenation of vegetable oils produces:
    a. butter   b. saturated fats   c. unsaturated fats

37. A common fused-ring aromatic compound is:
    a. benzene   b. naphthalene   c. toluene

38. In each case, select the property that is typical of organic rather than inorganic compounds.
    A.   a. melt below 200 °C              b. melt above 200 °C
    B.   a. water soluble                  b. water insoluble
    C.   a. ionic bonding                  b. covalent bonding
    D.   a. specific gravity less than 1   b. specific gravity greater than 1
    E.   a. flammable                      b. nonflammable

# Chapter 13 - Hydrocarbons

39. Which is not an appropriate name for $CH_2=CH-Cl$?
    a. chloroethylene
    b. chloroethene
    c. methylene chloride
    d. vinyl chloride

40. Which compound represents chloroform?

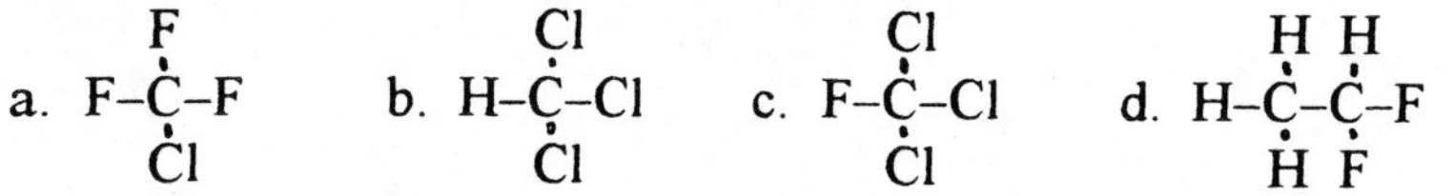

    a. F–C(F)(Cl)–F  b. H–C(Cl)(Cl)–Cl  c. F–C(Cl)(Cl)–Cl  d. H–C(H)(H)–C(F)(F)–F

41. What is the IUPAC name for the following compound?

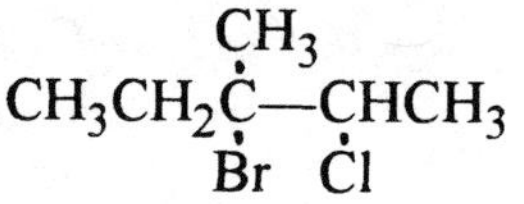

    $CH_3CH_2C(CH_3)(Br)-CH(Cl)CH_3$

    a. 3-bromo-4-chloro-3-methylpropane
    b. 2-chloro-3-bromo-3-hexane
    c. 3-bromo-2-chloro-3-methylpentane
    d. 3-bromo-4-chloro-3-methylpentane

42. Perchlorobenzene is:
    a. 1,4-dichlorobenzene
    b. 1,2-dichlorobenzene
    c. hexachlorobenzene
    d. dodecachlorocyclohexane

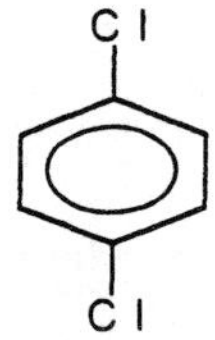

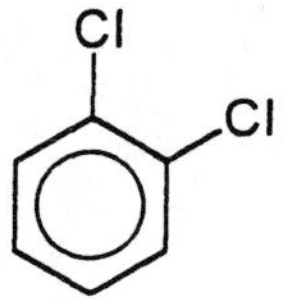

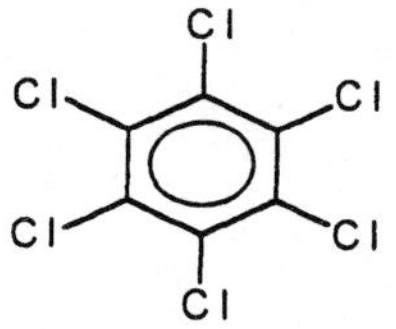

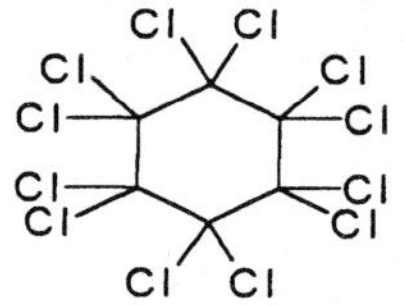

43. Which is not an alkyl halide?
    a. $CH_3CH_2Cl$   b. $CH_3CH_2CH_2Br$   c. $(CH_3)_3C-Cl$   d. chlorobenzene

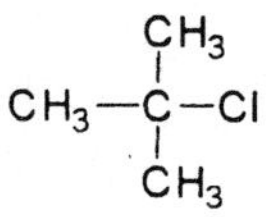

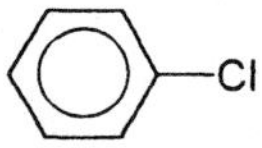

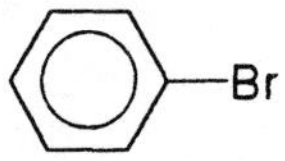

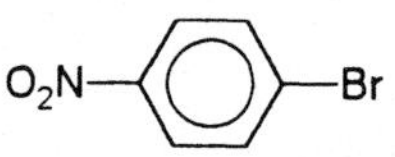

44. For which compound is the symbol Ar–Br not appropriate?
    a. cyclohexyl-Br   b. phenyl-Br   c. $O_2N$-phenyl-Br   d. naphthyl-Br

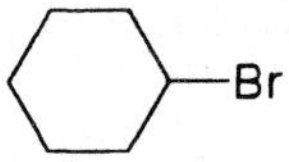

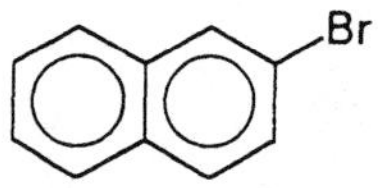

45. Which is not an aliphatic compound?
    a. $CH_3CH_2CH_2Br$   b. $Cl-CH=CH-Cl$   c.  cyclopentyl   d. $CHF_3$   e. 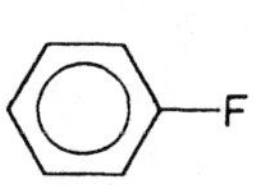 phenyl-F

46. Which is not a reason for discontinuing the use of carbon tetrachloride in fire extinguishers?
    a. The compound is flammable.
    b. Exposure to the compound itself can cause severe liver damage.
    c. The compound can react with water to form the extremely toxic gas phosgene.
    d. All of the above are valid reasons for discontinuing use of $CCl_4$ in fire extinguishers.

47. Which compound enjoyed widespread use as an anesthetic at one time?
    a. $CH_3Cl$   b. $CH_2Cl_2$   c. $CHCl_3$   d. $CCl_4$

48. The most important use for vinyl chloride is:
    a. as a dry-cleaning solvent
    b. as the starting material for the synthesis of vinyl plastics
    c. as a pesticide
    d. as an anesthetic
    e. as a plasticizer
49. If water and chloroform are mixed,
    a. the two miscible liquids form a clear solution
    b. a water layer floats on top of a chloroform layer
    c. a chloroform layer floats on top of a water layer
50. Which compound would propyl chloride most closely resemble in boiling point?
    a. ethane ($C_2H_6$)    b. pentane ($C_5H_{12}$)    c. heptane ($C_7H_{16}$)
51. The small-molecule starting materials from which macromolecules can be constructed are called:
    a. monomers        b. polymers        c. segmers
52. Which compound would not serve as a monomer in <u>addition</u> polymerization?
    a. $CH_2=CH-COOH$    b. $CH_2=CHCH_2OH$    c. $HO-CH_2-COOH$
53. The polymer formed from $CH_2=\overset{\cdot}{C}-CH_3$ is:
    $\phantom{CH_2=C-CH_3\ is:}\overset{\cdot}{Cl}$

    a. $-[-CH_2=\overset{CH_2Cl}{\underset{\cdot}{C}}-]_n-$    b. $-[-CH_2=\overset{Cl}{\underset{\cdot}{C}}-CH_2-]_n-$    c. $-[-CH_2-\overset{Cl}{\underset{CH_3}{\underset{\cdot}{C}}}-]_n-$    d. $-[-CH_2-\overset{CH_2Cl}{\underset{\cdot}{CH}}-]_n-$

54. If the monomer is $CH_3-CH=CH-Cl$, the polymer is:

    a. $[-CH_3CH=\overset{Cl}{\underset{\cdot}{CH}}-]_n$    b. $[-CH_3CH_2-\overset{Cl}{\underset{\cdot}{CH}}-]_n$    c. $[-CH_3CH=\overset{Cl}{\underset{\cdot}{CH}}-]_n$    d. $[-\overset{CH_3}{\underset{Cl}{\underset{\cdot}{CH}}}-CH-]_n$

55. Plasticizers are:
    a. polymers that show elastic properties
    b. polymers that soften on heating
    c. molecules that confer pliability on otherwise brittle polymers
56. The process in which large hydrocarbon molecules are heated in the absence of air and converted to smaller and more highly branched structures is called:
    a. combustion    b. cracking    c. hydration    d. hydrogenation
57. Tetraethyllead is a(n):
    a. gasoline additive    b. anesthetic    c. emollient
58. Hydrocarbons having from 5 to 12 carbons are called:
    a. gasolines    b. heating oil    c. paraffins    d. kerosenes
59. Hydrocarbons having from 12 to 16 carbons are called:
    a. gasolines    b. heating oil    c. paraffins    d. kerosenes
60. Hydrocarbons having from 15 to 18 carbons are called:
    a. gasolines    b. heating oil    c. paraffins    d. kerosenes

# Chapter 13 - Hydrocarbons

## ANSWERS

**Problems**: Nomenclature and Isomerism – Only the carbon skeleton of structural formulas is given.
(c & t = cis and trans isomers)

1. C–C–C–C–C–C–C  heptane

   C–C–C–C–C–C  2-methylhexane
     |
     C

   C–C–C–C–C–C  3-methylhexane
       |
       C

     C
     |
   C–C–C–C–C  2,2-dimethylpentane
     |
     C

     C C
     | |
   C–C–C–C–C  2,3-dimethylpentane

     C   C
     |   |
   C–C–C–C–C  2,4-dimethylpentane

       C
       |
   C–C–C–C–C  3,3-dimethylpentane
       |
       C

   C–C–C–C–C  3-ethylpentane
       |
       C–C

     C C
     | |
   C–C–C–C  2,2,3-trimethylbutane
     |
     C

2. C=C–C–C–C–C  1-hexene

   C–C=C–C–C–C  2-hexene (c & t)

   C–C–C=C–C–C  3-hexene (c & t)

      C
      |
   C=C–C–C–C  2-methyl-1-pentene

        C
        |
   C=C–C–C–C  3-methyl-1-pentene

          C
          |
   C=C–C–C–C  4-methyl-1-pentene

      C
      |
   C–C=C–C–C  2-methyl-2-pentene

   C–C=C–C–C  3-methyl-2-pentene
       |    (c & t)
       C

          C
          |
   C–C=C–C–C  4-methyl-2-pentene
             (c & t)

      C C
      | |
   C=C–C–C  2,3-dimethyl-1-butene

        C
        |
   C=C–C–C  3,3-dimethyl-1-butene
        |
        C

      C–C
      |
   C=C–C–C  2-ethyl-1-butene

     C C
     | |
   C–C=C–C  2,3-dimethyl-2-butene

3. C≡C–C–C–C–C  1-hexyne,  C–C≡C–C–C–C  2-hexyne,  C–C–C≡C–C–C  3-hexyne

       C
       |
   C≡C–C–C–C  3-methyl-1-pentyne,

         C
         |
   C≡C–C–C–C  4-methyl-1-pentyne

         C
         |
   C–C≡C–C–C  4-methyl-2-pentyne,

       C
       |
   C≡C–C–C  3,3-dimethyl-1-butyne
       |
       C

# Chapter 13 - Hydrocarbons

4.  1,2,3-trimethylbenzene   1,2,4-trimethylbenzene   1,3,5-trimethylbenzene

   (1,2-)   (1,3-)   (1,4-)
   *o*-, *m*-, and *p*-ethylmethylbenzene

   *n*-propyl- and isopropyl benzene

5.  1,1-dimethylcyclopropane   1,2-dimethylcyclopropane (cis & trans)

   ethylcyclopropane   methylcyclobutane   cyclopentane

**Self-Test**

| | | | | | |
|---|---|---|---|---|---|
| 1. e | 12. b | 21. a | 32. b | 39. c | 50. b |
| 2. c | 13. c | 22. d | 33. c | 40. b | 51. a |
| 3. b | 14. a | 23. d | 34. c | 41. c | 52. c |
| 4. b | 15. c | 24. c | 35. c | 42. c | 53. c |
| 5. a | 16. A. b | 25. d | 36. b | 43. d | 54. d |
| 6. b |     B. d | 26. b | 37. b | 44. a | 55. c |
| 7. c |     C. d | 27. c | 38. A. a | 45. e | 56. b |
| 8. a | 17. d | 28. a |     B. b | 46. a | 57. a |
| 9. a | 18. a | 29. b |     C. b | 47. c | 58. a |
| 10. a | 19. a | 30. a, d |     D. a | 48. b | 59. d |
| 11. c | 20. c | 31. d |     E. a | 49. b | 60. b |

# 14 Alcohols, Phenols, and Ethers

### KEY WORDS

| | | | | |
|---|---|---|---|---|
| *functional group* | *primary* | *Markovnikov's rule* | *carbolic acid* | *glycol* |
| *alcohol* | *secondary* | *fermentation* | *ether* | *glycerol* |
| *ethanol* | *tertiary* | *phenol* | *peroxides* | *dihydric* |
| *hydroxyl group* | *diol* | *triol* | *azeotrope* | $LD_{50}$ |
| *denatured* | *proof spirit* | *aldehyde* | *ketone* | *aryl* |
| *triglycerides* | *nitroglycerin* | *absolute alcohol* | *anesthesic* | *polyhydric* |

### SUMMARY

14.1 General Formulas and Functional Groups

   A. A *functional group* is a group of atoms that confers characteristic chemical and physical properties on a family of organic compounds. Alcohols, phenols and ethers are related to water, with one or both of its hydrogens replaced by an organic group. *Alcohols* have a *hydroxyl* (–OH) group attached to an aliphatic carbon, while *phenols* have an *aryl* group.

   B.
      *Alcohol*               *Phenol*               *Ether*
      R–OH               Ar–OH           –C–O–C–

     $CH_3CH_2OH$        C₆H₅–OH       $CH_3CH_2–O–CH_2–CH_3$
      (*ethanol*)           (phenol)         (diethyl ether)

14.2 Classification and Nomenclature of Alcohols

   A. Alcohols include molecules as diverse as ethanol (beverage alcohol), cholesterol, glucose, and vitamin A.

   B. Common names
      1. Alkyl group + alcohol: ethyl alcohol
      2. Alcohols are subdivided based on the number of C atoms attached to the carbon bearing the –OH group [primary (0 or 1), secondary (2), and tertiary (3)].

       –CH₂–OH           –C–CH–C–          –C–C–C–
                            OH                  OH
     *primary (1°)*        *secondary (2°)*     *tertiary (3°)*

   C. IUPAC system – identify the longest continuous carbon chain for the corresponding alkane that still contains the hydroxyl group, drop the "-*e*" and add "-*ol*." Identify the position of the hydroxyl group by the number of the carbon to which it is bonded. For "*diols*" and "*triols*," retain the "-*e*" of the parent alkane.

          OH                      OH OH OH
    $CH_3–CH–CH_2–CH_3$         $CH_2–CH–CH_2$
        2-Butanol             1,2,3-Propanetriol (glycerol)

# Chapter 14 - Alcohols, Phenols, and Ethers

### 14.3 Physical Properties of Alcohols
A. Alcohols can use the –OH group to form hydrogen bonds; thus they have higher boiling points than the corresponding hydrocarbons. Most common alcohols are liquids at room temperature.

$$CH_3-CH_2-\overset{..}{\underset{H}{O}}:-----H-\overset{..}{\underset{..}{O}}\diagdown CH_2-CH_3$$

B. The lower molecular weight alcohols are completely miscible with water, but the solubility decreases dramatically when the C:O ratio exceeds 4:1.

### 14.4 Preparation of Alcohols
A. Most alcohols are made by the hydration of the corresponding alkenes in the presence of acid.
  1. $CH_2=CH_2 + HOH \xrightarrow{H^+} CH_3-CH_2-OH$
  2. **Markovnikov's Rule** – "The rich get richer." When there is a difference, the hydrogen goes on the carbon atom of the double bond that already has more hydrogens bonded to it.

$$CH_3-CH=CH_2 + H-OH \rightarrow CH_3-\underset{OH}{CH}-CH_3 \quad \text{200 atm, 350 °C}$$

B. Methanol (wood alcohol) is produced commercially from: $CO + 2H_2 \xrightarrow{ZnO, Cr_2O_3} CH_3OH$

C. **Ethanol** is produced by **fermentation**.

$$\underset{\text{Starch}}{(C_6H_{10}O_5)_x} \longrightarrow \underset{\text{Glucose}}{C_6H_{12}O_6} \longrightarrow \underset{\text{Ethanol}}{2 C_2H_5OH + 2 CO_2}$$

  1. Source of sugar: grain, corn, molasses.
  2. Grind and cook the grain to produce mash.
  3. The enzyme diastase is added to convert the starch to maltose.
  4. Yeast is added to convert the maltose to glucose, and finally to ethanol (up to 18%).
  5. This is filtered and distilled to produce more concentrated liquors. The proof spirit value is twice the percent alcohol by volume, e.g., 50% alcohol in gin = 100 "proof."

### 14.5 Physiological Properties of Alcohols
A. Most alcohols are fairly poisonous, but ethanol and glycerol are less so than the others.
B. Methanol is a common industrial solvent. It is oxidized in the body to formaldehyde.

$$CH_3OH \xrightarrow{\text{liver enzymes}} \underset{H}{\overset{H}{C}}=O \text{ (formaldehyde)}$$

C. Ethanol is also toxic, but much less so than methanol. It too is oxidized in the liver, but to acetaldehyde, which can, in turn, be converted to harmless acetate and finally to $CO_2$.

$$\underset{\text{Ethanol}}{CH_3CH_2-OH} \longrightarrow \underset{\text{Acetaldehyde}}{H_3C-\overset{O}{\overset{\|}{C}}-H} \longrightarrow \underset{\text{Acetate}}{CH_3COO^-} \longrightarrow \underset{\text{Carbon dioxide}}{CO_2}$$

In laboratory tests, acetaldehyde is 27× less toxic ($LD_{50}$ = 1.9 g/kg vs. 0.07 g/kg; where $LD_{50}$ is the lethal dose for 50% of a population) to rats than is formaldehyde. However, prolonged intake of excessive amounts of ethanol can lead to fatty liver, followed by deterioration of the liver and other complications. Alcohol is a depressant of the central nervous system. In many states, a blood alcohol concentration (BAC) of 0.08% is legal evidence of intoxication. A BAC of 0.5% – 1% leads to coma and death.

D. Water forms a constant-boiling **azeotrope** containing 95% ethanol and 5% water. **Absolute alcohol** is 100% ethanol (200 proof). **Denatured** alcohol is ethanol with an additive that makes it unfit to drink.

# Chapter 14 - Alcohols, Phenols, and Ethers

E. A 70% isopropyl alcohol solution is common "rubbing alcohol."
F. Ethanol and methanol can be blended with gasoline to make "gasohol" to increase the octane rating of unleaded gasolines.

14.6 Chemical Properties of Alcohols
  A. Dehydration of alcohols
    1. Alkene formation

$$CH_3-CH_2-OH \xrightarrow[180\,°C]{conc.\ H_2SO_4} \underset{H\quad H}{\overset{H\quad H}{C=C}} + H_2O$$

    2. Ether formation

$$CH_3CH_2-OH + H-O-CH_2CH_3 \longrightarrow CH_3CH_2-O-CH_2CH_3 + H_2O$$

  B. Oxidation of alcohols
    1. Primary alcohols are oxidized to *aldehydes* (then carboxylic acids).

$$R-CH_2-OH \longrightarrow R-\overset{O}{\underset{H}{C}} \longrightarrow R-\overset{O}{\underset{OH}{C}}$$

    2. Secondary alcohols are oxidized to the corresponding *ketone*.

$$\underset{OH}{R-CH-CH_3} \xrightarrow[H^+]{K_2Cr_2O_7} \underset{O}{R-C-CH_3}$$

    3. Tertiary alcohols are resistant to such mild oxidation.

14.7 Multifunctional Alcohols: Glycols and Glycerol
  A. *Polyhydric* alcohols are alcohols with more than one hydroxyl group.
  B. *Glycols* are *dihydric* alcohols.
    1. Ethylene glycol or ethanediol – Main ingredient in permanent antifreeze; oxidized in the body to oxalic acid, $HO-CH_2CH_2-OH$, kidney stones are crystals (precipitate) of calcium oxalate.
    2. Propylene glycol – solvent for drugs; nontoxic since it is oxidized to pyruvic acid, which can enter our body's metabolism.
  C. *Glycerol* (or glycerin) is a *trihydric* alcohol (1,2,3-propanetriol). It is nontoxic and a product of the breakdown of fats (*triglycerides*). Glycerol reacts with nitric acid to form glycerol trinitrate (*nitroglycerin*), an oily explosive liquid used to make dynamite, and is also a drug used to relieve chest pain. Glycerol is also used as a lubricant, in hand lotions, and in the production of plastics and synthetic fibers.

14.8 Phenols
  A. *Phenols* are compounds with a hydroxyl group attached directly to an aromatic ring or *aryl* group.

  C$_6$H$_5$–OH     or simply  Ar–OH

    1. Compared with hydrocarbons, phenols have relatively high boiling points (most are solids at room temperature) because of intermolecular hydrogen bonding.
    2. By definition phenols have at least 6 carbon atoms, which is the borderline of water-solubility for compounds containing one oxygen. Phenols, in contrast to alcohols, are weakly acidic and thus most phenols are soluble in basic solutions.

$$Ar-OH + NaOH \longrightarrow Ar-O^-Na^+ + H_2O$$

# Chapter 14 - Alcohols, Phenols, and Ethers

B. **Phenol** (*carbolic acid*) was introduced by Joseph Lister in 1867 as the first widely used antiseptic.
C. Hexachlorophene (a chlorinated diphenol) was once widely used in germicidal cleaning solutions (pHisohex) and deodorant soaps, but was found to cause neurological disease.
D. Bakelite is a condensation polymer of formaldehyde and phenol.
E. Nomenclature: Simple compounds are named as derivatives of phenol, but many of the most interesting phenols are referred to by their common names.

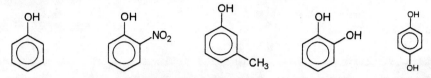

    Phenol     *o*-Nitrophenol  *m*-Cresol    Catechol    Hydroquinone

Other phenol compounds include picric acid, cresol, vanillin, resorcinol, and BHT.

## 14.9 Ethers

A. *Ether* compounds can be considered to be derivatives of water in which both hydrogens have been replaced by alkyl or aryl groups (e.g., R–O–R, R–O–Ar, Ar–O–Ar).
B. Ethers are named by naming the groups attached to the "–O–" and adding "ether."
    $CH_3$–O–$CH_3$ = dimethyl ether;    $CH_3CH_2$–O–$CH_2CH_3$ = diethyl ether
C. Physical properties
  1. Ethers have a boiling point comparable to hydrocarbons of similar molecular weight because ether molecules cannot hydrogen-bond with one another in pure form (no H-bond donors).
  2. Ethers are more soluble in water than corresponding hydrocarbons because ethers can interact with water molecules as acceptors in hydrogen bonding.
D. Chemically ethers are quite inert, more like alkanes, and often are used as an organic extraction medium. They are also very flammable and upon standing in air can form explosive *peroxides*.

$$-\underset{H}{\overset{|}{C}}-O-C- \;+\; O_2 \;\rightarrow\; -\underset{O-OH}{\overset{|}{C}}-O-C-$$

E. Anesthesia: An *anesthetic* acts to block pain. Many anesthetics are fat soluble and appear to work by dissolving in the fat-like membranes of nerve cells to depress conductivity of neurons.
  1. Diethyl ether is a well-known anesthetic that acts as a central nervous system depressant.
  2. Nitrous oxide (laughing gas, $N_2O$) was introduced in 1772.
  3. Chloroform ($CHCl_3$) was first used as a general anesthetic in 1847.
  4. Fluorine-containing compounds (halothane, enflurane) are relatively safe for the patient.

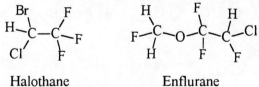

    Halothane         Enflurane

## DISCUSSION

Chapter 14 introduces the carbon-oxygen single bond (–C–O–) into our study of organic chemistry. Alcohols, phenols, and ethers all share this bonding pattern, yet have distinct physical and chemical properties. You should be able to classify compounds into each of these three categories on the basis of their chemical formulas. The physical properties can be correlated with the fact that alcohols and phenols can serve as donors in hydrogen-bond formation, while ethers cannot do so. The nomenclature in this chapter is complicated by the fact that many of these compounds are referred to by their common names. You should memorize the formulas of methanol, ethanol, glycerol, and phenol since you will use them frequently in subsequent chapters. The chemistry of alcohols is treated in more detail than the

# Chapter 14 - Alcohols, Phenols, and Ethers

reactions of phenols and ethers. In particular, take time to review the oxidation behavior of the various types of alcohols before attempting the **SELF-TEST**.

**SELF-TEST**

1. Which compound is an alcohol?
   a. $C_2H_5OH$  b. $CH_3-\overset{O}{\overset{\|}{C}}-CH_3$  c. $CH_3-O-CH_3$  d. $C_6H_6$

2. Which compound is an ether?
   a. $\underset{CH_2-CH-CH_2}{OH\ OH\ OH}$  b. $CH_3CH_2OH$  c. $CH_3-O-C_3H_7$  d. $CH_3-\overset{O}{\overset{\|}{C}}-CH_3$

3. A combination of atoms that confers chemical and physical properties on a compound is called a(n);
   a. ether     b. functional group     c. hydrogen bond

4. The presence of a hydroxyl group attached directly to a benzene ring makes the compound a(n):
   a. alcohol     b. ether     c. phenol     d. base     e. explosive

5. This compound is a(n);

   a. alcohol     b. aldehyde     c. benzene     d. ether     e. phenol

6. Which compound is a phenol?

   a. $CH_3CH_2OH$  b. $CH_3-O-CH_3$  c. $\underset{CH_2-CH_2}{OH\ OH}$  d. $CH_3-\overset{O}{\overset{\|}{C}}-H$  e. (benzene ring)-OH

7. For which compound is R–O–Ar an appropriate abbreviation?
   a. (benzene)-OH   b. $CH_3OH$   c. $CH_3-O-CH_3$   d. $CH_3-O-$(benzene)   e. (benzene)-O-(benzene)

8. Which compound is a trihydric alcohol?
   a. rubbing alcohol     b. propylene glycol     c. glycerol

9. Which is/are primary alcohols?
   a. $CH_3OCH_3$   b. $\underset{CH_3}{CH_3\overset{|}{C}H-CH_2OH}$   c. $\underset{OH}{CH_3CH_2\overset{|}{C}HCH_3}$   d. $\underset{CH_3}{CH_3\overset{OH}{\overset{|}{C}}CH_3}$

10. Which is/are secondary alcohols?
    a. $CH_3CH_2CH_2CH_2OH$  b. $\underset{CH3}{CH_3CH_2CH_2\overset{|}{C}HOH}$  c. $\underset{H}{CH_3\overset{CH_3}{\overset{|}{C}}-OH}$  d. $CH_3-O-CH_2CH_3$  e. $\overset{OH}{\overset{|}{C}H_2OH}$

11. Which compound is a tertiary alcohol?
    a. (benzene)-OH   b. (cyclohexane)-OH   c. (cyclohexane)-CH₂OH   d. (cyclohexane with CH₃ and OH)

12. The compound $\underset{CH_3}{CH_3CH_2\overset{|}{C}H-OH}$ is **not** properly called:
    a. 2-butanal     b. isobutyl alcohol     c. *sec*-butyl alcohol     d. 2-butanol

13. The correct name of $CH_3-O-CH_2CH_3$ is:
    a. diethyl ether     b. ethyl ether     c. methyl ethyl ether     d. ethyl methyl oxide

14. Wood alcohol is the same as:
    a. methanol     b. 2-propanol     c. glycerin     d. grain alcohol     e. rubbing alcohol

132

15. Phenol is:
   a.  [cyclohexyl-OH structure]   b. [benzene-OH structure]   c. CH₃CH₂CH₂–OH   d. CH₃CH₂CH₂CH₂CH₂OH

16. The formula of anesthetic "ether" is:
   a. CH₃OCH₃   b. C₂H₅OH   c. CH₃CH₂C(=O)CH₂CH₃   d. C₂H₅–O–C₂H₅

17. Which compound would have the highest boiling point?
   a. CH₃CH₃–O–CH₂CH₃   b. CH₃OCH₂CH₂OH   c. CH₃CH–CH₂ (OH OH)   d. CH₃CH₂CH₂CH₂OH

18. Which compound would be expected to have the lowest boiling point?
   a. CH₃CH₂CH₂OH   b. CH₃–O–CH₂CH₃   c. HO–CH₂CH₂–OH

19. The alcohol present in alcoholic beverages is:
   a. methyl alcohol   b. isopropanol   c. ethyl alcohol   d. denatured alcohol

20. Alcohols boil at appreciably higher temperatures than hydrocarbons of similar formula weight because:
   a. they have a higher molecular weight   b. alcohols are strongly acidic
   c. alcohols are ionic compounds   d. alcohols are soluble in water
   e. the molecules of the alcohol are associated through hydrogen bonding

21. Which alcohol is least soluble in water?
   a. CH₃OH   b. C₃H₇OH   c. C₆H₁₃OH   d. C₁₀H₂₁OH

22. Which compound would be most soluble in water?
   a. CH₂=CHCH₂CH₃   b. CH₃CH₂OCH₂CH₃   c. CH₃CH₂CH–CH₂OH (OH)

23. Which compound is not miscible with water?
   a. CH₃CH₂OH   b. CH₂–CH₂ (OH OH)   c. [phenol structure]

24. Which compound not miscible with water would be soluble in dilute base?
   a. CH₃CH₂CH₂CH₂CH₂CH₃   b. [phenol structure]   c. [benzene structure]

25. Fermentation of carbohydrates leads to the formation of:
   a. methyl alcohol   b. ethyl alcohol   c. glucose   d. 1–propanol

26. Which solution could not be obtained directly from the fermentation reaction?
   a. 10% ethyl alcohol in water
   b. 30% ethyl alcohol in water
   c. 10 proof ethyl alcohol
   d. 20 proof ethyl alcohol

27. Dehydration of alcohols does not produce:
   a. ethers   b. alkenes   c. aldehydes

28. Which term describes the reaction: CH₂=CH₂ + H₂O —H+→ CH₂–CH₂ (OH H)?
   a. combustion   b. dehydration   c. hydration   d. oxidation

29. Markovnikov's rule indicates that:
   a. when alcohols are dehydrated, they yield alkenes in preference to ethers.
   b. primary and secondary alcohols are readily oxidized, but tertiary alcohols are not.
   c. when water adds to alkenes, the hydrogen of water adds to the double-bonded carbon with the most hydrogens.

# Chapter 14 - Alcohols, Phenols, and Ethers

30. Which product would be expected from the hydration of 1-butene?
    a. $CH_3CH_2CH_2CH_2OH$   b. $CH_3\underset{OH}{CH}-CH_2CH_3$   c. $CH_3CH_2CH_2-O-CH_3$   d. $CH_3\underset{CH_3}{\overset{OH}{C}}-CH_3$

31. Which compound is *not* readily oxidized?
    a. $CH_3\underset{CH_3}{CH}-CH_2OH$   b. $CH_3CH_2\underset{OH}{CH}-CH_3$   c. $CH_3\underset{CH_3}{\overset{OH}{C}}-CH_3$   d. $CH_3CH_2\underset{H}{C}=O$

32. What is the product of the reaction: $CH_3\underset{}{\overset{OH}{CH}}-CH_3 \xrightarrow{K_2Cr_2O_7, H^+}$ ?
    a. $CH_3COOH$   b. $CH_3CH_2CH_2OH$   c. $CH_3CH_2\underset{H}{C}=O$   d. $CH_3CH_2COOH$   e. $CH_3\overset{O}{\overset{\|}{C}}-CH_3$

33. In chemical reactivity, ethers resemble:
    a. alcohols   b. alkanes   c. phenols

34. Air oxidation of ethers results in the formation of:
    a. aldehydes   b. carboxylic acids   c. ketones   d. peroxides

35. Denatured alcohol refers to:
    a. any alcohol not produced by fermentation
    b. grain alcohol that is highly taxed
    c. ethyl alcohol that has been treated with something to make it unfit to drink

36. The toxicity of wood alcohol results from its oxidation by liver enzymes to:
    a. carbon dioxide   b. formaldehyde   c. grain alcohol   d. methanol

37. Which is quite toxic when ingested by humans?
    a. ethylene glycol   b. propylene glycol   c. glycerol

38. Which compound is widely used as a general anesthetic?
    a. dimethyl ether   b. diethyl ether   c. bis(chloromethyl)ether (BCME)

39. What is the product when this compound undergoes oxidation?   [cyclohexyl-OH]
    a. [cyclohexyl-COOH]   b. [phenyl-OH]   c. [benzene]   d. [cyclohexanone]   e. cyclohexene

40. Select the response that gives the proper order for increasing solubility in water.
    a. butanol < hexane < diethyl ether
    b. diethyl ether < butanol < hexane
    c. hexane < butanol < diethyl ether
    d. diethyl ether < hexane < butanol
    e. hexane < diethyl ether < butanol

**ANSWERS**

| | | | | | | | | | |
|---|---|---|---|---|---|---|---|---|---|
| 1. | a | 9. | b | 17. | c | 25. | b | 33. | b |
| 2. | c | 10. | b and c | 18. | b | 26. | b | 34. | d |
| 3. | b | 11. | d | 19. | c | 27. | c | 35. | c |
| 4. | c | 12. | a and b | 20. | e | 28. | c | 36. | b |
| 5. | d | 13. | c | 21. | d | 29. | c | 37. | a |
| 6. | e | 14. | a | 22. | c | 30. | b | 38. | b |
| 7. | d | 15. | b | 23. | c | 31. | c | 39. | d |
| 8. | c | 16. | d | 24. | b | 32. | e | 40. | e |

# 15 Aldehydes and Ketones

## KEY WORDS
| | | | | |
|---|---|---|---|---|
| carbonyl | ketal | acetal | acetone | aldehyde |
| hemiacetal | keto-enol tautomerism | formaldehyde | ketone | hemiketal |
| hydrogen-bond | aldol condensation | Tollens's | hydrate | imine |

## SUMMARY

15.1 The Carbonyl Group: A Carbon-Oxygen Double Bond
   A. The *carbonyl* double bond is polar and tends to undergo addition reactions.
   B. Aldehydes and ketones – *Ketones* have two carbons attached to the carbonyl carbon; *aldehydes* have at least one hydrogen attached to the carbonyl carbon.

Aldehyde      Ketone

15.2 Names of Aldehydes
   A. Common names are frequently used for many aldehydes.
      1. *Formaldehyde* ($H_2C=O$) is a gas. Formalin, a 40% solution of formaldehyde, is a familiar biological preservative. Others in this series are:
         2 C – acetaldehyde      4 C – butyraldehyde
         3 C – propionaldehyde      5 C – valeraldehyde
      2. The positions of substituents are indicated by a Greek letter ($\alpha, \beta, \gamma$), where $\alpha$ corresponds to the C adjacent to the carbonyl group.
   B. The IUPAC rules for naming aldehydes are derived from the corresponding alkanes, dropping the "-e" and adding "-al." Note: No number is needed since the aldehyde carbonyl group is always on the end.

$$CH_3-\underset{\underset{Cl}{|}}{CH}-CH_2-\overset{\overset{O}{\|}}{C}\diagdown_H$$
               $\beta$-Chlorobutyraldehyde (Common name)
               3-Chlorobutanal     (IUPAC name)

15.3 Naming the Common Ketones
   A. Common names
      The simplest ketone has 3 carbons and is called *acetone*. Others are named like ethers using the alkyl groups in alphabetical order and the word ketone.

$$H_3C\overset{\overset{O}{\|}}{C}CH_3 \qquad CH_3-\overset{\overset{O}{\|}}{C}-CH_2-CH_3$$
Acetone          Ethyl methyl ketone

   B. The IUPAC rules for ketones use the longest continuous C chain containing the carbonyl group as the parent, the "-e" ending is dropped and "-one" added. The positions of the carbonyl group and of substituents are given by numbers.

$$CH_3-\overset{\overset{O}{\|}}{C}-CH_2-CH_2-CH_2-F \;=\; \text{5-Fluoro-2-pentanone}$$

# Chapter 15 - Aldehydes and Ketones

15.4 Physical Properties of Aldehydes and Ketones
  A. The physical properties of these compounds are greatly influenced by the marked polarity of the carbon-oxygen double bond, increasing the boiling point of aldehydes and ketones over the corresponding ethers.

$$\overset{\delta+}{\underset{}{\diagup}}\!\!\!\!\!\diagdown C = O^{\delta-}$$

  B. Pure aldehydes and ketones can not "*hydrogen bond*" because they are not hydrogen donors. However, they can participate as hydrogen-bond acceptors and thus behave like alcohols in their water solubility.

15.5 Preparation of Aldehydes and Ketones
  A. Aldehydes are made by oxidizing 1° alcohols.
  B. Ketones can be made by oxidizing 2° alcohols.

$$R-CH_2-OH \xrightarrow{[O]} R-\overset{O}{\underset{}{C}}-H \ ; \quad R-\underset{}{\overset{OH}{CH}}-R' \xrightarrow{K_2CrO_7,\ H_2SO_4} R-\overset{O}{\underset{}{C}}-R'$$

15.6 Chemical Properties of Aldehydes and Ketones
  A. Oxidation: Aldehydes are readily oxidized to carboxylic acids, while ketones resist mild oxidation.
    1. *Tollens's* reagent is a very gentle oxidant that is used to test for the presence of aldehydes. When Tollens's reagent oxidizes an aldehyde, silver ion is reduced to free silver, forming an easily recognizable mirror on the surface of the tube.
    $$R-CHO + 2\ Ag(NH_3)_2^+ + 2\ OH^- \rightarrow R-COO^-NH_4^+ + 2\ Ag^0 + 3\ NH_3 + H_2O$$
    Both aldehydes and ketones are also flammable and will undergo combustion.
    2. Benedict's and Fehling's tests use alkaline solutions of $Cu^{2+}$ (blue) which forms a red, solid precipitate of $Cu_2O$ when reduced in alkaline solutions.
  B. Reduction: Aldehydes are reduced to primary alcohols, and ketones are reduced to the corresponding secondary alcohols using $H_2$ and a Ni or Pt catalyst.
  C. Hydration of carbonyl compounds
    1. The lighter aldehydes readily dissolve in water to form unstable *hydrates*.
    2. The hydrate of trichloroacetaldehyde (chloral hydrate) is a stable solid. It is a powerful sedative and in a drink is called a "Mickey Finn."
  D. Addition of alcohols: hemiacetals and acetals
    1. Alcohols add to the carbonyl group of aldehydes (or ketones) to form *hemiacetals* (or *hemiketals*).

$$CH_3-\overset{O}{\underset{H\ (R)}{C}} + H-O-CH_3 \longrightarrow CH_3-\underset{H\ (R)}{\overset{OH}{C}}-OCH_3$$

   Aldehyde (ketone)                Hemiacetal (hemiketal)

  Hemiacetals are normally rather unstable. Sugars are polyhydroxy aldehydes or ketones that form cyclic, intramolecular hemiacetals that are fairly stable.
    2. Hemiacetals (or hemiketals) can be made to react further by a dehydration reaction with a second alcohol molecule to form a stable *acetal* (or *ketal*).

$$R-\underset{H}{\overset{OH}{C}}-O-R' + H-O-R'' \longrightarrow R-\underset{H\ \text{(an acetal)}}{\overset{O-R''}{C}}-O-R' + H_2O$$

  Acetals are frequently used to protect aldehyde functional groups because acetals are resistant to oxidation, whereas aldehydes are not.

E. Addition of ammonia and amines – *imine* formation (R–CHO + H–NH$_2$ → R–C=N–H + H$_2$O)
F. The hydrogen shift: tautomerism
1. The alpha hydrogens near a carbonyl group are slightly acidic. A carbonyl compound with alpha hydrogens exists in equilibrium with an isomeric form that has the hydrogen shifted to the carbonyl oxygen atom, an enol. This is referred to as ***keto–enol tautomerism***.
2. ***Aldol condensation***: The alpha hydrogens of a carbonyl compound are sufficiently acidic to be pulled off by a strong base. This negatively charged carbanion will add to the carbonyl carbon of another carbonyl molecule in an aldol condensation reaction. The 3 carbon sugars of glyceraldehyde and dihydroxyacetone combine to form the 6 carbon sugar, fructose, by aldol condensation.

$$2\ R-CH_2-\overset{O}{\underset{}{\overset{\|}{C}}}-H \quad \rightarrow \quad R-CH_2-\underset{H}{\overset{O-H}{\underset{}{\overset{|}{C}}}}-\underset{R}{\overset{}{\underset{}{\overset{|}{CH}}}}-\overset{O}{\underset{}{\overset{\|}{C}}}-H$$

(β-Hydroxy product)

15.7 Some Common Carbonyl Compounds
A. Formaldehyde – biological fixative (37% solution = formalin)
B. Acetaldehyde – fermentation of sugars, oxidation of ethanol
C. Acetone – solvent and one of the ketone bodies from lipid metabolism
D. Benzaldehyde – oil of bitter almond
E. Cinnamaldehyde – oil of cinnamon
F. Others – camphor, vanillin, muscone, progesterone, testosterone, etc.

## DISCUSSION

Tables 15.1 and 15.2 in Chapter 15 illustrate the nomenclature rules for aldehydes and ketones. You can use these tables to check your understanding of the common and IUPAC naming systems. If you have not already done so, answer the first twelve questions at the end of Chapter 15 for added practice.

By far, the greater part of this chapter is devoted to chemical reactions involving aldehydes and ketones. The selection of reactions considered is based on their significance in living systems. We will soon encounter aldol condensations, keto-enol tautomerism, and acetal formation in our study of the chemistry of carbohydrates. Carbohydrates contain many functional groups and, on first encounter, appear to be very complex molecules. This is why we are taking the time now to look at reactions that we'll consider again later. By looking at simple molecules, we can concentrate on the general pattern of these reactions. That same pattern will be followed when the reacting molecules are more complex. The reactions of Chapter 15 are gathered here for easy reference.

Chapter 15 - Aldehydes and Ketones

## Summary of Reactions of Aldehydes and Ketones

**I. Oxidation** – This is the reaction that most clearly distinguishes the aldehyde family from the ketone family.

   **A. Aldehydes:**

$$R-\overset{O}{\underset{H}{C}} \xrightarrow{[O]} R-\overset{O}{\underset{OH}{C}}$$

The usual oxidizing agents (like $K_2Cr_2O_7$) work, as do even weaker oxidizing agents (like $Ag^+$ in the form of Tollens's reagent).

   **B. Ketones:** Under ordinary conditions, ketones give no reaction.

**II. Addition reactions involving the carbonyl group –**

   **A. Addition of water (hydrate formation)**

$$-\overset{O}{\underset{}{C}}- \quad \overset{H}{\underset{}{O}}-H \longrightarrow -\overset{O-H}{\underset{}{C}}-O-H$$

   **B. Addition of alcohol (hemiacetal formation)**

    1. Hemiacetal formation

$$-\overset{O}{\underset{}{C}}- \quad \overset{H}{\underset{}{O}}-R \longrightarrow -\overset{O-H}{\underset{}{C}}-OR$$

    2. Acetal formation (accompanied by elimination of water)

$$-\overset{H\quad OH}{\underset{O\quad R'}{C}}-OR \xrightarrow{dry\ HCl} -\overset{H-OH}{\underset{O-R'}{C}}-OR$$

    3. Shortcut for determining acetal formed from one carbonyl and two alcohol units.

$$\overset{}{C}=O \quad \overset{H-OR}{\underset{H-OR}{}} \longrightarrow \overset{}{C}\overset{OR}{\underset{OR}{}} + H_2O$$

   **C. Addition of ammonia and derivatives**

    1. Imine formation (accompanied by elimination of water)

$$-\overset{O}{\underset{H}{C}}-\overset{H}{\underset{}{N}}-H \longrightarrow [-\overset{O-H}{\underset{H}{C}}-N-H] \longrightarrow -C=N-H + H_2O$$

    2. Phenylhydrazone formation (accompanied by elimination of water)

$$-\overset{O}{\underset{H}{C}}-\overset{H}{\underset{}{N}}-NHC_6H_5 \longrightarrow -\overset{O-H}{\underset{H}{C}}-N-NHC_6H_5 \longrightarrow -C=N-NHC_6H_5 + H_2O$$

    3. Shortcut for determining product from ammonia or derivative

$$\overset{}{C}=O \quad \overset{H}{\underset{H}{N}}- \longrightarrow \overset{}{C}=N- + H_2O$$

   **D. Addition of a second aldehyde or ketone (aldol condensation)**

$$-\overset{O}{\underset{}{C}}-\overset{H}{\underset{H}{C}}-\overset{O}{\underset{}{C}}- \longrightarrow -\overset{O-H}{\underset{}{C}}-\overset{}{C}-\overset{O}{\underset{}{C}}- \quad or \quad -\overset{OH}{\underset{}{C}}-\overset{}{C}=\overset{O}{\underset{}{C}}-$$

**III. Isomerism (keto-enol tautomerism)**

$$-\overset{H}{\underset{}{C}}-\overset{O}{\underset{}{C}}- \longrightarrow -\overset{H-O}{\underset{}{C}}=C- \quad or \quad -\overset{OH}{\underset{}{C}}=C-$$

# Chapter 15 - Aldehydes and Ketones

**Problems:** Draw the products of the reactions shown below:

1. 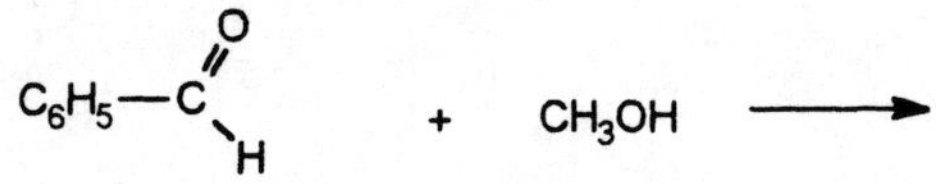 $C_6H_5-CHO$ + $CH_3OH$ ⟶

2. 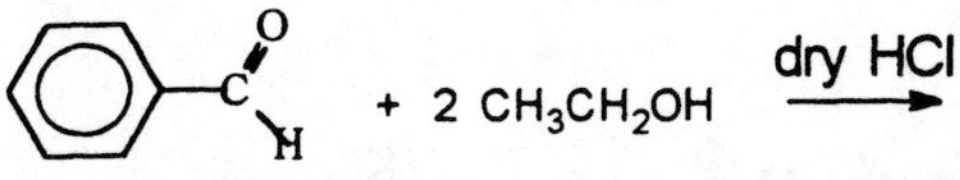 $C_6H_5-CHO$ + 2 $CH_3CH_2OH$ $\xrightarrow{\text{dry HCl}}$

3. 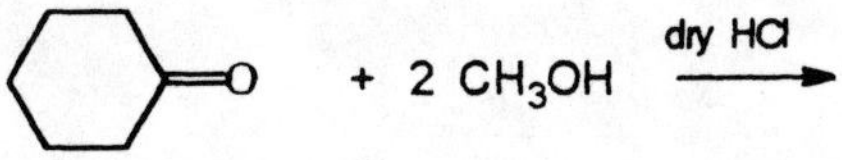 cyclohexanone + 2 $CH_3OH$ $\xrightarrow{\text{dry HCl}}$

4. 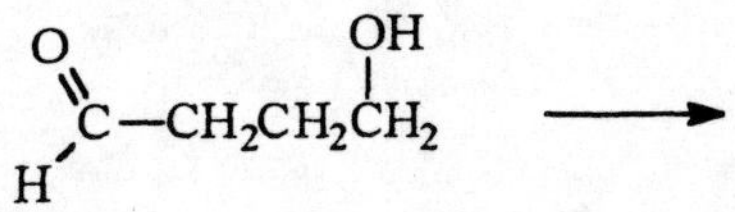 $OHC-CH_2CH_2CH_2-OH$ ⟶ a hemiacetal

5.  cyclohexanone $\xrightarrow{K_2Cr_2O_7, H^+}$

6. 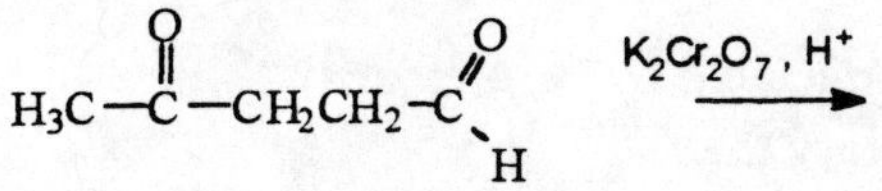 $H_3C-CO-CH_2CH_2-CHO$ $\xrightarrow{K_2Cr_2O_7, H^+}$

7. $C_6H_5-CHO$ + $NH_3$ ⟶

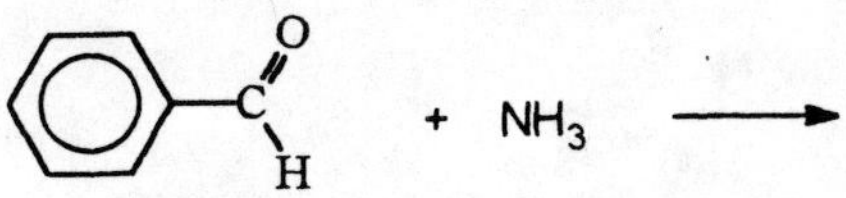

8. 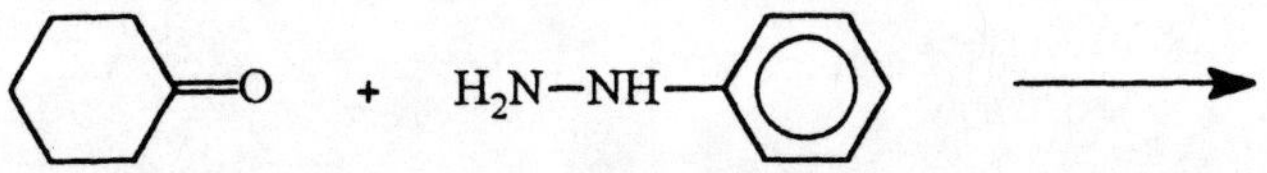 cyclohexanone + $H_2N-NH-C_6H_5$ ⟶

9. 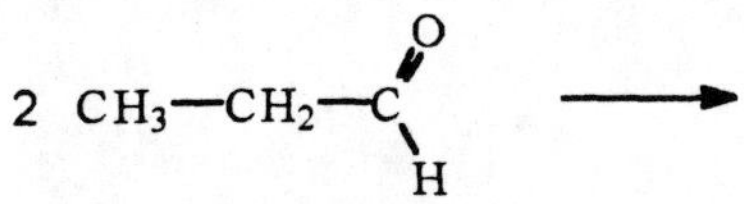 2 $CH_3-CH_2-CHO$ ⟶   Aldol condensation

## SELF-TEST

1. Which structural feature is possessed by aldehydes but not ketones?
   a. an alpha hydrogen      b. a hydrogen on the carbonyl carbon
   c. a hydroxyl group on the carbonyl carbon

2. A group that both aldehydes and ketones have in common is:
   a. –COOH    b. –C=O    c. –C=O    d. –OH    e. –O–
                     |              |
                     H

# Chapter 15 - Aldehydes and Ketones

3. The name of the functional group of aldehydes and ketones is:
   a. carbonyl group   b. carboxyl group   c. double bond   d. hydroxyl group
4. The precipitate that is produced in a positive Benedict's test is :
   a. Ag   b. AgCl   c. Cu   d. $Cu_2O$   e. CuO
5. The compound $CH_3COCH_3$ is:
   a. methyl alcohol   b. formic acid   c. formaldehyde   d. acetone   e. methanone
6. Benzaldehyde is:
   a. $CH_3CH_2CH_2CH=O$   b.   c.   d.
7. The name of $CH_3CH_2\underset{O}{\overset{..}{C}}-CH_3$ is:
   a. methyl propyl ketone   b. 2-pentanone   c. 3-butanone   d. 2-butanone
8. $CH_3-\underset{O}{\overset{..}{C}}-CH_3$ is not:
   a. acetone   b. dimethyl aldehyde   c. dimethyl ketone   d. propanone
9. The compound 3-chlorobutanal is also properly called:
   a. 3-chlorobutanol   b. 3-chlorobutyraldehyde   c. α-chlorobutanal
   d. α-chlorobutyraldehyde   e. β-chlorobutyraldehyde
10. An aqueous solution of formaldehyde is called:
    a. aldol   b. acetone   c. formalin   d. formic acid   e. methanal
11. In general, aldehydes and ketones exhibit lower water solubility than:
    a. alcohols   b. alkenes   c. ethers
12. Which compound has the lowest boiling point?
    a. $CH_3CH_2OH$   b. $CH_3CH=O$   c. $CH_3-O-CH_3$
13. Tollens's reagent will oxidize a(n):
    a. aldehyde   b. alcohol   c. ketone   d. carboxylic acid
14. When an aldehyde is oxidized, the product is a(n):
    a. alcohol   b. ketone   c. carboxylic acid   d. aldehyde
15. Which structure would give a positive Tollens's test?
    a. $CH_3CH=O$   b.   c.   d. $CH_3COOH$
16. An aldehyde can be distinguished from a ketone by means of:
    a. the Tollens's test
    b. reaction with phenylhydrazine
    c. Both contain carbonyl groups and cannot be distinguished by the above tests.
17. A ketone can be distinguished from an alcohol by means of:
    a. the Tollens's test
    b. reaction with phenylhydrazine
    c. cannot be distinguished by the above tests
18. Ketones are prepared by the oxidation of:
    a. primary alcohols   b. secondary alcohols   c. tertiary alcohol
    d. carboxylic acids   e. cannot be prepared by oxidation reactions
19. Which reaction does not involve addition to the aldehyde carbonyl oxygen?
    a. hydrate formation   b. hemiacetal formation   c. imine formation   d. oxidation

## Chapter 15 - Aldehydes and Ketones

20. Both tautomerism and the aldol condensation depend on the relative acidity of the:
    a. hydrogen alpha to the carbonyl group     b. hydrogen on the carbonyl carbon atom
    c. carbon of the carbonyl group     d. oxygen of the carbonyl group

21. In general, which is the most stable type of compound?
    a. hydrate     b. hemiacetal     c. acetal

22. Which organic compound would be isolated from the reaction:

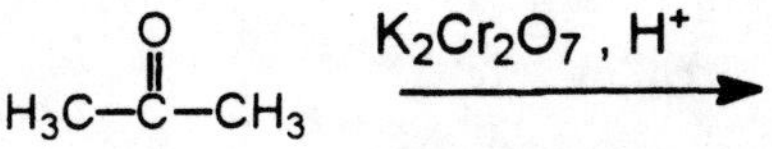

$$H_3C-\overset{O}{\underset{}{C}}-CH_3 \xrightarrow{K_2Cr_2O_7,\ H^+}$$

   a. $CH_3CH_2COOH$   b. $CH_3\overset{COOH}{\underset{}{C}}HCH_3$   c. $CH_3\overset{O}{\underset{}{C}}CH_3$   d. $CH_3\overset{O}{\underset{}{C}}COOH$

23. What must be added to acetaldehyde to form $CH_3-\overset{H}{\underset{OCH_3}{C}}-OH$?

   a. $CH_3CH_2OH$   b. $CH_3-\overset{O}{\underset{}{C}}-H$   c. $CH_3OH$   d. $CH_2=O$

24. Which compound represents the tautomer of $CH_3-\overset{O}{\underset{}{C}}-CH_3$?

   a. $CH_3CH_2\overset{O}{\underset{}{C}}-H$   b. $CH_3CH=\overset{OH}{\underset{}{C}}H$   c. $CH_3\overset{OH}{\underset{}{C}}HCH_3$   d. $CH_2=\overset{OH}{\underset{}{C}}-CH_3$

25. Which of the following reagents is required to accomplish each of the transformations shown below:
    a. $H^+, H_2O$   b. dry HCl   c. NaOH   d. $K_2Cr_2O_7, H^+$   e. no additional reagent required

   A.  $CH_3-\overset{O}{\underset{}{C}}-H \longrightarrow CH_3-\overset{O}{\underset{}{C}}-OH$

   B.  $CH_3-\overset{O}{\underset{}{C}}-H + 2\ CH_3OH \longrightarrow CH_3-\overset{O-CH_3}{\underset{H}{C}}-O-CH_3$

   C.  $CH_3-\overset{O}{\underset{}{C}}-H + H_2O \longrightarrow CH_3-\overset{OH}{\underset{H}{C}}-OH$

   D.  $2\ CH_3-\overset{O}{\underset{}{C}}-H \longrightarrow CH_3-\overset{OH}{\underset{H}{C}}-CH_2-\overset{O}{\underset{}{C}}-H$

26. Oxidation of wood alcohol by liver enzymes produces the toxic substance:
    a. ethanol     b. formaldehyde     c. methanol     d. nicotine

27. Addition of ammonia to an aldehyde produces water and a(n):
    a. alcohol     b. ketal     c. hydrate     d. N-acetal     e. imine

28. Isopropanol could be made from the reduction of which compound?
    a. propionic acid     b. ethyl methyl ketone     c. acetone     d. acetaldehyde

29. A β-hydroxy product is formed when an aldehyde reacts with a(n) _____.
    a. alcohol     b. aldehyde     c. ketone     d. ammonia

30. Which of these compounds would be most soluble in water?
    a. $CH_3CH_2CH_2CH_2CH_3$         b. $CH_3CH_2CH_2-O-CH_3$
    c. $CH_3CH_2CH_2CH_2OH$           d. $CH_3CH_2CH_2-\underset{H}{\overset{}{C}}=O$

## Chapter 15 - Aldehydes and Ketones

## ANSWERS

### Problems

1. $C_6H_5-CHO + CH_3OH \longrightarrow C_6H_5-CH(OCH_3)(OH)$

2. $C_6H_5-CHO + 2\ CH_3CH_2OH \xrightarrow{dry\ HCl} C_6H_5-CH(OCH_2CH_3)_2$

3. cyclohexanone $+ 2\ CH_3OH \xrightarrow{dry\ HCl}$ 1,1-dimethoxycyclohexane

4. $OHC-CH_2CH_2CH_2-OH \longrightarrow$ a hemiacetal (2-hydroxytetrahydrofuran)

5. cyclohexanone $\xrightarrow{K_2Cr_2O_7,\ H^+}$ No reaction

6. $H_3C-CO-CH_2CH_2-CHO \xrightarrow{K_2Cr_2O_7,\ H^+} CH_3-CO-CH_2CH_2-COOH$

7. $C_6H_5-CHO + NH_3 \longrightarrow C_6H_5-CH=N-H$

8. cyclohexanone $+ H_2N-NH-C_6H_5 \longrightarrow$ cyclohexanone phenylhydrazone

9. $2\ CH_3-CH_2-CHO \longrightarrow CH_3CH_2-CH(OH)-CH(CH_3)-CHO$

### Self-Test

| | | | | | |
|---|---|---|---|---|---|
| 1. b | 6. d | 11. a | 16. a | 21. c | 26. b |
| 2. c | 7. d | 12. c | 17. b | 22. c | 27. e |
| 3. a | 8. b | 13. a | 18. b | 23. c | 28. c |
| 4. d | 9. e | 14. c | 19. d | 24. d | 29. b, c |
| 5. d | 10. c | 15. a | 20. a | 25. A. d, B. b, C. e, D. c | 30. c |

# 16 Carboxylic Acids and Derivatives

## KEY WORDS

| | | | | |
|---|---|---|---|---|
| carboxylic acid | acetic acid | anhydrides | amide | carboxyl |
| acetate | esterification | anilides | hydrolysis | ester |
| saponification | carbonyl | hydroxyl | polyester | polyamide |
| condensation polymerization | | dimer | | |

## SUMMARY

**16.1 Carboxylic Acids and Their Derivatives: The Functional Groups**

A. The *carbonyl* group is also found in carboxylic acids and their derivatives, but comprises only a part of the functional groups of these families. The *carboxyl* group is a combination of the *carbonyl* and *hydroxyl* groups. This new combination now has acidic properties.

B. *Amides* and *esters* are derived from carboxylic acids by replacing the –OH group with either an amine (amide) or an alcohol (ester), and named accordingly.

$$-\overset{O}{\underset{}{C}}-OH\ (-COOH) \qquad -\overset{O}{\underset{}{C}}-NH- \qquad -\overset{O}{\underset{}{C}}-O-\overset{}{C}-$$
$$\text{carboxyl} \qquad\qquad \text{amide} \qquad\qquad \text{ester}$$

$$CH_3-\overset{O}{\underset{}{C}}-OH \qquad CH_3-\overset{O}{\underset{}{C}}-NH_2 \qquad CH_3-\overset{O}{\underset{}{C}}-OCH_3$$
$$\text{Acetic acid} \qquad\qquad \text{Acetamide} \qquad\qquad \text{Methyl acetate}$$

**16.2 Some Common Carboxylic Acids: Structures and Names**

A. Organic or *carboxylic acids* are weak acids, often having pungent odors. Many have very old histories and go by their common names, using Greek letters to indicate the positions of substituents.

$$-\underset{\gamma}{C}-\underset{\beta}{C}-\underset{\alpha}{C}-COOH$$

B. The IUPAC rules for naming carboxylic acids use the name of the corresponding alkane, but drop the "-*e*" ending and add "-*oic* acid." Substitutents are numbered using the carboxyl carbon as carbon "1."

$$-\underset{4}{C}-\underset{3}{C}-\underset{2}{C}-\underset{1}{COOH}$$

C. Common carboxylic acids, R–COOH:

| Formula | Name | Found in |
|---|---|---|
| HCOOH | Formic acid (methanoic) | Ant bites |
| CH$_3$COOH | *Acetic acid* (ethanoic) | Vinegar |
| CH$_3$–(CH$_2$)$_{14}$–COOH | Palmitic acid | Animal fat |
| CH$_3$–(CH$_2$)$_{16}$–COOH | Stearic acid | Animal fat |
| C$_6$H$_5$–COOH | Benzoic acid | |

**16.3 Preparation of Carboxylic Acids**

A. Oxidation of primary alcohols: Ethanol —[O]→ Acetaldehyde —[O]→ Acetic acid

B. *Hydrolysis* of fats: Triglycerides —[H$_2$O, H$^+$]→ Glycerol + Fatty acids

# Chapter 16 - Carboxylic Acids and Derivatives

16.4. Physical Properties of Carboxylic Acids
  A. The carboxyl group is very polar and is capable of acting as both a hydrogen-bond donor and acceptor. This gives rise to strong intermolecular forces, and high boiling points. Methanoic (C1) through nonanoic (C9) are colorless liquids with foul odors.
  B. Hydrogen-bonded dimers of carboxylic acids are found even in the vapor phase. Those carboxylic acids containing 4 or fewer carbons are completely miscible with water.

$$CH_3-C\underset{OH\ -----\ O}{\overset{O\ -----\ HO}{\diagup\diagdown}}C-CH_3$$

  Acetic acid dimer

  C. Pure *acetic acid* freezes at 16.6 °C, just below room temperature, and has been observed to freeze on the laboratory shelf. The common name for pure acetic acid is "glacial acetic acid."

16.5 Chemical Properties of Carboxylic Acids: Neutralization
  $$RCOOH\ +\ NaOH\ \longrightarrow\ RCOO^-Na^+\ +\ H_2O$$
  A. Carboxylic acids are weak acids that ionize slightly; ($pK_a$ 3~5).
  $HCl, H_2SO_4, HNO_3 > H_3PO_4 > RCOOH > H_2CO_3 > ArOH > H_2O > ROH > RH$
  (strongest) ←←←←←←←←←←←←←←←←←←←←←←←←←← (weakest)
  B. Carboxylic acids turn blue litmus red and react with bases to form salts and water. The salts are named by naming the cation first, then naming the anion by changing the "-ic" ending to "-ate."
   $CH_3COOH$     acetic acid
   $CH_3COO^-Na^+$  sodium *acetate*
  C. Propionate, benzoate, and sorbate salts are frequently used as food preservatives.

16.6 An Ester by Any Other Name . . .
  A. *Esters* are produced when a carboxylic acid reacts with an alcohol with loss of water. Many esters have very pleasant odors. *Polyesters* are important in biology and textiles.
  B. Esters are named much the same way that carboxylate salts are named except that the alkyl (or aryl) group from the alcohol is named first in place of the cation.

$$CH_3\overset{O}{\overset{\|}{C}}OH\ +\ HOCH_3\ \longrightarrow\ CH_3\overset{O}{\overset{\|}{C}}-O-CH_3\ +\ H_2O$$

  Acetic acid  +  Methanol                    Methyl acetate

16.7 Physical Properties of Esters
  A. Many fragrances of fruits and flowers are esters.
  B. Esters are polar, but pure esters are <u>not</u> capable of intramolecular hydrogen bonding because they are not hydrogen donors.
  C. Esters have lower boiling points and somewhat lower solubility in water than do the isomeric carboxylic acids. Ethyl acetate is a common organic solvent.

16.8 Preparation of Esters: *Esterification*
  A. Esters are prepared by heating the corresponding acid and alcohol in the presence of a mineral acid catalyst. To obtain high yields of the ester, it is usually necessary to supply one of the reactants in excess or to remove one of the products as it forms.

$$R-COOH\ +\ HO-R'\ \longrightarrow\ R-\overset{O}{\overset{\|}{C}}-O-R'\ +\ H_2O$$

## Chapter 16 - Carboxylic Acids and Derivatives

B. Esters are more commonly prepared in the laboratory by the reaction of acyl chlorides and alcohols.

$$R-\overset{O}{\underset{\|}{C}}-Cl + HO-R' \longrightarrow R-\overset{O}{\underset{\|}{C}}-O-R' + HCl$$

C. Acetate esters are usually prepared from acetic anhydride. (*Anhydrides* are prepared by removal of water from two acid molecules.)

$$CH_3-\overset{O}{\underset{\|}{C}}-O-\overset{O}{\underset{\|}{C}}-CH_3 + HO-CH_3 \longrightarrow CH_3-\overset{O}{\underset{\|}{C}}-OCH_3 + CH_3-COOH$$

D. *Polyesters* (Dacron) are polymers made from the condensation of diols with dicarboxylic acids.

HO—CH₂CH₂CH₂CH₂CH₂CH₂—OH + [terephthalic acid] ⟶

### 16.9 Chemical Properties of Esters: *Hydrolysis*
A. Esters are neutral compounds, neither acidic nor basic.
B. Hydrolysis (splitting with water)
   1. Acid hydrolysis is the reverse of esterification.

   $$\text{Ester} \xrightarrow{H^+} \text{Carboxylic acid + Alcohol}$$

   2. Basic hydrolysis produces the salt of the acid and the corresponding alcohol. Alkaline hydrolysis of fats and oils is called *saponification* (to make soap).

   $$R-\overset{O}{\underset{\|}{C}}-O-R' + NaOH \longrightarrow R-\overset{O}{\underset{\|}{C}}-O^-Na^+ + HO-R'$$

### 16.10 Esters of Phosphoric Acid
A. Esters can also be made from the reaction of inorganic acids and alcohols. Nitroglycerin is a powerful explosive (dynamite), but it is also used to relieve chest pains by relaxing smooth heart muscles. Alfred Nobel (from Nobel prizes) discovered dynamite in 1866.
B. Important esters of inorganic acids include the diphosphate, also called pyrophosphoric acid, which is found in adenosine diphosphate (ADP) and the triphosphate found in adenosine triphosphate (ATP). Phosphate esters of sugars such as glucose-6-phosphate are important in carbohydrate metabolism.

ATP            Glucose-6-phosphate

### 16.11 Amides: Structures and Names
A. *Amides* are prepared when a carboxylic acid reacts with ammonia or its derivatives (amines) with the loss of water.
B. Simple amides are named by dropping the "*-ic*" suffix and adding "*-amide.*" Substituted amides are named by using a capital *N* to indicate that the substituent is on the nitrogen. Phenyl substituents are referred to as *anilides*.

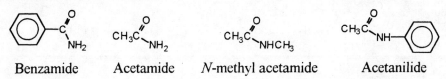

Benzamide    Acetamide    *N*-methyl acetamide    Acetanilide

# Chapter 16 - Carboxylic Acids and Derivatives

16.12 Physical Properties of Amides
  A. Most amides are solids at room temperature, indicating the presence of strong intermolecular forces.
  B. The hydrogen bonds in polyamides are very important in determining the three-dimensional structure of proteins.

16.13 Synthesis of Amides
  A. First step: preparation of acyl chloride

$$R-\overset{O}{\underset{}{C}}-OH + SOCl_2 \longrightarrow R-\overset{O}{\underset{}{C}}-Cl + SO_2 + HCl$$

  B. Second step: treatment of acyl chloride with ammonia (or amine)

$$R-\overset{O}{\underset{}{C}}-Cl + 2\,NH_3 \longrightarrow R-\overset{O}{\underset{}{C}}-NH_2 + NH_4Cl$$

  C. Nylon and proteins are examples of *polyamides*.

16.14 Chemical Properties of Amides: *Hydrolysis*
  A. Amides are neutral compounds. The lone pair of electrons of the basic ammonia have been delocalized in forming a partial double bond with the carbonyl C.

$$R-C\underset{NH_2}{\overset{\displaystyle O}{\diagup\!\!\!\diagdown}} \longrightarrow R-C\underset{\underset{H}{N}-H}{\overset{\displaystyle O}{\diagup\!\!\!\diagdown}} \quad \text{Planar amide group}$$

  B. Amides are fairly stable in water but hydrolyze in acidic and basic solutions. Hydrolysis of amide bonds occurs during the digestion of proteins.

## DISCUSSION

We have included here a summary of all the nomenclature rules and the chemical properties of carboxylic acids and their derivatives presented in Chapter 16.

| Nomenclature of Carboxylic Acids and Derivatives | | | |
|---|---|---|---|
| Type of compound | Common Name | IUPAC Name | Structure |
| Carboxylic acids | Valeric acid | Pentanoic acid | $CH_3CH_2CH_2CH_2\overset{O}{\underset{}{C}}-OH$ |
| Salts of acids | Sodium valerate | Sodium pentanoate | $CH_3CH_2CH_2CH_2\overset{O}{\underset{}{C}}-O^-Na^+$ |
| Esters | Ethyl valerate | Ethyl pentanoate | $CH_3CH_2CH_2CH_2\overset{O}{\underset{}{C}}OCH_2CH_3$ |
| Simple amide | Valeramide | Pentanamide | $CH_3CH_2CH_2CH_2\overset{O}{\underset{}{C}}-NH_2$ |
| Substituted amide | *N,N*-Dimethylvaleramide | *N,N*-Dimethylpentanamide | $CH_3CH_2CH_2CH_2\overset{O}{\underset{}{C}}-\underset{CH_3}{N}-CH_3$ |
| Anilides | Valeranilide | Pentananilide or *N*-phenylpentanamide | $CH_3CH_2CH_2CH_2\overset{O}{\underset{}{C}}-\underset{H}{N}-Ar$ |

# Chapter 16 - Carboxylic Acids and Derivatives

## Reactions of Carboxylic Acids and Derivatives

**Preparation of acids** (oxidation of primary alcohols or aldehydes)

$$RCH_2OH \xrightarrow{K_2Cr_2O_7,\ H^+} RCH=O \longrightarrow RCHOOH$$

**Salt formation**

$$RCOOH + NaOH \longrightarrow RCOO^-Na^+ + H_2O$$

$$RCOOH + NaHCO_3 \longrightarrow RCOO^-Na^+ + H_2O + CO_2$$

$$2\ RCOOH + Na_2CO_3 \longrightarrow 2\ RCOO^-Na^+ + H_2O + CO_2$$

**Ester formation**

$$R\overset{O}{\underset{}{C}}-OH + HO-R' \xrightarrow{H^+} R-\overset{O}{\underset{}{C}}-O-R' + H_2O$$

$$CH_3\overset{O}{\underset{}{C}}-O-\overset{O}{\underset{}{C}}CH_3 + HOR \longrightarrow \underset{\text{ester}}{CH_3\overset{O}{\underset{}{C}}-O-R} + \underset{\text{acid byproduct}}{CH_3\overset{O}{\underset{}{C}}-OH}$$

**Ester hydrolysis**

Acid-catalyzed: $\quad R-\overset{O}{\underset{}{C}}-O-R' + H_2O \xrightarrow{H^+} R-\overset{O}{\underset{}{C}}-OH + HOR'$

Base-catalyzed: $\quad R-\overset{O}{\underset{}{C}}-O-R' + H_2O \xrightarrow{OH^-} R-\overset{O}{\underset{}{C}}-O^- + HOR'$

**Amide formation**

Simple amide: $\quad R-\overset{O}{\underset{}{C}}-OH \xrightarrow{SOCl_2} R-\overset{O}{\underset{}{C}}-Cl \xrightarrow{NH_3} R-\overset{O}{\underset{}{C}}-NH_2$

Substituted amide: $\quad R-\overset{O}{\underset{}{C}}-OH \xrightarrow{SOCl_2} R-\overset{O}{\underset{}{C}}-Cl \xrightarrow{H_2N-R'} R-\overset{O}{\underset{}{C}}-\underset{H}{N}-R'$

$\quad R-\overset{O}{\underset{}{C}}-OH \xrightarrow{SOCl_2} R-\overset{O}{\underset{}{C}}-Cl \xrightarrow{HNR'_2} R-\overset{O}{\underset{}{C}}-\underset{R'}{N}-R'$

**Amide hydrolysis**

Acid-catalyzed: $\quad R-\overset{O}{\underset{}{C}}-NH_2 \xrightarrow{H^+,\ H_2O} R-\overset{O}{\underset{}{C}}-OH + NH_4^+$

Base-catalyzed: $\quad R-\overset{O}{\underset{}{C}}-NH_2 \xrightarrow{OH^-,\ H_2O} R-\overset{O}{\underset{}{C}}-O^- + NH_3$

(Substituted amides give the corresponding substituted ammonia derivatives, e.g., $RNH_2$ or $RNH_3^+$)

# Chapter 16 - Carboxylic Acids and Derivatives

**Problems:** The following problems will test your understanding of the chemistry of acids and acid derivatives. Remember – no matter how complicated individual structures may appear, the chemistry is determined by the functional group and follows the patterns summarized previously. Identify the functional groups in each of the following compounds. There may be more than one functional group in a single compound. Also, the same functional group may appear in more than one compound.

| Functional Groups | | Compound | Functional Groups Present |
|---|---|---|---|
| a. alcohol | 1. | OH O<br>CH$_3$–CH–C–O–CH$_3$ | a b c d e f g h |
| b. aldehyde | 2. | CH$_3$CH$_2$–O–CH$_2$CH$_2$OH | a b c d e f g h |
| c. amide | 3. | CH$_3$–O–CH$_2$C–OH<br>‖<br>O | a b c d e f g h |
| d. carboxylic acid | | | |
| e. ester | 4. | CH$_3$–C–O–CH$_2$CH$_2$–OH<br>‖<br>O | a b c d e f g h |
| f. ether | | | |
| g. ketone | 5. | CH$_3$–C–O–CH$_2$C–H<br>‖ ‖<br>O  O | a b c d e f g h |
| h. phenol | | | |
| | 6. | CH$_3$–O–C–CH$_2$–C–NH$_2$<br>‖ ‖<br>O  O | a b c d e f g h |
| | 7. | HO–CH$_2$CH$_2$–C–OH<br>‖<br>O | a b c d e f g h |
| | 8. | (benzene ring with HO and COOH substituents) | a b c d e f g h |
| | 9. | CH$_3$–O–(benzene ring)–NHCH(=O) | a b c d e f g h |
| | 10. | (steroid structure with O=C–CH$_3$ ester group and ketone) | a b c d e f g h |

# Chapter 16 - Carboxylic Acids and Derivatives

**Reactions:** Draw the principal organic products of the reactions.

1. C$_6$H$_5$COOH + CH$_3$CH(OH)CH$_3$ $\xrightarrow{H^+}$

2. (CH$_3$CO)$_2$O + C$_6$H$_5$OH $\longrightarrow$

3. HOOC-CH$_2$CH$_2$-COOH + 2 CH$_3$OH $\xrightarrow{H^+}$

4. CH$_3$-CO-Cl + C$_6$H$_5$-NH$_2$ $\longrightarrow$

5. C$_6$H$_5$-CO-Cl + NH$_3$ $\longrightarrow$

6. (CH$_3$)$_2$CH-CO-Cl + CH$_3$-NH-CH$_3$ $\longrightarrow$

7. CH$_3$CH$_2$-COOH + KOH $\longrightarrow$

8. o-C$_6$H$_4$(COOH)$_2$ + Na$_2$CO$_3$ $\longrightarrow$

9. CH$_3$CH$_2$-CO-OCH$_2$CH$_3$ + H$_2$O $\xrightarrow{OH^-}$

10. CH$_3$CH$_2$-CO-OCH$_2$CH$_3$ + H$_2$O $\xrightarrow{H^+}$

11. HO-C$_6$H$_4$-NH-CO-CH$_3$ + H$_2$O $\xrightarrow{H^+}$

12. (3-pyridyl)-CO-NH$_2$ + H$_2$O $\xrightarrow{OH^-}$

# Chapter 16 - Carboxylic Acids and Derivatives

**SELF-TEST**

1. Which group bestows functional group properties of an amide?
   a. –CH=O   b. –NH$_2$   c. C$_6$H$_5$–   d. –CH$_2$OH   e. $-\underset{\underset{O}{\|}}{C}-NH_2$   f. $-\underset{\underset{O}{\|}}{C}-OH$

2. The –COOH group is called a(n):
   a. carboxyl group   b. carbonyl group   c. aldehyde group   d. hydroxyl group

3. Which is not a mineral acid?
   a. HNO$_3$   b. HCl   c. HCOOH   d. H$_2$SO$_4$

4. Which is the acid found in vinegar?
   a. nitric acid   b. acetic acid   c. valeric acid   d. formic acid

5. The product of the reaction between an amine and an acid is known as a(n):
   a. acid anhydride   b. amide   c. ester   d. ether   e. salt

6. The correct name of the compound CH$_3$CH$_2$CH$_2$CH$_2$$\underset{\underset{O}{\|}}{C}$–OH is:

   a. pentanoic acid   b. caproic acid   c. succinic acid   d. butanonic acid

7. $\underset{}{CH_3}$
   CH$_3$$\overset{|}{C}$HCH$_2$–$\underset{\underset{O}{\|}}{C}$–OH is:
   a. α-methylbutyric acid   b. β-methylbutyric acid
   c. 2-pentanoic acid   d. 1-methylbutanoic acid

8. Ammonium acetate is:
   a. CH$_3$$\underset{\underset{O}{\|}}{C}$–O$^-$ NH$_4^+$   b. CH$_3$$\underset{\underset{O}{\|}}{C}$–NH$_2$   c. CH$_3$$\underset{\underset{O}{\|}}{C}$–O$^-$Na$^+$   d. NH$_4^+$Cl$^-$

9. The name of H–$\underset{\underset{O}{\|}}{C}$–O$^-$ Na$^+$ is not:
   a. sodium carbonate   b. sodium formate   c. sodium methanoate

10. The correct name of CH$_3$CH$_2$O–$\underset{\underset{O}{\|}}{C}$–CH$_3$ is:
    a. ethyl acetate   b. ethyl formate   c. ethyl methyl ketone
    d. methyl acetate   e. ethyl methyl ester

11. Which acid is palmitic acid?
    a. CH$_3$(CH$_2$)$_{12}$COOH   b. CH$_3$(CH$_2$)$_{10}$COOH   c. CH$_3$(CH$_2$)$_{14}$COOH   d. CH$_3$(CH$_2$)$_{16}$COOH

12. Methyl acetate is:
    a. C$_2$H$_5$COOCH$_3$   b. CH$_3$COOC$_2$H$_5$   c. CH$_3$COOCH$_3$   d. C$_2$H$_5$COOC$_2$H$_5$

13. Butyramide is:
    a. CH$_3$CH$_2$CH$_2$$\underset{\underset{O}{\|}}{C}$–O–NH$_4^+$   b. $\overset{NH_2}{\underset{|}{C}}$H$_2$CH$_2$CH$_2$$\underset{\underset{O}{\|}}{C}$–OH   c. CH$_3$CH$_2$CH$_2$$\underset{\underset{O}{\|}}{C}$–NH$_2$

14. CH$_3$$\underset{\underset{O}{\|}}{C}$–NHCH$_3$ is:
    a. N-methylamide   b. methylformamide   c. N-methylethanamide   d. N-methylmethanamide

15. The compound is:

    (C$_6$H$_5$)–NH–C(=O)–CH$_3$

    a. methylbenzamide   b. N-methylethanamide   c. methananilide   d. acetanilide

16. Dacron is a synthetic fiber representative of which type of compound?
    a. amide   b. ether   c. polyester   d. ester   e. polyamide

150

17. Proteins are representative of which type of compound?
    a. amide    b. ether    c. polyester    d. ester    e. polyamide
18. Glacial acetic acid is:
    a. a frozen solution of acetic acid
    b. a mixture of acetic acid and water
    c. pure acetic acid
19. Which compound is the weakest acid?
    a. $CF_3COOH$    b. [phenol structure]    c. $CH_3C(O)-OH$    d. $H_2CO_3$

20. Which compound is the strongest acid?
    a. $CF_3COOH$    b. [phenol structure]    c. $CH_3C(O)-OH$    d. $H_2CO_3$

21. Which compound is oxalic acid?
    a. [benzoic acid structure]    b. HOOC–COOH    c. HOOC–$CH_2$COOH    d. [phthalic acid structure]

22. For which pure compound is hydrogen bonding not possible?
    a. $CH_3C(O)-O-CH_3$    b. $CH_3C(O)-OH$    c. $CH_3C(O)-NHCH_3$    d. $CH_3CH_2OH$
23. Which compound(s) would produce a nearly neutral solution in water?
    a. amide    b. carboxylic acid    c. phenol    d. ester    e. amine
24. Which compound has the highest boiling point?
    a. $CH_3CH_2O-CH_2CH_3$    b. $CH_3C(O)-O-CH_3$    c. $CH_3C(O)CH_2CH_3$
    d. $CH_3CH_2CH_2CH_2OH$    e. $CH_3CH_2C(O)-OH$

25. Which compound has the lowest boiling point?
    a. HO–$CH_2CH_2CH_2$OH    b. $CH_3CH_2C(O)-OH$    c. $CH_3C(O)-O-CH_3$
26. The odors associated with the smaller members of this class of compounds are distinctly unpleasant.
    a. amides    b. alcohols    c. esters
27. The aromas associated with this class of compounds are regarded, in general, as pleasant.
    a. amides    b. carboxylic acids    c. esters
28. Alkaline hydrolysis of large, fatty esters is called:
    a. solvation    b. saponification    c. sublimation    d. hydration
29. If $CH_3CH_2C(O)-O-CH_2CH_3$ is hydrolyzed in base, which set of products is formed?
    a. $CH_3CH_2OH$    b. $CH_3CH_2O^-Na^+$    c. $CH_3CH_2OH$    d. $CH_3CH_2O^-Na^+$
       $CH_3CH_2COOH$       $CH_3CH_2COOH$       $CH_3CH_2COO^-Na^+$    $CH_3CH_2COO^-Na^+$
30. Esters can be synthesized from carboxylic acids and alcohols in the presence of mineral acids. Which alcohol is needed to synthesize this ester?
    $$CH_3CH_2CH_2CH_2O-C(O)CH_3$$
    a. ethanol    b. propanol    c. methanol    d. pentanol    e. butanol

## Chapter 16 - Carboxylic Acids and Derivatives

### ANSWERS

### Problems

| | | | | |
|---|---|---|---|---|
| 1. a,e | 3. d,f | 5. b,e | 7. a,d | 9. c,f |
| 2. a,f | 4. a,e | 6. c,e | 8. d,h | 10. e,g |

### Reactions

1. PhCOOH + (CH$_3$)$_2$CHOH $\xrightarrow{H^+}$ PhCOOCH(CH$_3$)$_2$ + H$_2$O

2. (CH$_3$CO)$_2$O + PhOH $\longrightarrow$ PhOCOCH$_3$ + CH$_3$COOH

3. HOOC–CH$_2$CH$_2$–COOH + 2 CH$_3$OH $\xrightarrow{H^+}$ CH$_3$OOC–CH$_2$CH$_2$–COOCH$_3$ + 2 H$_2$O

4. CH$_3$COCl + PhNH$_2$ $\longrightarrow$ PhNHCOCH$_3$ + HCl

5. PhCOCl + NH$_3$ $\longrightarrow$ PhCONH$_2$ + HCl

6. (CH$_3$)$_2$CHCOCl + (CH$_3$)$_2$NH $\longrightarrow$ (CH$_3$)$_2$CHCON(CH$_3$)$_2$ + HCl

7. CH$_3$CH$_2$COOH + KOH $\longrightarrow$ CH$_3$CH$_2$COO$^-$K$^+$

8. o-C$_6$H$_4$(COOH)$_2$ + Na$_2$CO$_3$ $\longrightarrow$ o-C$_6$H$_4$(COO$^-$Na$^+$)$_2$

9. CH$_3$CH$_2$COOCH$_2$CH$_3$ + H$_2$O $\xrightarrow{OH^-}$ CH$_3$CH$_2$COO$^-$ + CH$_3$CH$_2$OH

10. CH$_3$CH$_2$COOCH$_2$CH$_3$ + H$_2$O $\xrightarrow{H^+}$ CH$_3$CH$_2$COOH + CH$_3$CH$_2$OH

## Chapter 16 - Carboxylic Acids and Derivatives

11. HO—C₆H₄—NH—C(=O)—CH₃ + H₂O →(H⁺) HO—C₆H₄—NH₃⁺ + CH₃COOH

12. (pyridine-3-carboxamide) + H₂O →(OH⁻) (pyridine-3-carboxylate) + NH₃

**Self-Test**

| | | | | |
|---|---|---|---|---|
| 1. e | 7. b | 13. c | 19. b | 25. c |
| 2. a | 8. a | 14. c | 20. a | 26. a |
| 3. c | 9. a | 15. d | 21. b | 27. c |
| 4. b | 10. a | 16. c | 22. a | 28. b |
| 5. b | 11. c | 17. e | 23. a, d | 29. c |
| 6. a | 12. c | 18. c | 24. e | 30. e |

# B Drugs: Some Carboxylic Acids, Esters, and Amides

**KEY WORDS**

*analgesic*      *agonist*         *opiates*         *hallucinogen*     *aspirin*
*alkaloid*       *antipyretic*     *antagonist*      *narcotic*         *salicylate*
*lysergic acid*  *THC*             *ibuprofen*       *prostaglandins*   *endorphin*
*enkephalin*     *morphine*        *heroin*          *codeine*          *LSD*
*acetaminophen*  *anticoagulant*   *drug*            *marijuana*        *NSAIDs*

**SUMMARY**

B.1 Aspirin and Other Salicylates
  A. "*Aspirin*" (acetylsalicylic acid) was introduced in 1899 as one of the first successful synthetic *analgesics* (pain relievers) and has become the largest-selling drug in the world. A *drug* is any substance that affects an individual in a manner that brings about a physiological, emotional, or behavioral change.
    1. Willow bark was a "home remedy" for reducing fever.
    2. Salicylic acid was isolated from willow bark in 1860. It was found to be a good analgesic (pain killer) and antipyretic (fever reducer) but is sour and irritating to take by mouth.
    3. Sodium *salicylate* (1875), phenyl salicylate (1866) and acetyl salicylate (1899) are chemical modifications of this natural drug introduced as having removed some of the undesirable effects of salicylic acid while retaining its desirable properties.
  B. *Aspirin* is synthesized in the laboratory by the treatment of salicylic acid with acetic anhydride to make the acetate ester. "Bufferin" is aspirin that contains antacids, but it is not truly buffered. Ordinary aspirin contains ~325 mg/tablet, while "extra strength" aspirin contains ~500 mg/tablet.

Salicylic acid    Acetic anhydride         Aspirin           Acetic acid

  C. Aspirin relieves minor aches and suppresses inflammation. It is an example of a nonsteroidal anti-inflammatory (*NSAIDs*) drugs, and is often used to treat arthritis. Aspirin acts by inhibiting the enzyme (Chapter 22) cyclomonooxygenase, which is needed for the synthesis of *prostaglandins* that affect blood pressure and inflammation.
  D. Aspirin is also an *anticoagulant*. This property may help aspirin reduce the risk of a heart attack or stroke. It can also cause bleeding in the stomach. Using aspirin to treat children with fevers has been associated with Reye's syndrome.
  E. Hazards: Some people are allergic to aspirin.

B.2 Aspirin Substitutes and Combination Pain Relievers
  A. Many aspirin substitutes include the amide **acetaminophen** (*p*-hydroxyacetanilide) or a combination of aspirin, acetaminophen, caffeine, and buffers. (Excedrin = aspirin, acetaminophen, and caffeine.) Overuse of acetaminophen is linked to liver and kidney damage in those who drink a lot of alcohol.

# Selected Topic B - Drugs

B. *Ibuprofen* is an anti-inflammatory drug in common use (Advil, Nuprin, Motrin).

Acetaminophen     Ibuprofen     Caffeine

## B.3 Opium Alkaloids

A. *Alkaloids* are physiologically active, N–containing compounds, such as nicotine and caffeine, which are isolated from plants. *Opiates* are obtained from the "poppy" plant. *Morphine* is a *narcotic* producing sedation and analgesia, but it is strongly addictive. It is a natural alkaloid from opium. Morphine was first isolated in 1805 and remains a standard for narcotic analgesics in medicine.

B. The human brain receptors used by the opiates are designed for our bodies' natural pain killers, a family of peptides called *endorphins*. The two most characteristic of these compounds are pentapeptides called *enkephalins*: Leu–enkephalin (Tyr–Gly–Gly–Phe–Leu), Met–enkephalin (Tyr–Gly–Gly–Phe–Met).

C. Endorphins are released by extreme trauma, acupuncture, and strenuous physical activity (the "high" of the long-distance runner).

D. *Codeine* is very similar to morphine, with –OCH₃ in place of the –OH group. It is less potent and less addictive than morphine.

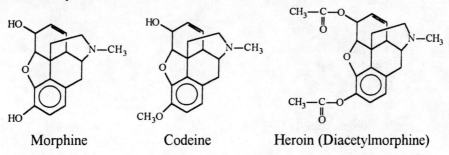

Morphine     Codeine     Heroin (Diacetylmorphine)

## B.4 Semisynthetic Opiates

A. *Heroin* is the common name of diacetylmorphine, introduced by the Bayer company of Germany in 1874 as a synthetic analgesic and antidote for morphine addiction. Heroin produces a strong feeling of euphoria but is strongly addictive and illegal in the United States.

B. Deaths from heroin are usually the result of overdose, since street samples of the drug vary widely.

## B.5 Synthetic Narcotics: Analgesia and Addiction

A. Thousands of morphine analogs have been synthesized to find a nonaddictive analgesic.
1. Meperidine (Demerol) is less addictive, but also less effective, than morphine.
2. Methadone is also highly addictive but does not cause the sleepy stupor of heroin so that the addict can still hold a job and function to care for himself or herself.

B. *Agonists* are molecules with drug-like action. *Antagonists* are drugs that block the action of other drugs.

## B.6 LSD: A Hallucinogenic Drug

A. *LSD* (*N,N*–diethylamide of lysergic acid) was discovered by Albert Hofmann in 1943. It is a *hallucinogenic* drug related to *lysergic acid* and other ergot alkaloids. As little as 10 µg of this powerful drug can bring on colorful hallucinations. It was the popular "acid" of the 1960's, but its

## Selected Topic B - Drugs

use has declined after reports that LSD damages chromosomes. It is an illegal drug in the United States and Great Britain.

B. The LSD structure resembles that of serotonin. LSD probably acts as a serotonin agonist. Lysergic acid is obtained from the ergot fungus that grows on rye. Consumption of spoiled rye stores by villagers in the Middle Ages could account for reports of mass hallucinations.

LSD

### B.7 Marijuana: Some Chemistry of *Cannabis*

A. The plant *Cannabis sativa* has long been used to provide tough fibers for making rope and to supply drugs for ceremonies. *Marijuana* ("pot") refers to a preparation made by drying the leaves, flowers, and seeds of *Cannabis*.

B. *THC* or tetrahydrocannabinol is the main active ingredient in marijuana and represents ~ 0.1 to 1% of the "pot." Hashish or hash can have a THC content of 5 to 12%.

C. Smoking "pot" impairs complex motor skills, distorts one's sense of time, and may cause brain damage and have "feminizing" properties.

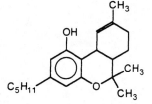
Tetrahydrocannabinol (THC)

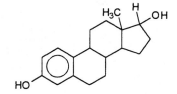
Estradiol

### B.8 Thalidomide: Revival of a Discarded Drug

A. Thalidomide was introduced as a tranquilizer in the 1950's. It was considered safe and often prescribed for morning sickness for pregnant women. The drug was not approved in the United States, but in Germany thalidomide was available without a prescription. Unfortunately, women who took thalidomide during the first three months of pregnancy had an increased risk of having babies with shortened or absent arms and legs. The drug was banned worldwide as a tranquilizer in 1962.

Thalidomide

B. However, the drug has been found useful for its anti-inflammatory effects and ability to modify the immune response. In 1998 the drug was approved for use in treating leprosy and is being tested for treatment of other serious diseases such as tuberculosis, certain sarcomas, autoimmune diseases, and AIDS-related ulcers.

## Selected Topic B - Drugs

## DISCUSSION

In this unit we have introduced some compounds that look rather unusual or perhaps we should say "unnatural." That is because these compounds are drugs, compounds that are not normally found in the body. Note the unusual ring systems that are found in heroin and thalidomide. We will see in later chapters on biochemistry that such compounds are not normal substrates for our bodies' metabolic machinery; thus such compounds can have unexpected and undesirable side effects.

Several derivatives of carboxylic acids have found use as drugs to relieve pain. Willow bark has been used as a "home remedy" for thousands of years. The following SELF-TEST will help you check your understanding of this material.

## SELF-TEST

1. Aspirin is:
    a. diacetylmorphine     b. acetylsalicylic acid     c. acetaminophen
2. Which of the following are valid pharmacological classifications of aspirin?
    a. antipyretic     b. tranquilizer     c. analgesic     d. hallucinogen
3. Which compound is salicylic acid?
    a.     b.     c.     d.
4. Aspirin acts to inhibit the enzyme cyclomonooxygenase which is needed for the synthesis of:
    a. acetic anhydride     b. caffeine     c. prostaglandins     d. morphine
5. Molecules that mimic the action of a drug are called:
    a. agonists     b. opiates     c. caffeine     d. antagonists
6. Alkaloids are:
    a. buffered aspirin products that dissolve with a bubbling action
    b. $N$-containing drugs of all sorts
    c. $N$-containing, physiologically active compounds obtained from plants
    d. forms of THC (tetrahydrocannabinol)
7. Which of these compounds was introduced by Bayer Company of Germany and was once thought to be an antidote for morphine addiction?
    a. acetominophen     b. salicylamide     c. heroin     d. thalidomide
8. Which of the following compounds has been used extensively as an aspirin substitute for those who are allergic to aspirin?
    a. acetominophen     b. LSD     c. thalidomide     d. morphine
9. Which of the following compounds was marketed as a safe tranquilizer but found to cause birth defects when taken by expectant mothers during the first trimester of pregnancy?
    a. acetominophen     b. LSD     c. thalidomide     d. morphine
10. Which compound is classified as a hallucinogen?
    a. acetominophen     b. $N,N$-diethyllysergamide     c. morphine     d. thalidomide
11. Which of the following common drugs is not an alkaloid?
    a. morphine     b. heroin     c. THC (pot)     d. LSD
12. Natural, small pentapeptides that mimic some of the actions of opiates are called:
    a. heroins     b. enkephalins     c. endorphins     d. analgesics     e. thalidomides

**Selected Topic B - Drugs**

**ANSWERS**

1. b
2. a, c
3. b
4. c
5. a
6. c
7. c
8. a
9. c
10. b
11. c
12. b

# 17 Amines and Derivatives

**KEY WORDS**

| 1° amine | 2° amine | 3° amine | 4° quaternary | imidazole |
| aniline | ninhydrin | nitroso | amide | purine |
| pyrimidine | indole | pyridine | alkaloid | heterocyclic |
| pyrrole | nylon | amino group | | |

## SUMMARY

### 17.1 Structure and Classification of Amines
A. **Amines** are basic compounds ("the bases of life") related to ammonia, where one or more hydrogens have been replaced by alkyl or aryl groups.
B. Amines are classified according to the number of C atoms directly bonded to the N atom.

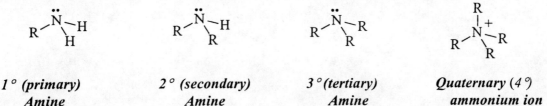

1° (primary) Amine    2° (secondary) Amine    3° (tertiary) Amine    Quaternary (4°) ammonium ion

### 17.2 Naming Amines
A. Simple aliphatic amines are named by specifying the alkyl groups and adding the suffix "-amine."
B. The simplest aromatic amine is called **aniline**.
C. Substituted ammonium ions are named by regarding the alkyl groups as substituents on the parent species, the ammonium ion.

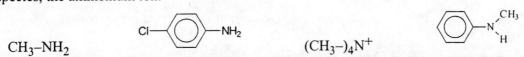

CH$_3$–NH$_2$    p-Chloroaniline    (CH$_3$–)$_4$N$^+$    N-Methylaniline
Methyl amine                  Tetramethylammonium ion

D. Sometimes the amino group is named as a substituent (H$_2$N–CH$_2$–CH$_2$–OH = 2-aminoethanol).

### 17.3 Physical Properties of Amines
A. Amines are polar compounds. Pure 1° and 2° amines are capable of forming intermolecular hydrogen bonds. All three classes of amines can hydrogen-bond to water and are soluble as long as the C:N ratio does not exceed 5:1.
B. Many amines smell bad; two diamines are particularly well known for this.

    H$_2$N–CH$_2$–CH$_2$–CH$_2$–CH$_2$–NH$_2$      H$_2$N–CH$_2$–CH$_2$–CH$_2$–CH$_2$–CH$_2$–NH$_2$
              Putrescine                                     Cadaverine

C. Many aromatic amines are toxic or are potential carcinogens.

### 17.4 Amines as Bases
A. Both ammonia and amines contain a nitrogen with an unshared pair of electrons that can accept a proton, and thus both are weak bases.

         CH$_3$–NH$_2$ + H$_2$O $\rightleftharpoons$ CH$_3$–NH$_3^+$ + OH$^-$

# Chapter 17 - Amines and Derivatives

Aromatic amines are much weaker bases than ammonia because of the electron-withdrawing property of the aromatic ring.

B. An amine and its salt serve as a c.b.–c.a. pair and can be used as effective buffers. The $pK_a$ for most substituted ammonium ions is in the range of 7 to 9, so these systems buffer in the alkaline pH range.

C. Most amines (even the very weak bases) can react with strong acids to form water-soluble salts. These cations are named as substituted ammonium ions.

## 17.5 Other Chemical Properties of Amines

A. *Amide* formation: 1° or 2° amines react with acid chlorides or acid anhydrides to form substituted amides. Proteins are polyamides.

$$R-\overset{O}{\underset{}{C}}-Cl \quad + \quad H-\underset{H}{N}-R' \quad \rightarrow \quad R-\overset{O}{\underset{}{C}}-\underset{H}{N}-R' \quad + \quad H_2O$$

Amide formation is also important in the synthesis of plastics like *Nylon* and also proteins.

$$Cl-\overset{O}{\underset{}{C}}-CH_2-CH_2-CH_2-CH_2-\overset{O}{\underset{}{C}}-Cl \quad + \quad H-\underset{H}{N}-CH_2-CH_2-CH_2-CH_2-CH_2-CH_2-\underset{H}{N}-H$$

$$\longrightarrow \quad -\overset{O}{\underset{}{C}}-CH_2-CH_2-CH_2-CH_2-\overset{O}{\underset{}{C}}-\underset{H}{N}-CH_2-CH_2-CH_2-CH_2-CH_2-CH_2-\underset{H}{N}-$$

B. Nitrous acid reactions:

1. 1° Amines react with nitrous acid to give a quantitative yield of nitrogen gas.
   $$R-NH_2 \; + \; HNO_2 \; \rightarrow \; N_2(g) + \text{other products}$$

2. 2° Amines react with nitrous acid to form oily, dangerous *N-nitroso* compounds.
   $$R-\underset{R}{N}-H + HO-N=O \; \rightarrow \; R-\underset{R'}{N}-N=O \; + \; H_2O$$

   These compounds may be the cause of the high rates of stomach cancer associated with nitrites in food.

3. 3° Amines react to form a salt.

C. *Ninhydrin* reaction: Many amines react with ninhydrin to form a purple product. Ninhydrin solution is commonly sprayed on paper chromatograms to identify the positions of the separated amino acids.

## 17.6 Heterocyclic Amines

A. *Heterocyclic* compounds are ring structures containing N, O, S, or some other element in addition to carbon.

B. Several heterocyclic N-ring systems are important in biochemistry.

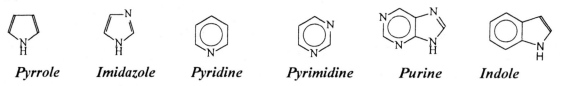

*Pyrrole*   *Imidazole*   *Pyridine*   *Pyrimidine*   *Purine*   *Indole*

C. **Alkaloids** are naturally occurring, N-containing compounds with pharmacological activity. Many alkaloids are heterocyclic amines with striking physiological properties; e.g., morphine, LSD, etc.

## DISCUSSION

Amines are organic nitrogen–containing compounds related to ammonia. It is important to remember that amines are also weak bases, the "bases of life." The nomenclature rules for amines are few and are presented in Section 17.2. The chemistry of amines is summarized here for your convenience.

---

**Chemistry of Amines**

**I. Synthesis of amines**

    A. Chloride substitution: $\quad R-Cl + NH_3 \rightarrow R-NH_2 + HCl$

    B. Reductive amination: $\quad R-\overset{O}{\underset{\|}{C}}-R \xrightarrow{NH_3, H_2, Ni} R-\overset{NH_2}{\underset{|}{C}H}-R$

**II. Reactions of amines**

    A. Salt formation:

        All classes of amines (1°, 2°, 3°) react with acids to produce salts.

$$R-NH_2 \xrightarrow{HCl} RNH_3^+ \; Cl^-$$
$$R_2NH \xrightarrow{HCl} R_2NH_2^+ \; Cl^-$$
$$R_3N \xrightarrow{HCl} R_3NH^+ \; Cl^-$$

    B. Amide formation: (1° and 2° amines only)

        1. From acid chlorides: $\quad RNH_2 + Cl-\overset{O}{\underset{\|}{C}}-R' \rightarrow R-NH-\overset{O}{\underset{\|}{C}}-R'$

$$R_2NH + Cl-\overset{O}{\underset{\|}{C}}-R' \rightarrow R-\underset{\underset{R}{|}}{N}-\overset{O}{\underset{\|}{C}}-R'$$

        2. From acid anhydrides: $\quad RNH_2 + CH_3\overset{O}{\underset{\|}{C}}-O-\overset{O}{\underset{\|}{C}}CH_3 \rightarrow R-NH-\overset{O}{\underset{\|}{C}}-CH_3$

$$R_2NH + CH_3\overset{O}{\underset{\|}{C}}-O-\overset{O}{\underset{\|}{C}}CH_3 \rightarrow R-\underset{\underset{R}{|}}{N}-\overset{O}{\underset{\|}{C}}-CH_3$$

        3. Unsubstituted amides are commercially prepared by heating the ammonium salts of carboxylic acids.

$$R-\overset{O}{\underset{\|}{C}}-OH + NH_3 \rightarrow R-\overset{O}{\underset{\|}{C}}-O^- NH_4^+ \xrightarrow{heat} R-\overset{O}{\underset{\|}{C}}-NH_2$$

    C. **Reaction with nitrous acid:** Each class of amines reacts with nitrous acid in its own way.

        1. 1° Amines yield nitrogen gas and a variety of organic products.
It is the nitrogen gas that is measured in the Van Slyke method for the quantitative determination of primary amino groups.

        2. 2° Amines produce *N*-nitroso compounds: $\quad R_2NH \xrightarrow{HO-N=O} R_2N-N=O$

        3. 3° Amines may undergo a variety of reactions, but salt formation is most important.

# Chapter 17 - Amines and Derivatives

The last section of the chapter presented heterocyclic amines. Included in this group are pyrrole, imidazole, indole, the purines and pyrimidines, and a heterogeneous class of naturally occurring plant heterocyclic amines called alkaloids. We will encounter the imidazole and indole ring systems again in our study of the amino acids histidine and tryptophan (Chapter 21, Table 21.1); the purines and pyrimidines in our study of nucleic acids (Chapter 23). And we referred to the alkaloids with our study on drugs (Selected Topic B).

The questions at the end of the chapter serve as a review for the chemistry and nomenclature of amines. The **SELF-TEST** will further check your understanding of this material.

## SELF-TEST

1. Amines are compounds that would be classified as:
   a. neutral salts   b. weak acids   c. weak bases   d. strong bases   e. strong acids
2. The group that identifies a primary amine is represented by:
   a. R–NH–R   b. R=N–O   c. $R_4N^+Cl^-$   d. R–NH$_2$
3. Predict how many isomers are possible for amines with the molecular formula $C_4H_{11}N$.
   a. 5   b. 6   c. 7   d. 8   e. 9
4. An example of a secondary amine is:
   a. $C_2H_5NH_2$   b. $(C_2H_5)_2NH$   c. $(C_2H_5)_3N$   d. $(C_2H_5)_4N^+$
5. Which compound is aniline?

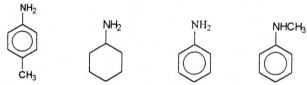

   a.         b.         c.         d.

6. Which compound contains a primary amine and a secondary alcohol group?
   $$\text{a. } \overset{OH}{C}H_2\overset{NH_2}{C}H-CH_2CH_3 \qquad \text{b. } \overset{NHCH_3}{C}H_2CH_2\overset{OH}{C}HCH_3 \qquad \text{c. } CH_3\overset{OH}{C}H-\overset{NH_2}{C}H-CH_3$$

7. $CH_3-\overset{NH_2}{\underset{}{C}H}-\underset{\underset{O}{\|}}{C}-OH$  is a(n):

   a. amide   b. primary amine   c. secondary amine   d. tertiary amine

8. $CH_3-\overset{H}{\underset{}{N}}-CH_2CH_3$ is:
   a. 2-aminopropane   b. ethylmethylamine
   c. isopropylamine   d. methyldimethylamine

9. The compound (imidazole structure) is:
   a. pyrrole   b. imidazole   c. pyridine   d. pyrimidine   e. indole

10. The compound (pyrrole structure) is:
    a. pyrrole   b. imidazole   c. pyridine   d. pyrimidine   e. indole

# Chapter 17 - Amines and Derivatives

11. The compound  is:

    a. pyrrole  b. imidazole  c. pyridine  d. pyrimidine  e. indole

12.  is:

    a. isopentylamine  b. ethyldimethylamine
    c. 2-aminoisopentane  d. 2-amino-2-methylbutane

13.  is:

    a. *o*-chloroaniline  b. aniline hydrochloride  c. anilinium chloride

14. Aniline hydrobromide is:

    a.  b.  c.  d.

15. Which is a quaternary ammonium ion?

    a. $[CH_3-\overset{H}{\underset{H}{N}}-H]^+$  b. $[CH_3-\overset{H}{\underset{CH_3}{N}}-H]^+$  c. $[CH_3-\overset{H}{\underset{CH_3}{N}}-CH_3]^+$  d. $[CH_3-\overset{CH_3}{\underset{CH_3}{N}}-CH_3]^+$

16.  is a(n):

    a. aniline  b. primary amine  c. secondary amine  d. tertiary amine

17.  is:

    a. benzene  b. purine  c. pyridine  d. pyrimidine

18. Alkaloids can best be generally classified with the:

    a. alcohols  b. acids  c. amines  c. amides  e. esters

19. Which of the following compounds when pure is not associated through hydrogen bonding?

    a. $CH_3CH_2-OH$  b. $CH_3-\underset{O}{\overset{\parallel}{C}}-OH$  c. $CH_3-\underset{O}{\overset{\parallel}{C}}-NH_2$  d. $CH_3-NH-CH_3$  e. $CH_3-\underset{CH_3}{N}-CH_3$

20. Which has the highest boiling point?

    a. $CH_3CH_2CH_2OH$  b. $CH_3CH_2CH_2NH_2$  c. $CH_3NH-CH_2CH_3$  d. $CH_3\underset{CH_3}{N}-CH_3$

21. Which of the compounds shown in question 20 has the lowest boiling point?

    a.  b.  c.  d.

22. Some drugs that contain amino groups are converted to their salts to increase their:

    a. acidity  b. basicity  c. solubility

23. Amines react with strong acids to form:

    a. salts  b. esters  c. amino acids  d. bases

## Chapter 17 - Amines and Derivatives

24. Which compound does not exhibit the properties of a base?

   a. $CH_3CH_2CH_2NH_2$   b. $CH_3\overset{.}{C}HCH_3$   c. $CH_3\overset{..}{C}NH_2$   d. $CH_3NCH_3$   e. 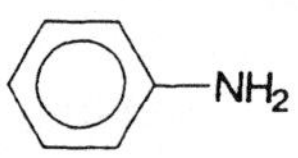 ⌬—$NH_2$
                              $NH_2$              $O$                $CH_3$

25. If a carboxylic acid and an amine react to form a salt, which of the following is the product?
   a. $RCOO^-RNH_3^+$   b. $RCOO^+RNH_3^-$   c. $RCOOH^+RNH_2^-$   d. $RCOOH^+RNH_2^-$

26. The product of the reaction between isopropyl chloride and ammonia is:

   a. $CH_3-\underset{NH_2}{\overset{Cl}{C}}-CH_3$   b. $CH_3-\underset{}{\overset{NH_2}{CH}}-CH_3$   c. $CH_3-\overset{O}{C}-NH_2$

27. The reaction between acetic anhydride and aniline produces:

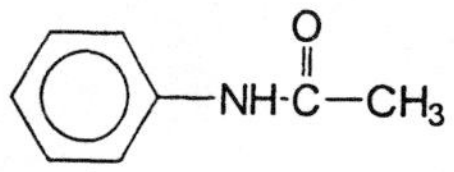

   a. 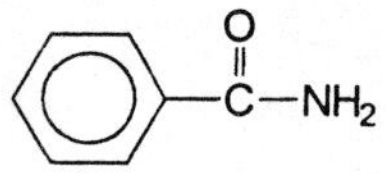 ⌬—C(=O)—$NH_2$   b. ⌬—NH-C(=O)—$CH_3$   c. $CH_3-\overset{O}{C}-NH_2$   d. $CH_3-\overset{O}{C}-O^-NH_4^+$

28. What is the product of the following reaction?

   $CH_3\overset{O}{C}-CH_3 \xrightarrow{H_2, NH_3, Ni}$ ?

   a. $CH_3-\overset{O}{C}-NH_2$   b. $CH_3-\overset{O}{C}-NH-CH_3$   d. $CH_3CH_2-\overset{O}{C}-NH_2$   d. $CH_3-\underset{NH_2}{CH}-CH_3$

29. Which compound would not react with an acid chloride to form an amide?

   a. $NH_3$   b. $CH_3NH_2$   c. $CH_3-NH-CH_3$   d. $CH_3\overset{.}{N}-CH_3$   e. 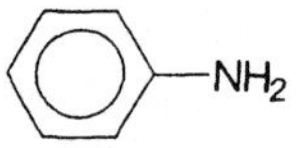 ⌬—$NH_2$
                                                            $CH_3$

30. Which reaction is called reductive amination?

   a. $CH_3CH_2Cl \xrightarrow{NH_3} CH_3CH_2NH_2$

   b. $CH_3\overset{O}{C}-CH_3 \xrightarrow{H_2, NH_3, Ni} CH_3\underset{NH_2}{CH}-CH_3$

   c. $CH_3-NH-CH_3 \xrightarrow{HNO_2} CH_3-\underset{CH_3}{N}-N=O$

31. Which reagent is used to detect the presence of amino acids through a color reaction?
    a. ninhydrin   b. pyrimidine   c. epinephrine

32. In the Van Slyke method, primary amino groups are detected by:
    a. the development of a blue color
    b. the change of litmus paper from red to blue
    c. the evolution of nitrogen gas

33. Members of which class of compounds have not been identified as carcinogens?
    a. N-nitroso compounds   b. aromatic amines   c. amino acids

34. These amines react with nitrous acid to form dangerous N-nitroso compounds.
    a. primary   b. secondary   c. tertiary   d. quaternary

# Chapter 17 - Amines and Derivatives

## ANSWERS

| | | | | | |
|---|---|---|---|---|---|
| 1. c | 7. b | 13. a | 19. e | 25. a | 31. a |
| 2. d | 8. b | 14. d | 20. a | 26. b | 32. c |
| 3. d | 9. b | 15. d | 21. d | 27. b | 33. c |
| 4. b | 10. a | 16. d | 22. c | 28. d | 34. b |
| 5. c | 11. e | 17. d | 23. d | 29. d | |
| 6. c | 12. d | 18. c | 24. c | 30. b | |

# C  Brain Amines and Related Drugs

**KEY WORDS**

| | | | | |
|---|---|---|---|---|
| axon | adrenalin | synapse | neurotransmitter | phenethylamine |
| amphetamines | stimulants | alkaloid | anesthetic | barbiturate |
| tranquilizers | dopamine | antihistamines | synergism | histamine |
| serotonin | catecholamines | acetylcholine | GABA | MAO |
| nitric oxide | benzodiazepine | cocaine | caffeine | nicotine |

**SUMMARY**

C.1 Some Chemistry of the Nervous System
   A. Nerve cells carry messages between the brain and other parts of the body. Although the *axons* of a nerve cell can be quite long, the nerve impulse must be transmitted across short gaps (*synapses*) via chemical messengers called *neurotransmitters*. Each type of neurotransmitter binds to a specific type of receptor site, completing the intended action.
   B. Many drugs (and poisons) act by either blocking or mimicking the action of these natural neurotransmitters.
   C. Since many neurotransmitters are amines, it is not surprising that many drugs also contain nitrogen.

C.2 Brain Amines
   A. *Catecholamines* are adrenergic transmitters. The neurotransmitter norepinephrine is chemically related to the hormone *adrenaline* (epinephrine). It is synthesized from the amino acid tyrosine. High levels of norepinephrine are associated with a state of elation. Drugs that block its action cause depression, while those that mimic its action act as stimulants.

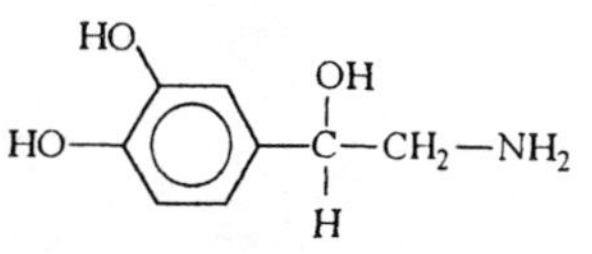

Norepinephrine 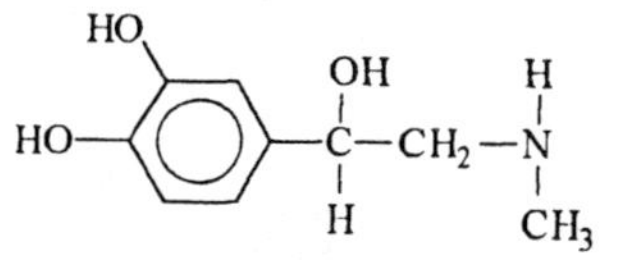     Adrenaline (epinephrine) 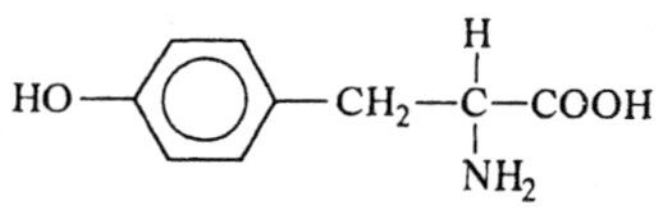     Tyrosine (an amino acid)

   B. *Serotonin* is a neurotransmitter that seems to play a role in mental illness. Serotonin is synthesized from the amino acid tryptophan. Serotonin agonists are used to treat depression and anxiety.

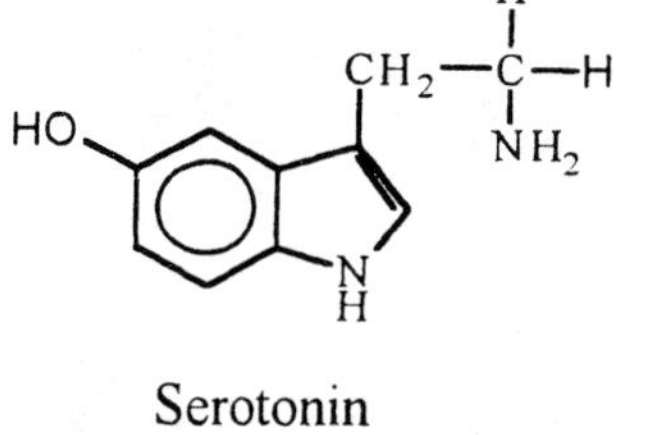

Serotonin 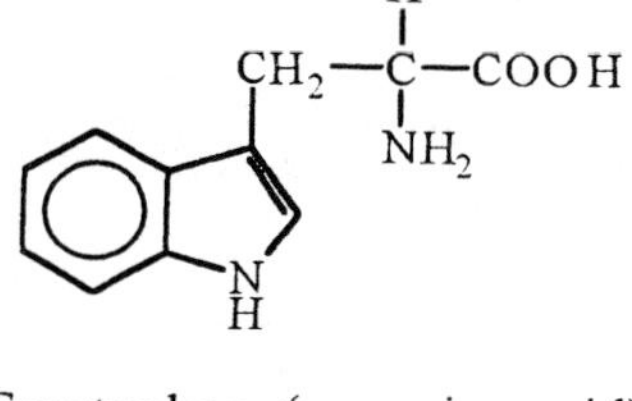     Tryptophan (an amino acid)

   C. *Histamine* is formed by the decarboxylation of the amino acid histidine. It is found in the central nervous system (CNS) but also mediates allergic reactions, inflammation, and acid production in the stomach (ulcers). Products such as Cope, Compoz, and Vanquish contain aspirin plus an antihistamine. *Antihistamines* inhibit the release of histamine.

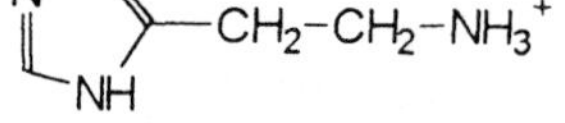

Histamine

### Selected Topic C - Brain Amines and Related Drugs

D. Amino acids can act as excitatory (glutamate, aspartate) or inhibitory (gamma–aminobutyric acid = *GABA*) transmitters.

$$\begin{array}{c} ^+H_3N-CH-COO^- \\ ^-OOC-CH_2-CH_2 \\ \text{Glutamate} \end{array} \quad \rightarrow \quad \begin{array}{c} ^+H_3N-CH \\ ^-OOC-CH_2-CH_2 \\ \text{GABA} \end{array}$$

E. *Acetylcholine* is wide spread in the CNS. Acetylcholine may be involved in short-term memory. Acetylcholine levels appear to be diminished in patients with Alzheimer's disease.
F. Neurotransmitters and mental illness – psychoactive drugs.
     antipychotic / antidepressants / antianxiety
G. Brain cells also make *nitric oxide* (NO), which serves as a chemical messenger.

C. 3 Antianxiety Agents and Sedatives: GABA Modulators
  A. The *benzodiazepines* contain a 7-member heterocyclic ring system as found in Valium. After 20 years of use, benzodiazepines were found to be addictive.

    Diazepam (Valium)      Chlordiazepoxide (Librium)

  B. Buspirone (Buspar) inhibits serotonin activity but enhances dopamine and noradrenaline activity.
  C. Barbiturates enhance GABA activity and decrease oxidative metabolism in the brain.
    1. Barbituric acid was synthesized from urea and malonic acid in 1864 by von Baeyer. Since then, thousands of *barbiturates* have been synthesized. They are generally sedatives (sleeping pills) but can be lethal in larger doses. Barbiturates are particularly dangerous when ingested with alcohol. The *synergistic* effect of these two depressants can enhance the effect of the barbiturate by factors up to 200-fold. Barbiturates are strongly addictive, and withdrawal, hazardous.
    2. Barbiturates are cyclic amides that resemble pyrimidine and may act by substituting for pyrimidine bases (thymine) in nucleic acids (Chapter 23).

    Barbital    Phenobarbital (Luminal)    Sodium pentothal    Thymine

  D. Ethanol is the most widely used tranquilizer, as in a drink to "unwind" at the end of the day.

C.4 Antidepressants: Monoamine Enhancers
  A. Depression is a widespread mood disorder. The biogenic amine theory of depression relates depression to low levels of monoamines; thus the goal is to increase the level of monoamines by inhibiting their breakdown by *monoamine oxidase* (*MAO*). The major classes of antidepressant drugs are MAO inhibitors. Imipramine and amitriptyline are closely related to Promazine, but act as antidepressants instead of tranquilizers.

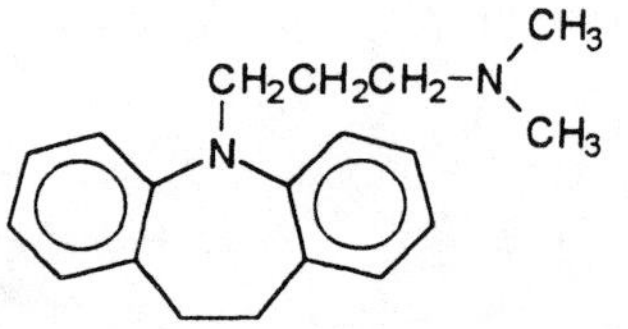

### Selected Topic C - Brain Amines and Related Drugs

**Imipramine**
(Tofranil)

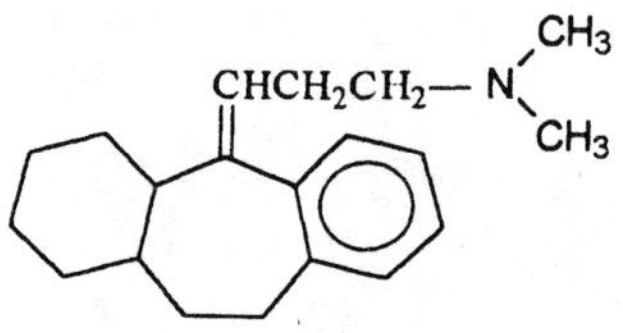

**Amitriptylene**
(Elavil)

B. The most popular "new" antidepressant is fluoxetine (Prozac), with 1993 sales of over $3 billion. Prozac appears to block the reabsorption of serotonin, enhancing its effect. It has been prescribed to help people cope with obesity, fears, and PMS.

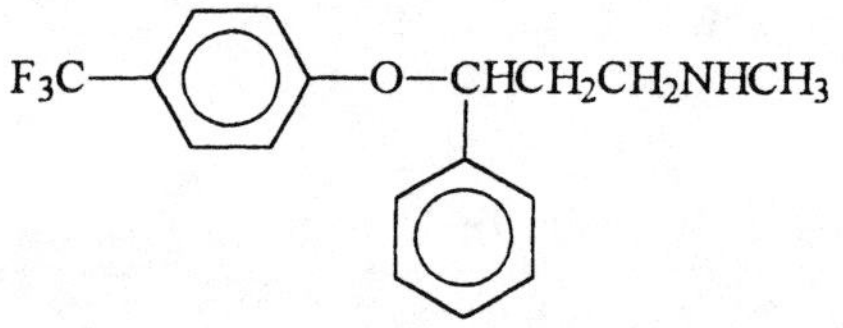

**Fluoxetine (Prozac)**

C. Lithium carbonate is a widely used drug for bipolar (manic–depressive) illness.

### C.5 Antipsychotic Agents: Dopamine Antagonists

A. Reserpine is the active alkaloid in the snakeroot plant (*Rauwolfia serpentina*) used by people of India to treat fever, snakebite, and maniacal forms of mental illness. It was found to reduce blood pressure and bring about sedation. By 1953 it replaced electroshock therapy for many psychotic patients.

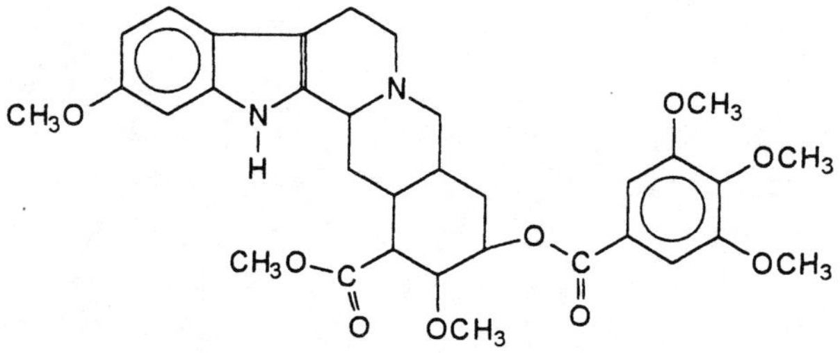

**Reserpine**

B. Slight structural changes can result in a profound change in properties. Promazine is a tranquilizer; imipramine (see below) is a psychic energizer. These are phenothiazines that act as dopamine antagonists to block the action of *dopamine* and to relieve the symptoms of schizophernia. Today, more than 90% of all schizophrenics no longer need hospitalization.

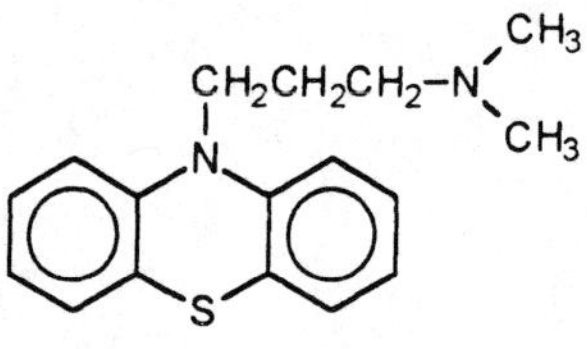

**Promazine**
(tranquilizer)

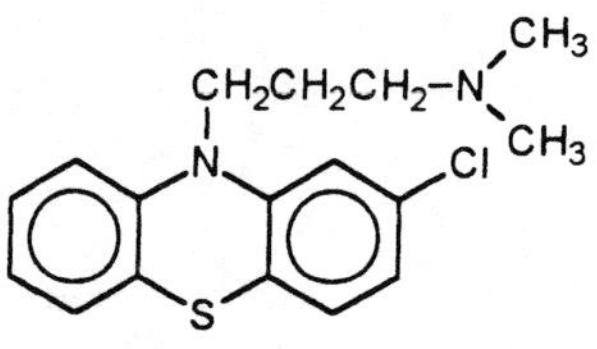

**Chlorpromazine**
(or Thorazine)

# Selected Topic C - Brain Amines and Related Drugs

C.6 Stimulant Drugs: Dopamine Enhancers
  A. The *phenethylamines* (*amphetamines*) form a group of synthetic *stimulants* that may act by mimicking the natural brain amine, norepinephrine.

  Phenethylamine      Amphetamine (Benzedrine)      Methamphetamine (Methedrine)

  1. Amphetamine has been used for weight reduction. Methamphetamine is the "speed" of many drug users. Phenylpropanolamine is used as an appetite suppressant.
  2. Designer drugs are synthetic analogs of compounds that have proven pharmacological activity. Angel dust = PCP, or phencyclidine; china white = α-methyl fentanyl (similar to heroin); Ecstasy = MDMA, or 3,4-methylene dioxymethamphetamine.

  B. *Cocaine* was first isolated in 1860 from the leaves of the coca plant. Cocaine was once used as a local anesthetic, but it is quite toxic. It is also a stimulant, increasing stamina and reducing fatigue by increasing the level of dopamines. It was the "sniff" of the jet-set and it is now widely used. The stimulant effects are short-lived, followed by depression. Overdoses may cause death.

  C. *Caffeine* is the most widely used stimulant. It is an *alkaloid* found in coffee, soft drinks, and "No-Doz" tablets (~100 mg caffeine /tablet vs. ~ 200 mg/cup of strong coffee).

  D. *Nicotine* is an alkaloid found in smoking tobacco. The lethal dose for humans is estimated at about 50 mg.

  Cocaine      Caffeine      Nicotine

C.7 Anesthesia Revisited
  A. Local anesthetics are drugs that block the nerve signals to the brain from the treated tissue. Many local *anesthetics* are ester derivatives of, or related to, *p*-aminobenzoic acid.

  *p*-Aminobenzoic acid      Novocaine (Procaine-HCl)      Benzocaine

  1. Lidocaine or mepivacaine is often the local anesthetic of choice today.
  2. A general anesthetic acts on the brain to produce unconsciousness. Two of the first general anesthetics used in surgical practice were diethylether (1846) and chloroform (1847). General anesthetics appear to act by dissolving fatty membranes of nerve cells.

  B. Dissociative anesthetics: ketamine and PCP
  1. Ketamine is an anesthetic that also produces "near-death" type hallucinations.
  2. PCP (phencyclidine) is also known as "angel dust" or "crystal." It is a dangerous drug but has found use as an animal tranquilizer. It also appears to depress the immune system.

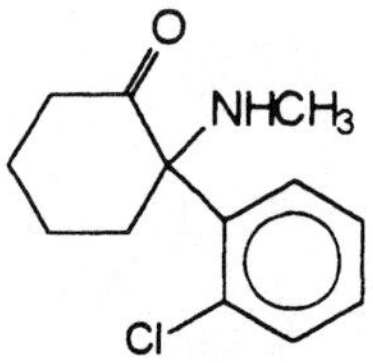

## Selected Topic C - Brain Amines and Related Drugs

Ketamine

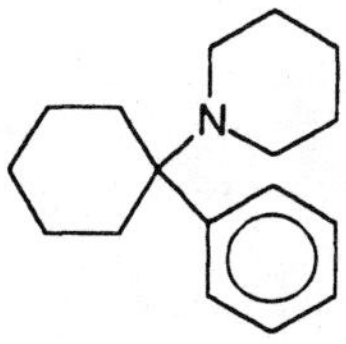

Phencyclidine (PCP)

## DISCUSSION

This Selected Topic is devoted to physiologically important amines and drugs that affect our mental state. It is important that you realize that our moods, our sense of feeling "up" or "down", our sense of and tolerance to pain, etc., are all influenced by the presence or absence of chemical molecules that help regulate our body functions. We now know that many of the drugs that have been used and abused for centuries are capable of eliciting their effects because they somehow mimic a natural neurotransmitter or body regulator molecule. This usually involves some structural similarity that allows the drug molecule to bind to a receptor site intended for the normal body regulator molecule.

The listing below offers a brief summary of this material. The alkaloids are included here again but were first introduced in Selected Topic B.

### Physiologically Active Compounds

| Compound | Common Structural Feature | Some Effects |
|---|---|---|
| **Mood Hormones:** epinephrine (adrenaline) norepinephrine serotonin | See Section C.2 | Balances between compounds produce states ranging from aggression to depression |
| **Amphetamines** amphetamine methamphetamine | 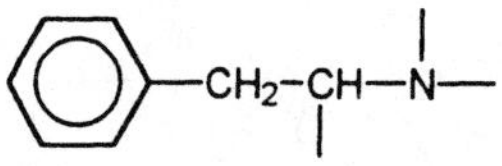 Ph-CH$_2$-CH(-)-N(-) | Stimulants; can cause insomnia, tremors, hallucinations |
| **Barbiturates** barbital pentobarbital secobarbital phenobarbital amobarbital thiopental | 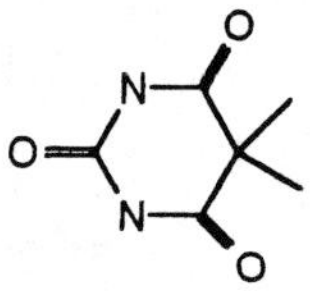 (barbiturate ring structure) | Sedatives and soporifics are strongly addictive |

## Selected Topic C - Brain Amines and Related Drugs

**Alkaloids**
Opium Alkaloids and Derivatives  —  Narcotics; some are
 (morphine, codeine, heroin)  strongly addictive
coniine (from hemlock)  The alkaloids are  Nausea, paralysis and
 nitrogen-containing  death
 heterocyclic
caffeine (from coffee beans,  compounds of  Stimulant
 tea leaves, cola nuts)  varying degrees
 of complexity.

nicotine (from tobacco)  Stimulant
cocaine (from coca leaves)  Powerful stimulant

**Local Anesthetics**  Localized insensitivity
benzocaine  to pain
tetracaine
procaine

**Tranquilizers**  Calmatives and mild
Promazines  soporifics
 promazine
 chlorpromazine
 thioridazine

The questions in the text serve as an excellent review of this material. The following SELF-TEST will further check your understanding of this material.

## SELF-TEST

1. Adrenaline is structurally related to the amino acid:
    a. phenylalanine    b. alanine    c. tyrosine    d. methioine    e. tryptophan
2. Which compound is commonly referred to as adrenaline?
    a. serotonin    b. epinephrine    c. norepinephrine
3. The molecular theory of mental illness postulates an imbalance between two biochemicals. An excess of which compound produces mental depression?
    a. serotonin    b. epinephrine    c. norepinephrine
4. Which class of compounds are related to β-phenylethylamine?
    a. alkaloids    b. amphetamines    c. barbiturates    d. tranquilizers
5. Which compound is classified as a "downer"?
    a. amphetamine    b. barbital    c. cocaine

### Selected Topic C - Brain Amines and Related Drugs

6. The amphetamines are structurally related to:
   a. the natural stimulants epinephrine and norepinephrine
   b. the alkaloids, cocaine and caffeine
   c. the tranquilizers Valium and Librium
7. Which molecular group is common to all barbiturates?
   a.                         b.                         c.    $-O-\overset{O}{\overset{\|}{C}}-N-$

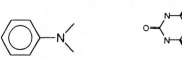

8. Caffeine, nicotine, and cocaine are all classified as:
        a. indole compounds     b. amphetamines     c. barbiturates     d. alkaloids
9. Esters of *p*-aminobenzoic acid are used widely as:
        a. "uppers"          b. "downers"          c. hallucinogens          d. local anesthetics
10. This compound was used by people of India to treat snakebite and is used today to reduce blood pressure and to bring about sedation.
         a. serotonin     b. adrenaline     c. nicotine     d. caffeine     e. reserpine
11. Valium and Librium are two common drugs that are widely used as:
    a. antianxiety drugs      b. anesthetics
    c. sleeping pills          d. major tranquilizers
12. Reserpine, lithium carbonate, and the promazines are all used:
    a. in the treatment of mental illnesses
    b. as local anesthetics
    c. as general anesthetics
13. This popular antidepressant drug is thought to act by blocking the readsorption of serotonin.
         a. Valium      b. Promazine     c. Reserpine     d. Prozac     e. Novocaine

### ANSWERS

| | | | | | | |
|---|---|---|---|---|---|---|
| 1. c | 3. a | 5. b | 7. b | 9. d | 11. a, c | 13. d |
| 2. b | 4. b | 6. a | 8. d | 10. e | 12. a | |

# 18 Stereoisomerism

## KEY WORDS

isomers, structural formula, molecular formula, enzymes, polarized light, levorotatory, stereoisomer, diastereomers, optically active, chirality, enantiomers, meso compound, specific rotation, asymmetric carbon, nonsuperimposable, racemic mixture, dextrorotatory, geometric isomer, cis-, trans-isomer, unpolarized light, achiral, chiral center, Fischer projection, pheromone, polarimeter, structural isomer, van't Hoff's rule

## SUMMARY

*Isomers* are different compounds represented by the same *molecular formula*. *Structural isomers* include positional isomers (1-propanol vs. 2-propanol) and functional-group isomers (ethanol vs. dimethyl ether). In this chapter we are concerned with *stereoisomers*, or space isomerism, which can be distinguished by how the compounds interact with plane-polarized light.

18.1 Polarized Light and Optical Activity
   A. Light is a wave with an oscillating electric vector. *Unpolarized light* can be pictured as a bundle of waves vibrating in all directions. *Polarized light* vibrates in only a single plane.
   B. Certain substances can act on polarized light causing the plane of polarization to rotate. Such substances are said to be optically active. A *polarimeter* employs two polarizing lenses, one before and one after the solution being measured, to measure the degree of *optical activity*.
   C. Optical activity is dependent on the substance, the concentration of the solution, the length of the tube, wavelength of the light, and the temperature of the solution. However, when these things are normalized, a *specific rotation* can be calculated, which is a characteristic property of that substance.
      1. *Dextrorotatory* (+) – right or clockwise rotation of plane-polarized light
      2. *Levorotatory* (-) – left or counter clockwise rotation of plane-polarized light

18.2 Chiral Centers
   A. Optical activity arises from "handedness," or "*chirality*," made possible by the tetrahedral bonding by carbon atoms in the molecules. Organic compounds can be optically active if they have *chiral centers*, carbons with four different groups attached. Chiral carbon centers are often referred to as *asymmetric carbons*. *Achiral* objects are superimposable upon their mirror images.

Mirror plane

   B. Terms
      1. *Stereoisomers* are isomers that differ only in the orientation of atoms in space. Stereoisomers not only have the same *molecular formula*, they have the same *structural formula*, but with a different configuration in space.

# Chapter 18 - Stereoisomerism

2. *Enantiomers* are stereoisomers that are *nonsuperimposable* mirror images of each other. *Fischer projection* formulas assume that the vertical bonds lie behind the page and the horizontal bonds lie above the plane of the paper.

3. A *racemic mixture* is optically inactive because it contains equal amounts of both (+,−) enantiomers.

C. Enantiomers have identical physical properties except that they rotate plane-polarized light in exactly opposite directions.

D. Enantiomers have the same chemical properties when reacting with molecules where handedness is not important. However, they can have drastically different chemical properties when reacting with a chiral molecule such as an enzyme or receptor site. Enantiomers may have different tastes and smells, or the one isomer might be physiologically active and the other not.

18.3 Multiple Chiral Centers

A. Molecules may have several chiral centers. If a molecule has $n$ chiral centers, there are $2^n$ possible stereoisomers of that compound (*van't Hoff's rule*; first Nobel Prize in chemistry in 1901). For example, a molecule with three chiral carbons will have $2^3 = 8$ stereoisomers, or 4 sets of enantiomeric pairs. Those that are not enantiomers will be related as *diastereomers*. Diastereomers can have different physical and chemical properties.

B. A *meso compound* has at least two optically active chiral carbon atoms but contains an internal mirror plane that renders the molecule as a whole optically inactive. A meso compound is superimposable on its own mirror image.

18.4 Geometric Isomerism (Cis–Trans Isomerism)

A. *Geometric isomers* are compounds that have different configurations due to a restraint in the structure such as a cyclic ring structure or double bonds.

B. Nomenclature: Use *cis-* for compounds when the similar groups are on the same side, and use *trans-* when the similar groups are on opposite sides of the ring or double bond.

Examples:   cis-2-Butene   trans-2-Butene   trans-1,2-Dichlorocyclopropane

18.5 Biochemical Significance

A. Chemical reactions in living organisms are catalyzed and controlled by protein molecules called *enzymes*. These are chiral molecules and usually will operate only on specific chiral substrates.

B. Foodstuffs and medicines must have the proper "handedness" to be used by these enzymes and thus be beneficial. (Most natural sugars are D-sugars and most natural amino acids are L-amino acids.)

C. *Pheromones* are chemicals used for communication between insects. The chiral properties of pheromones is important for their recognition.

## DISCUSSION

We have introduced the study of optical isomers and the problem of "handedness" in three dimensional molecules. Stereoisomers are molecules with the same structural formula but which differ in

# Chapter 18 - Stereoisomerism

the way the bonds of the molecule are oriented in space. This is a very important distinction, however, because we will find that most biochemical processes are specific for only one "hand" of a given molecule. We will learn in Chapter 19 that carbohydrates found in nature are mostly D-sugars and in Chapter 21 that proteins are composed of L-amino acids. For now, we want you to be able to recognize chiral molecules, to master the terminology used to describe these isomers, and to work with the various ways of representing the formulas of such compounds. We have gathered some of the important terms of this chapter here for easy reference.

> **Isomers** – different compounds that have the same chemical formula or composition.
> **Stereoisomers** - isomers that have the same structural formula but differ in the spatial arrangement of the atoms.
> **Chirality** – a "handedness" property associated with organic compounds which have one or more carbon atoms with four different groups attached. If there are $n$ chiral carbon centers, then there are $2^n$ stereoisomers possible.
> **Enantiomers** – stereoisomers that are mirror images of each other, i.e., D-glucose and L-glucose.
> **Diastereomers** – stereoisomers that are not mirror-images of one another, i.e., glucose and galactose.
> **Racemic modification** – an equal mixture of a pair of enantiomers. A racemic mixture is not optically active, because it contains equal amounts of the (+) and (-) enantiomorphic forms.
> **Meso form** – a molecule containing chiral carbon centers but which is not itself optically active because of an internal mirror plane.

The remaining problems at the end of the chapter and the **SELF-TEST** given here will help you check your understanding of these concepts.

## SELF-TEST

1. How many chiral carbon centers are in the molecule shown?

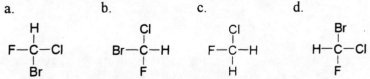

   a. 1   b. 2   c. 3   d. 4   e. 5
2. How many stereoisomers are possible for the compound shown in Question 1?
   a. 1   b. 2   c. 3   d. 4   e. 5   f. 8   g. 10   h. 16   i. 24   j. 32
3. Which of the following one-carbon compounds is not chiral?

   a.              b.              c.              d.

   F—C—Cl with H up, Br down      Br—C—H with Cl up, F down      F—C—H with Cl up, H down      H—C—Cl with Br up, F down

4. Which of the following properties would be different for each of two enantiomers?
   a. solubility in water
   b. density
   c. interaction with plane-polarized light

# Chapter 18 - Stereoisomerism

5. The extent of optical activity of a solution depends on which of the following?
   a. concentration of the solution
   b. temperature
   c. wavelength of light used in the experiment
   d. length of the polarimeter light path
   e. all of the above

6. Substances that rotate polarized light to the right, or in a counter-clockwise manner, are said to be:
   a. dextrorotatory   b. levorotatory

7. Which compound would be expected to exist as a pair of mirror-image isomers?

   a.                  b. 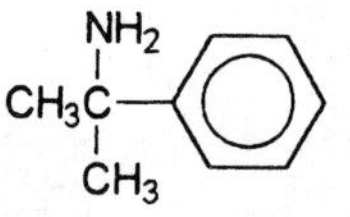                  c. 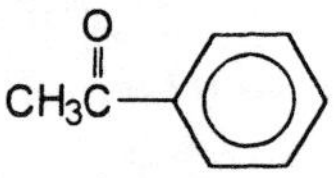                  d.

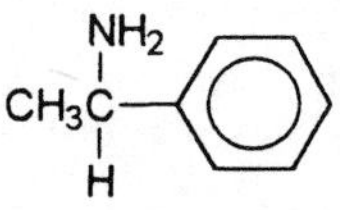

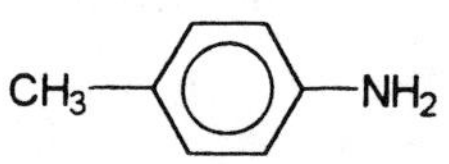

   $CH_3\overset{NH_2}{\underset{H}{C}}$—⬡       $CH_3\overset{NH_2}{\underset{CH_3}{C}}$—⬡       $CH_3\overset{O}{\overset{\|}{C}}$—⬡       $CH_3$—⬡—$NH_2$

The remaining questions refer to the following compounds.

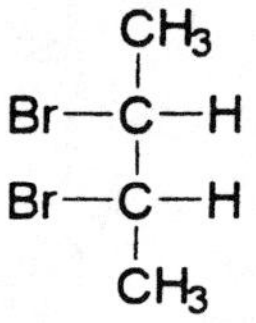

```
    CH3            CH3            CH3            CH3
    |              |              |              |
Br—C—H         H—C—Br         H—C—Br         Br—C—H
    |              |              |              |
Br—C—H         H—C—Br         Br—C—H         H—C—Br
    |              |              |              |
    CH3            CH3            CH3            CH3
     I              II             III            IV
```

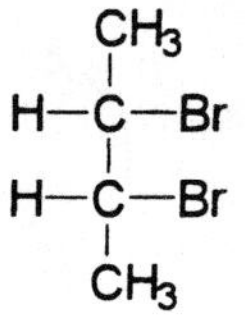

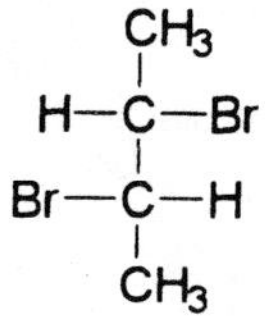

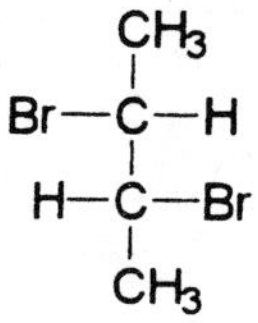

8. Which of the above compounds represent enantiomers?
   I   II   III   IV
9. Which of the above compounds represent diastereomers?
   I   II   III   IV
10. Which of the above compounds represent a *meso*-form?
    I   II   III   IV
11. Which of the following solutions would rotate polarized light?
    a. 0.2 M of compound I
    b. 0.2 M II
    c. 0.2 M III
    d. 0.2 M IV
    e. 0.2 M III plus 0.2 M IV
    f. 0.2 M II plus 0.2 M III

## ANSWERS

1. d
2. h
3. c
4. c
5. e
6. b
7. a
8. III and IV
9. I and III, I and IV
10. I or II (I = II)
11. c, d, f

# D  Chemistry of the Senses

## KEY WORDS
rhodopsin   cones   opsin   photochemical isomerization   retinal
olfactory   odorant   rods   visual pigments

## SUMMARY
We perceive the world around us through our five senses: sight, sound, smell, taste, and touch. Here we look at the molecules for seeing and smelling.

D.1 Vision: Cis-Trans Isomerism
  A. The retina of the human eye contains two kinds of receptor cells, *rods* and *cones*. Cone cells in the eye contain *rhodopsin*, which consists of a protein (*opsin*) in complex with a small organic molecule derived from vitamin A (11-*cis-retinal*).
  B. The vision cycle (*photochemical isomerization*)
    1. Light is absorbed by 11-*cis* retinal which converts the molecule to its trans isomer, 11-*trans-retinal*. This isomerization is initiated in a few picoseconds.
    2. The structural change from the bent *cis*-isomer to the more linear trans-isomer acts like a switch to trigger a nerve impulse in the optic nerve that is sent to the brain. Also, during this stage the rhodopsin complex splits into the protein opsin and free retinal.
    3. Enzymes are used to re-form the 11-*cis* isomer.
    4. The *cis* isomer complexes with opsin to reform rhodopsin so the cycle can start again.

11-*cis*-Retinal    ~ $10^{-12}$ sec →    11-*trans*-Retinal

D.2 Smell and Taste
  A. The process of olfaction (smelling) involves the binding of an odorous molecule (*odorant*) to a receptor cell of the *olfactory* system (nose).
  B. The receptors in the nose have are divided into odor types. Molecules that can bind to multiple receptor sites will send a "mixed" odor signal to the brain.
  C. The theory of taste (gustation) by the taste buds is very similar to that proposed here for smell, e.g. there may be small number of distinct taste receptors that are mixed and matched to produce a broad spectrum of different tastes.

## SELF-TEST
1. The protein part of the visual receptor in cone cells is called:
    a. retina   b. opsin   c. rhodopsin   d. retinal   e. sightase
2. The pigment part of the visual receptor in cone cells is called:
    a. retina   b. opsin   c. rhodopsin   d. retinal   e. sightase
3. The adsoption of light causes a(n) _____ of the receptor molecules.
    a. oxidation   b. reduction   c. *cis-trans* isomerization   d. cleavage

ANSWERS   1. b   2. d   3. c

# 19 Carbohydrates

**KEY WORDS**

| | | | | |
|---|---|---|---|---|
| sugar | ketose | lactose | acetal | monosaccharide |
| aldose | sucrose | α-D-glucose | disaccharide | D-glucose |
| maltose | β-D-glucose | polysaccharide | mannose | starch |
| amylopectin | "reducing" sugar | galactose | glycogen | mutarotation |
| hemiacetal | fructose | amylose | cellulose | Tollens' reagent |
| Fischer | lactose intolerance | triose | pentose | hexose |
| Haworth | glycosidic linkage | invert sugar | non-reducing | Fehling's reagent |
| Benedict's reagent | anomeric carbon | anomers | carbohydrate | galactosemia |

**SUMMARY**

Many carbohydrates have the molecular formula $(C \cdot H_2O)_n$, but they are not hydrates of carbon as their name implies. **Carbohydrates** are polyhydroxy aldehydes (aldoses) or ketones (ketoses). The terms carbohydrate, saccharide, and **sugar** are used interchangeably. **Sugars** are also classified as "*reducing*" or "*non-reducing*" sugars, depending on whether the sugar will reduce **Tollens's** or related reagents that are mild oxidants. Most carbohydrates are crystalline solids at room temperature, with high melting points. Carbohydrates readily form hydrogen bonds and are very soluble in water (e.g., 100 g glucose/100 mL $H_2O$ at 25 °C).

*Monosaccharides* – simple sugar molecules

*Disaccharides* – carbohydrates that can be hydrolyzed to two monosaccharides

*Polysaccharides* – carbohydrates made up of many monosaccharides

19.1 General Terminology and Stereochemistry
  A. Monosaccharides (*-ose* ending) are classified according to whether the carbonyl is present as an aldehyde (*aldoses*) or ketone (*ketoses*).
  B. They are also classified by the number of C atoms
     (3, tri*ose*; 4, tetr*ose*; 5, pent*ose*; 6, hex*ose*).
  C. The simplest *sugars* are triose oxidation products of glycerol. There are 2 aldotrioses because the glyceraldehyde molecule has one chiral carbon center, but only one ketotriose since it has no chiral center. Usually carbohydrates found in nature are D-sugars. By convention, all D-type sugars are drawn with the "-OH" on the right at the asymmetric C atom furthest from the carbonyl carbon (usually the penultimate carbon) as shown in the *Fischer* projections below.

```
        O                O
        ‖                ‖
        C—H              C—H              CH₂OH
        |                |                |
   OH—C—H           H—C—OH              C=O
        |                |                |
       CH₂OH            CH₂OH           CH₂OH
        L-               D-
          Glyceraldehyde                Dihydroxyacetone
```

  D. D- and L-glyceraldehyde are optical isomers and differ in the way they rotate plane polarized light. D(+) indicates that a D-sugar rotates plane-polarized light in a clockwise (+), or dextrorotatory direction, and L(−) indicates a counter-clockwise (−), or levorotatory, direction.

E. Major monosaccharides in biochemistry
  1. *Trioses*: glyceraldehyde, dihydroxyacetone
  2. *Pentoses*: D-Ribose is found in ribonucleic acids (RNAs), and 2-deoxyribose occurs in deoxyribonucleic acids (DNAs).

```
        CHO                    CHO
     H—C—OH                 H—C—H
     H—C—OH                 H—C—OH
     H—C—OH                 H—C—OH
       CH₂OH                  CH₂OH
      D-Ribose              D-2-Deoxyribose
```

  3. *Hexoses*: Glucose is an aldohexose and the most abundant sugar occurring in nature.

```
     CHO            CHO            CHO           CH₂OH
   H—C—OH        H—C—OH         HO—C—H           C=O
   HO—C—H        HO—C—H         HO—C—H         HO—C—H
   H—C—OH        HO—C—H         H—C—OH         H—C—OH
   H—C—OH        H—C—OH         H—C—OH         H—C—OH
     CH₂OH         CH₂OH          CH₂OH           CH₂OH
   D(+)-Glucose  D(+)-Galactose  D(+)-Mannose   D(-)-Fructose
```

## 19.2 Hexoses

A. **D-Glucose** is the normal "blood sugar" and the sugar our cells use directly for the production of energy. D-Glucose (or dextrose) is the monomer of *starch* and *cellulose,* and is also found in the familiar disaccharides *lactose* and *sucrose*.

B. D-Glucose has 4 asymmetric carbon centers; thus there are $2^4$, or 16, possible stereoisomers of D-glucose. However, only 2 of the other aldohexoses are very common in nature.
  1. *Mannose* – differs from glucose only at the second C atom; a component of the polysaccharide mannan.
  2. *Galactose* – differs from glucose only at the fourth C atom; it is bound with glucose to form the disaccharide, lactose (milk sugar), and is a component of galactans and glycolipids of nerve and brain tissue called cerebrosides and gangliosides.

C. **D-Fructose** (also called levulose) is a ketohexose. Fructose and glucose combine to make the common disaccharide, *sucrose*. It is found in honey and is the sweetest of all natural sugars.

D. Aspartame, saccharin, and sodium cyclamate are popular artificial sweeteners, 100x–300x as sweet as fructose. Aspartame (Nutrasweet) is an ester of a dipeptide and breaks down into aspartic acid, phenylalanine, and methanol. Individuals sensitive to phenylalanine should regulate their consumption of Nutrasweet. Acesulfame K has recently been approved as an artificial sweetener. It is reported to be more heat-stable than Nutrasweet and can survive the cooking process.

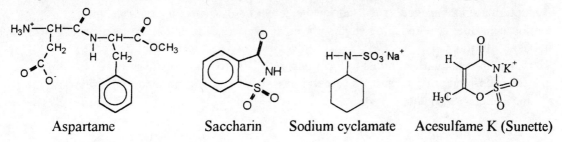

Aspartame        Saccharin        Sodium cyclamate        Acesulfame K (Sunette)

# Chapter 19 - Carbohydrates

19.3 Cyclic Structures of Monosaccharides
  A. Monosaccharides are polyhydroxy aldehydes or ketones. Aldehydes and ketones can react with alcohols to form hemiacetals or hemiketals. Pentoses and hexoses exist in solution as internal *hemiacetals*, forming stable 5- or 6-membered rings.

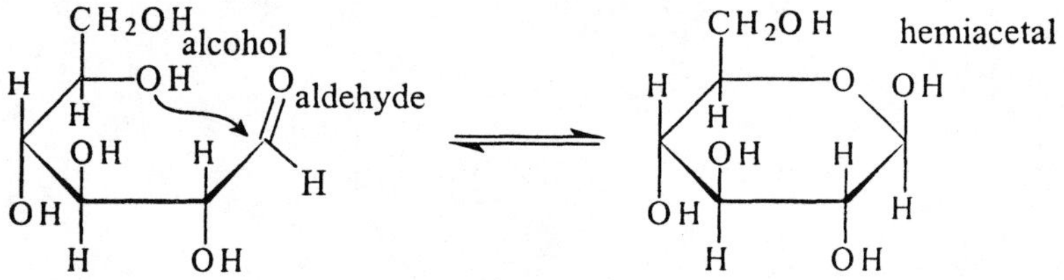

  B. α- and β-D-Glucose: The internal hemiacetal reaction creates a new asymmetric carbon (*anomeric carbon*) at the C-1 position of glucose. These two different forms of glucose (*anomers*) are called α-D-glucose and β-D-glucose, shown here as *Haworth* formulas. The β form always has the new -OH group on the same side of the ring as the terminal -CH$_2$OH group. These compounds are not enantiomers and have slightly different physical properties as well as different chemical properties.

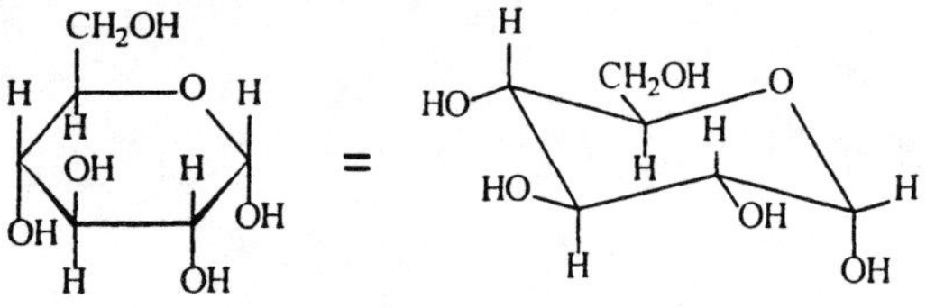

  *α-D-Glucose* (m.p. 146 °C) 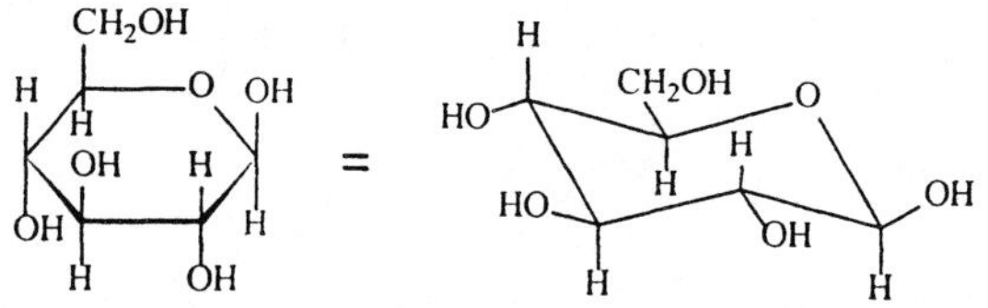    *β-D-Glucose* (m.p. 150 °C)

  C. The β form of D-Glucose is more stable than the α form, but both forms are always in equilibrium with a very small amount of the free aldehyde. The interconversion of α- and β-isomers via the reopening and reclosing of the hemiacetal form is called *mutarotation*.

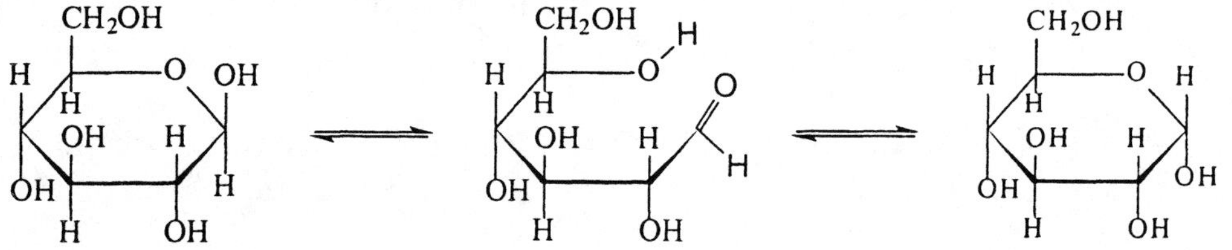

  β-D-Glucose (64%)    D-Glucose (aldehyde form) (<0.02%)    α-D-Glucose (36%)

19.4 Properties of Monosaccharides
  A. Most sugars are crystalline solids at room temperature and are water soluble.
  B. Chemical properties
    1. The -OH groups on sugars behave like typical alcohols, reacting to form esters and ethers.
    2. Mild oxidation of the aldehyde by *Tollens's* (reduction of Ag$^+$), *Benedict's*, or *Fehling's* (reduction of Cu$^{2+}$) *reagents* is used for the detection of *reducing sugars*. Note: all monosaccharides are "reducing sugars" because even ketoses exhibit some aldehyde character in basic solution.
    3. Hemiacetals react with alcohols to form *acetals*. This reaction is important in forming disaccharides and polysaccharides. Acetals are stable and will not undergo mutarotation or reduce Tollens's reagent.

# Chapter 19 - Carbohydrates

19.5 Disaccharides

A. Disaccharides are distinguished by their composition and linkage. The three major disaccharides are sucrose, maltose, and lactose.

B. *Maltose* is the disaccharide produced during the partial hydrolysis of starch. It is composed of two D-glucose units, connected "head to tail" by an α (1→4) glycosidic linkage. It is classified as a reducing sugar because it still also possesses a hemiacetal on one end.

C. *Lactose,* or milk sugar, is composed of a β-D-galactose unit with a β (1→4) glycosidic *acetal* linkage to C 4 of D-glucose. Lactose is also a reducing sugar because of the presence of the hemiacetal group in the glucose ring. Most infants have a "lactase" enzyme and can digest lactose more readily than adults. Over 70% of the world's adult population suffers from *lactose intolerance,* characterized by cramps and diarrhea when milk lactose is left undigested. *Galactosemia* is a genetic disease resulting from the inability to convert galactose to glucose.

*Maltose*        *Lactose*

D. *Sucrose* is the common table sugar, cane sugar, beet sugar, or just *sugar*. It is composed of α-D-glucose and β-D-fructose with an α (1→2) *glycosidic linkage* from the C-1 carbon of glucose to the β C-2 carbon of fructose. Sucrose is a nonreducing sugar since the latent carbonyl groups of both units are involved in the acetyl linkage. *Invert sugar* is a 1:1 mixture of glucose and fructose and has properties that differ from sucrose.

*Sucrose*    or

19.6 Polysaccharides

A. Polysaccharides are large polymers of sugars that are used as food/energy reserves (starch) or structural components (cellulose). Starch, glycogen, and cellulose are all homopolymers comprised from just one sugar, D-glucose.

B. *Starch* is a polymer of α-D-glucose. Plants make glucose from carbon dioxide by photosynthesis and then store it in the form of starch. Starch can be separated into two fractions.
   1. *Amylose* (~20%): Unbranched chains of 60 to 300 glucose units with α (1→4) linkages; water soluble.

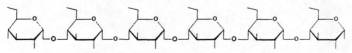

## Chapter 19 - Carbohydrates

2. *Amylopectin* (~80%): A branched-chain polymer of 300 to 6000 glucose units; the chain segments have α (1→4) linkages as in amylose, but also are connected by α (1→6) acetal linkages at the branch points; water insoluble.

C. *Glycogen* is the equivalent of starch in animals. Glycogen is structurally a branched-chain polymer of D-glucose similar to amylopectin but even more highly branched. Glucose reserves are stored as glycogen in liver and muscle tissues.

D. *Cellulose* is a linear polymer of D-glucose similar to amylose but with β-1,4-glysosidic (acetal) linkages. The β linkage is more stable and confers a different three-dimensional structure on cellulose than that found for starch. Man and most animals cannot digest cellulose, but grazing animals and termites can because of the action of the enzymes in the bacteria of their digestive tracts. Although it has no nutritive value for man, cellulose makes up the greater part of dietary fiber. Cotton fibrils are nearly pure cellulose. Chemical modifications of the -OH groups of cellulose have produced a wide variety of synthetic polymers (Celluloid).

## DISCUSSION

In Chapter 19 the terminology and molecular architecture associated with carbohydrates were presented. Our study of carbohydrates will continue in Chapters 24 and 25 when we trace the critical biochemical role of these compounds. To do that it is essential that you familiarize yourself now with carbohydrate structures. You should aim at being able to read carbohydrate structures as well as you now read the structures of simpler compounds. For example, when we say "the hydroxyl group of ethanol," with only the slightest pause, you should think "$CH_3-CH_2-OH$." After studying the material in this chapter, you should also be able to look at a carbohydrate structure and, with a pause only a bit longer than before, locate the acetal linkage or determine whether the compound is glucose or fructose or mannose, and whether or not it is a reducing sugar.

Before we present the **SELF-TEST**, let's establish a way of drawing saccharides in abbreviated form. For the open-chain form, we'll adopt this shorthand notation.

The branches on the stem of the abbreviated formula give the orientation of the secondary hydroxyl groups. The same simplification can be used for ring forms.

# Chapter 19 - Carbohydrates

We'll use the simplified structures in the **SELF-TEST**. Be sure you're able to translate these forms to the more familiar ones before you proceed.

## SELF-TEST

Refer to the following structures in answering Questions 1 through 6.

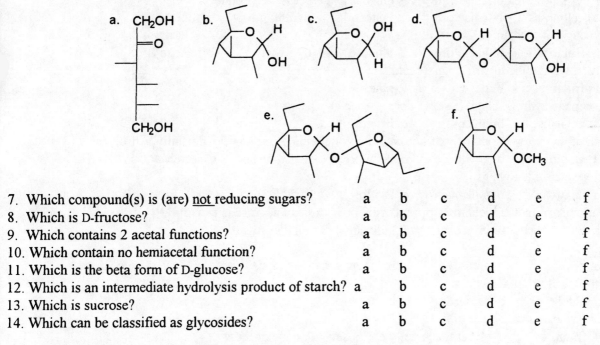

1. Which of the compounds is not a sugar?   a   b   c   d   e
2. Which is not a member of the D-family of sugars?   a   b   c   d   e
3. Which is D-glyceraldehyde?   a   b   c   d   e
4. Which is the product of oxidation of D-glyceraldehyde?   a   b   c   d   e
5. Which is a ketotetrose?   a   b   c   d   e
6. Which is (are) tetroses?   a   b   c   d   e

Refer to the following structures in answering Questions 7 through 14.

7. Which compound(s) is (are) <u>not</u> reducing sugars?   a   b   c   d   e   f
8. Which is D-fructose?   a   b   c   d   e   f
9. Which contains 2 acetal functions?   a   b   c   d   e   f
10. Which contain no hemiacetal function?   a   b   c   d   e   f
11. Which is the beta form of D-glucose?   a   b   c   d   e   f
12. Which is an intermediate hydrolysis product of starch?   a   b   c   d   e   f
13. Which is sucrose?   a   b   c   d   e   f
14. Which can be classified as glycosides?   a   b   c   d   e   f

Refer to the following structure in answering Questions 15 through 22.

15. The monosaccharide unit on the right is:
    a. fructose   b. galactose   c. glucose   d. lactose   e. maltose
    f. mannose   g. sucrose   h. none of these

183

# Chapter 19 - Carbohydrates

16. The monosaccharide unit on the left is:
    a. fructose   b. galactose   c. glucose   d. lactose   e. maltose
    f. mannose   g. sucrose   h. none of these
17. The disaccharide is:
    a. fructose   b. galactose   c. glucose   d. lactose   e. maltose
    f. mannose   g. sucrose   h. none of these
18. The hemiacetal group is:
    a. alpha   b. beta
19. The acetal group is:
    a. alpha   b. beta
20. The rings are attached through positions:
    a. 1,4   b. 1,6   c. ortho   d. meta   e. para
21. The compound is a:
    a. reducing sugar   b. nonreducing sugar
22. On hydrolysis, the compound yields:
    a. aldohexoses   b. aldopentoses   c. ketohexoses   d. ketopentoses
23. Which of the following yields a sugar other than glucose upon complete hydrolysis?
    a. amylopectin   b. cellulose   c. lactose   d. maltose   e. starch
24. Common table sugar is more formally described as:
    a. glucose   b. lactose   c. maltose   d. sucrose
25. Hydrolysis of which disaccharide gives glucose and fructose as products?
    a. cellulose   b. galactose   c. lactose   d. maltose   e. sucrose
26. Blood sugar is the same as:
    a. fructose   b. galactose   c. glucose   d. glycogen   e. lactose   f. sucrose
27. On hydrolysis, milk sugar does not yield:
    a. fructose   b. galactose   c. glucose
28. On complete hydrolysis, table sugar does not yield:
    a. galactose   b. glucose   c. fructose
29. Which of the following carbohydrates would not yield maltose if it were partially hydrolyzed?
    a. amylose   b. amylopectin   c. cellulose   d. glycogen   e. starch
30. Fructose is also known as:
    a. dextrose   b. milk sugar   c. levulose   d. table sugar
31. The mixture of monosaccharides produced by hydrolysis of sucrose is called:
    a. blood sugar   b. invert sugar   c. milk sugar   d. table sugar
32. Saccharin is a(n):
    a. hexose   b. aldose   c. reducing sugar   d. glycoside   e. none of these
33. Fructose is a(n):
    a. aldohexose   b. aldopentose   c. ketohexose   d. ketopentose   e. triose
34. 2-Deoxyribose is an isomer of:
    a. fructose   b. glyceraldehyde   c. ribose   d. none of these
35. Dihydroxyacetone is an isomer of:
    a. glyceraldehyde   b. glucose   c. ribose   d. saccharin   e. none of these
36. Ribose is a(n):
    a. aldopentose   b. ketohexose   c. ketotriose   d. aldotetrose   e. none of these
37. Cellulose is a(n):
    a. monosaccharide   b. disaccharide   c. polysaccharide   d. none of these
38. In which of the following are monosaccharide units not joined by alpha linkages?
    a. maltose   b. amylose   c. glycogen   d. cellulose

39. Which shorthand formula correctly shows the orientation of the secondary hydroxy groups in D-mannose?

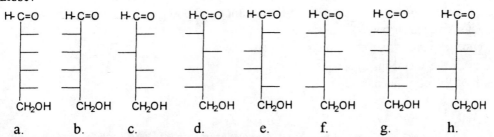

    a.      b.      c.      d.      e.      f.      g.      h.

40. Which compound contains a β-acetal linkage?
    a. amylose    b. glycogen    c. lactose    d. maltose    e. starch

41. Which compound does not show 1,6 branching?
    a. amylopectin    b. amylose    c. glycogen

42. Which compound contains α-acetal linkages and 1,6 branching?
    a. amylose    b. cellulose    c. glycogen    d. sucrose

43. Fructose is not a(n):
    a. alcohol    b. aldehyde    c. carbohydrate    d. sugar    e. saccharide

44. Which is not a reducing sugar?
    a. fructose    b. galactose    c. glucose    d. lactose
    e. maltose    f. sucrose

45. If a carbohydrate gives a positive Fehling's test, the carbohydrate:
    a. is a reducing sugar    b. is reduced    c. will give a negative Tollens's test

46. Which reagent will accomplish the following reaction?

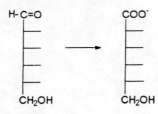

    a. Tollens's reagent    b. Fehling's reagent    c. Benedict's reagent
    d. All of these reagents will yield the product shown.
    e. None of these reagents will yield the product shown.

47. Which compound will not be oxidized by Fehling's reagent?

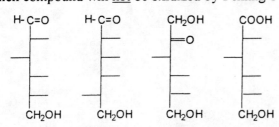

    a.      b.      c.      d.      e. All will be oxidized.

48. Is the following compound a reducing sugar?

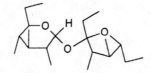

    a. yes    b. no

Chapter 19 - Carbohydrates

49. Which compound cannot be metabolized by human beings?
   a. amylopectin   b. amylose   c. cellulose   d. maltose   e. starch
50. Which structural feature prevents the digestion of the compound described in Question 49?
   a. the branches formed through 1,6 linkages
   b. the presence of galactose units in the polysaccharide
   c. the β form of the acetal linkages
   d. the length of the polysaccharide

**ANSWERS**

| | | | | |
|---|---|---|---|---|
| 1. c | 11. c | 21. a | 31. b | 41. b |
| 2. d | 12. d | 22. a | 32. e | 42. c |
| 3. a | 13. e | 23. c | 33. c | 43. b |
| 4. c | 14. d, e, f | 24. d | 34. d | 44. f |
| 5. e | 15. f | 25. e | 35. a | 45. a |
| 6. b, e | 16. c | 26. c | 36. a | 46. d |
| 7. e, f | 17. h | 27. a | 37. c | 47. d |
| 8. a | 18. b | 28. a | 38. d | 48. b |
| 9. e | 19. a | 29. c | 39. g | 49. c |
| 10. a, e, f | 20. a | 30. c | 40. c | 50. c |

# 20 Lipids

**KEY WORDS**

| | | | | |
|---|---|---|---|---|
| lipid | fat | "hard" water | glycolipids | fatty acid |
| oils | phosphatides | steroid | saturated | triglyceride |
| lecithin | cholesterol | monounsaturated | saponification | cephalin |
| polyunsaturated | iodine number | soap | sphingolipid | membranes |
| hydrogenation | detergent | choline | ethanolamine | integral |
| rancidity | antioxidant | hydrophobic | hydrophilic | peripheral |
| cerebroside | semipermeable | eutrophication | biodegradable | micelle |
| facilitated diffusion | active transport | bile salts | waxes | fluid mosaic |
| arteriosclerosis | ganglioside | bilayer | gallstones | heart attack |
| nonsaponifiable | stroke | cis-double bond | LDL, VLDL | HDL |
| cytoplasm | emulsion | glycolipid | lipoprotein | phospholipid |
| sphingosine | sphingomyelin | | | |

**SUMMARY**

*Lipids* are a diverse class of biomolecules defined on the basis of being more soluble in nonpolar solvents like diethyl ether than in water. Lipids represent an important energy source, play a major role in membranes, and serve as vitamins and hormones. Types of compounds included as lipids are:
1. fatty acids   2. triglycerides   3. waxes   4. phospholipids
5. glycolipids   6. steroids   7. prostaglandins   8. fat-soluble vitamins

## 20.1 Fatty Acids

A. *Fatty acids* are long-chain carboxylic acids. Most natural fatty acids contain an even number of carbon atoms and are components of fats and oils.

B. Fatty acids are divided into two subclasses:
1. *Saturated* – the hydrocarbon "tail" of a saturated fatty acid is fairly straight, permitting close dispersion forces resulting in higher melting points; for example, *fats.*
2. *Unsaturated* – most natural unsaturated fatty acids have *cis-double bonds* that cause the hydrocarbon portion to bend, thus lessening the strength of intermolecular forces and resulting in lower melting points; for example, *oils.*

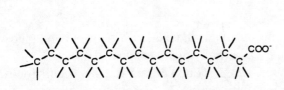

Saturated fatty acid          Unsaturated fatty acid (*cis* double bond)

C. Some common fatty acids: (Δ-9 refers to the position of a double bond after C9, etc.)

*Saturated* fatty acids      $CH_3(CH_2)_{12}COOH$   $C_{14}$   Myristic acid
     $CH_3(CH_2)_{14}COOH$   $C_{16}$   Palmitic acid
     $CH_3(CH_2)_{16}COOH$   $C_{18}$   Stearic acid

# Chapter 20 - Lipids

| | | | |
|---|---|---|---|
| *Monounsaturated* fatty acids | $C_{18}$ | $\Delta\,9$ | Oleic acid |
| *Polyunsaturated* fatty acids | $C_{18}$ | $\Delta\,9, 12$ | Linoleic acid |
| | $C_{18}$ | $\Delta\,9, 12, 15$ | Linolenic acid |
| | $C_{20}$ | $\Delta\,5, 8, 11, 14$ | Arachidonic acid |

20.2 Fats and Oils
  A. Most *"fats"* are fatty acid esters of glycerol, called *triglycerides*. Animal fats are usually solids at room temperature because their triglycerides contain mostly saturated fatty acids. Fats can be broken down by basic hydrolysis (*saponification*) to yield glycerol and the salts of the fatty acids (*soaps*).

  B. Oils are fats that are liquids at room temperature. Their lower melting points are due to a higher proportion of unsaturated fatty acid units. They are obtained principally from vegetable sources. *Polyunsaturated* fatty acids have two or more double bonds. Some polyunsaturated fatty acids are "essential" fatty acids that must be supplied in the diet for good health.
  C. Saturated fats and cholesterol have been implicated in *arteriosclerosis* (hardening of the arteries).
  D. Olestra is a fat-based fat replacer. Olestra is a sucrose core with 6–8 fatty acids attached. Because of its larger size, Olestra is too large to be hydrolyzed by lipases and thus adds no calories and no fat. However, it does reduce the absorption of the fat-soluble vitamins (A, D, E, and K).
  D. Properties of fats and oils
    1. Physical properties: Fats and oils are lighter than water (0.8 g/cm$^3$). Most are colorless, odorless, and tasteless.
    2. Chemical properties
      a. Hydrolysis – Basic hydrolysis = soap making = saponification
      b. *Iodine Number*: The degree of unsaturation is measured in terms of the "iodine number" which is defined as the number of grams of iodine that will be consumed by 100 g of fat or oil. The larger the iodine number, the higher the degree of unsaturation.

        | Fat or Oil | Iodine Number |
        |---|---|
        | Butter | 25–40 |
        | Lard | 47–70 |
        | Sunflower oil | 130–140 |

      c. *Hydrogenation* of oils can produce solids or semisolids. Margarine is a butter substitute consisting of vegetable oils that have been partially hydrogenated.
      d. *Rancidity* (production of odorous breakdown products) of fats and oils occurs by hydrolysis of ester bonds and oxidation of unsaturated fatty acids. *Antioxidants* are added to increase the shelf life of foods.

20.3 Soaps
  A. Natural *soaps* are made by the saponification (basic hydrolysis) of fats. The sodium or potassium salts of the released fatty acids have an ionic head group and a nonpolar or "oily" tail. The ionic end (*hydrophilic*) keeps it dissolved in water, while the *hydrophobic* hydrocarbon tail helps disperse

# Chapter 20 - Lipids

oils. The oil and water form an *emulsion*, which cleans away the dirt and oil. Many liquid soaps and shampoos are potassium salts that produce a finer lather and are more soluble.
B. Problems with natural soaps
   1. Acidic conditions – natural soaps are converted to insoluble fatty acids under acidic conditions, precipitating out as "greasy scum."
   2. *"Hard" water* – refers to the presence of calcium, magnesium, or iron ions that form insoluble salts with the fatty acid anions that precipitate as "bathtub ring."
C. Synthetic *detergents* have soluble calcium salts so that they are still effective in "hard" water. Today there is great emphasis on detergents that must be *biodegradable* and also not contribute to the *eutrophication* of our lakes and rivers.
D. A *wax* is an ester formed from a long-chain fatty acid and a long-chain monohydric alcohol.

Waxes serve as the protective coatings on fruits and are used in many cosmetics and ointments.
   1. Beeswax is a by-product of honey production.
   2. Spermaceti is obtained from the oil taken from the head of the sperm whale.
   3. Lanolin is a wax from sheep's wool. It is a mixture of esters and polyesters.

## 20.4 Membrane Lipids
A. *Phosphatidates* are lipid esters of glycerol containing two fatty acid groups and one phosphoric acid residue and one alcohol. Such *phospholipids* are a major component of cell *membranes*.
   1. *Cephalins* are phosphatides containing *ethanolamine*. They are found in brain and nerve tissue.
   2. *Lecithins* are phosphatides containing *choline*. Lecithins are found in brain and nerve tissue and are widely used in foods as emulsifying agents. Lecithin is plentiful in egg yolk and is a common emulsifying agent in foods.

    Cephalins                    Lecithins

B. *Sphingolipids* are based on the unsaturated amino alcohol, *sphingosine*, rather than glycerol. The sphingosine molecule contributes one fatty acid-like group and binds a second in an amide linkage. *Sphingomyelin* is an important constituent of the myelin sheath that surrounds the axis of a nerve cell.

                    Sphingomyelin

C. Glycolipids
   1. *Glycolipids* are distinguished by the presence of a sugar unit, and no phosphoric acid unit.

# Chapter 20 - Lipids

2. Glycolipids composed of galactose, a fatty acid, and sphingosine are called *cerebrosides*. They are important constituents of the membranes of nerve and brain cells. Gaucher disease results from the substitution of glucose for galactose in these glycolipids.

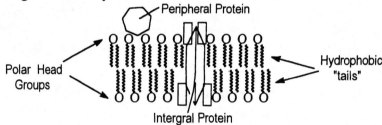

3. *Gangliosides* are similar to cerebrosides but contain a more complex oligosaccharide in place of galactose.

## 20.5 Cell Membranes

A. Polar lipids form *micelles*, monolayers, and bilayers. The *bilayers* have the polar head groups interacting with the polar aqueous media and have the nonpolar hydrophobic tails of the fatty acids interacting with each other on the inside (see figure of membrane below).

B. Major lipid classes present in cell membranes are phospholipids, glycolipids, and cholesterol.

C. Cell *membranes* are phospholipid bilayers but also contain several proteins. *Peripheral* proteins are associated with the inside or the outside of the lipid bilayer, while *integral* proteins completely penetrate the cell membrane. In the *"fluid mosaic"* model of cell membranes, the proteins are viewed as moving in a sea of lipids.

D. Cell membranes are *semipermeable*, allowing only certain materials to pass across the membranes. Many proteins in the membrane are receptors and involved in the active transport of materials into the *cytoplasm* of the cell.

1. If the direction of transport is dictated by a favorable concentration gradient, this is called *"facilitated diffusion."*
2. If the movement takes place against a concentration gradient, this requires energy and is called *"active transport."*

## 20.6 Steroids: Cholesterol and Bile Salts

A. *Steroids* are a group of *nonsaponifiable* lipids that include the bile salts, cholesterol, and the sex hormones. All steroids share a common structural unit – four fused rings (three 6-membered rings and one 5-membered ring).

B. *Cholesterol*, a steroidal alcohol, is the most abundant steroid. It is a common component of animal cell membranes and represents about 10% of the brain, but does not occur in plants. Cholesterol is the precursor of cholic acid and some hormones and vitamins.

*Cholesterol*

C. **Bile salts** are emulsifying agents released from the gallbladder into the digestive tract to aid in the digestion of fats. They are derived from the steroid, cholic acid. About 500 mL of yellowish-green bile liquid is secreted each day by the gallbladder. *Gallstones* are solid precipitates of cholesterol.

20.7 Cholesterol and Cardiovascular Disease
  A. Animal food products (meat, eggs, cheese, etc.) are normally high in cholesterol. A single egg contains about 200 mg of cholesterol.
  B. The liver will reduce its production of cholesterol if dietary cholesterol is high, but individuals with diets high in cholesterol tend to have elevated cholesterol levels in the blood. Lower cholesterol levels appear to reduce atherosclerosis and the risk of coronary artery disease.
  C. Cholesterol and other lipids form complexes with water-soluble proteins and are transported as lipoproteins. *Lipoproteins* are classified according to their density and composition.
    1. *VLDLs* (very-low-density lipoproteins) mainly transport triglycerides.
    2. *LDLs* (low-density lipoproteins) are the carriers of cholesterol.
    3. *HDLs* (high-density lipoproteins) also carry cholesterol.
  High levels of LDLs (low-density lipoproteins), which are rich in cholesterol, have been associated with increased heart disease.

| Lipoprotein | Density | % Protein | % Triglyceride | % Cholesterol (free) | % Cholesterol (ester) | % Phospholipid |
|---|---|---|---|---|---|---|
| Chylomicrons | <0.94 | 1–2 | 85–95 | 1–3 | 2–4 | 3–6 |
| VLDL | 0.94–1.006 | 6–10 | 50–65 | 4–8 | 16–22 | 15–20 |
| LDL | 1.006–1.063 | 18–22 | 4–8 | 6–8 | 45–50 | 18–24 |
| HDL | 1.063–1.21 | 45–55 | 2–7 | 3–5 | 15–20 | 26–32 |

  D. *Heart attacks* occur when a coronary artery is blocked. The cells deprived of nutrients soon die. A *stroke* occurs when a clot or a plaque deposit obstructs an artery supplying the brain. The "normal fasting levels" of lipids in blood plasma are given below. Abnormally high levels of *triglycerides* and *cholesterol* are associated with hardening of the arteries and an increased potential for stroke or heart attack.
    1. Total cholesterol:     120–250 mg/100mL
    2. Triglycerides:         25–260 mg/100mL
    3. Total phospholipids:   150–250 mg/100mL
    4. Total lipids:          400–700 mg/100mL

## DISCUSSION

The SUMMARY provides you with the structures of all the saponifiable lipids discussed in Chapter 20. The compounds have been drawn without hydrogens to emphasize similarities in the structures and to make it easier to compare the structures. You should be able to draw from memory a typical triglyceride structure. You should be able to recognize the others. That's not as hard as it sounds, particularly if you have a good grasp of the triglyceride structure. After all – if you know "*glyco*" comes from glycose, which means sugar, then identifying a glycolipid should be a snap because we've just spent

## Chapter 20 - Lipids

some time looking at sugars in Chapter 19. Identifying a phospholipid should be even easier. Look for phosphorus. The same can be said of sphingosine-based lipids. If you know what the glycerol unit in a triglyceride looks like, you know that the sphingosine unit in sphingosine-based lipids is obviously different and identifiable.

Note that phospholipid is a more general term than phosphatide. All phosphatides are phospholipids, but not all phospholipids are phosphatides. The phosphatides are those compounds that contain both phosphorus and the glycerol unit. Names like lecithin and cerebroside give no hint of the corresponding structure (in contrast to designations like phospholipid or glycolipid). Therefore, associating these names with structural features involves some straight memorization. For example, if you want to be able to recognize a cerebroside, you'll have to mentally translate that to "a sphingosine-based glycolipid" and then look for a sphingosine unit and a sugar unit.

The nonsaponifiable steroids are very easy to recognize because of their characteristic fused-ring structure. Actually writing out the structure frequently impresses it much more firmly on one's mind, and we advise you to do this. Why all this talk about structure? Because as we begin discussing bigger and more complicated molecules, the very appearance of those molecules gets in our way. When such a structure appears on a page, a message seems to travel from eye to brain and says, "WARNING...COMPLICATED STRUCTURE...YOU'LL NEVER UNDERSTAND WHAT THEY'RE TALKING ABOUT!" By spending the time to familiarize ourselves with the structures, we are shutting off (or at least turning down) that alarm system.

It isn't important that you be able to draw out perfectly the structure of a phospholipid. But you should be able to recognize a significant difference between a phospholipid and a triglyceride. One incorporates an ionized group; the other doesn't. Is that important? Well, one (the triglyceride) is actually stored in the body as droplets of fat within special cells (that's what adipose tissue, or "fat," is). The other, because it has a polar end (the ionized group) and long, nonpolar hydrocarbon chains, can organize its molecules into bilayers and is the structural material of membranes. You can see what very different roles these molecules play in the body because of their structure. Be sure to go through the problems at the end of the chapter in the text before attempting the **SELF-TEST**.

## SELF-TEST

1. Which of these compounds would <u>not</u> be classified as a lipid?
    a. prostaglandins   b. steroids   c. triglycerides   d. glycolipids   e. glycogen
2. Which is a <u>not</u> a major role of lipids?
    a. membranes   b. vitamins   c. enzymes   d. hormones   e. energy storage
3. Animal fats and vegetable oils are natural:
    a. polymers   b. soaps   c. esters   d. hydrocarbons   e. carboxylic acids
4. The body's principal energy reserves are in the form of:
    a. alcohol   b. carbohydrates   c. enzymes   d. fats   e. proteins
5. The double bond in an unsaturated fatty acid causes a bend in the hydrocarbon chain. Because of this, these compounds _____ compared to saturated fatty acids:
    a. experience weaker intermolecular attractive forces
    b. have higher melting points
    c. do not react as strongly with glycerol
6. Mixed triglycerides contain:
    a. at least two different fatty acid units
    b. a phosphate unit and fatty acid units
    c. choline and ethanolamine
7. Oils typically have a greater percentage of _____ fatty acids.
    a. saturated   b. unsaturated   c. steroidal

# Chapter 20 - Lipids

8. In general, a vegetable triglyceride would not be hydrogenated to produce:
    a. margarine   b. shortening   c. cooking oil
9. The iodine number of a sample of a butter is 27; the iodine number of three samples of margarine are 72–(I), 80–(II), and 92–(III). Which sample is the more highly unsaturated fat?
    a. butter   b. margarine I   c. margarine II   d. margarine III
10. Hydrogenation of arachidonic acid yields:
    a. lauric acid   b. linoleic acid   c. linolenic acid   d. stearic acid   e. none of these
11. When butter turns rancid, which chemical reaction is not involved?
    a. hydrogenation   b. hydrolysis   c. oxidation
12. Basic hydrolysis of a triglyceride is called:
    a. esterification   b. hydrogenation   c. polymerization   d. saponification
13. Sodium salts of long-chain fatty acids are called:
    a. lecithins   b. soaps   c. micelles   d. synthetic detergents
14. Complete saponification of a fat yields:
    a. salts of fatty acids and glycerol
    b. salts of fatty acids and a glyceride
    c. fatty acids and glycerol
    d. fatty acids and a glyceride
15. Modern synthetic detergents are not:
    a. biodegradable   b. soluble in hard water   c. soaps
16. Saponification refers to the reaction of lipids with:
    a. an enzyme   b. hydrogen   c. phosphate   d. sodium hydroxide
17. A wax is:
    a. a solid fat
    b. any solid lipid
    c. the ester of a long-chain fatty acid and a long-chain alcohol
18. Which is a wax according to the chemical definition of that term?
    a. carbowax   b. carnauba wax   c. paraffin wax
19. Spermaceti is a wax obtained from:
    a. honeycombs   b. palm leaves   c. whales
20. Which compound would not be expected to exhibit detergent action?
    a. $CH_3CH_2CH_2CH_2CH_2CH_2CH_2CH_2CH_2CH_2CH_2CH_2CH_2CH_2CH_2COO^-Na^+$
    b. $CH_3CH_2CH_2CH_2CH_2CH_2CH_2CH_2CH_2CH_2CH_2CH_2CH_2CH_2CH_2COOH$
    c. $CH_3CH_2CH_2CH_2CH_2CH_2CH_2CH_2CH_2CH_2CH_2CH_2CH_2CH_2CH_2OSO_3^-Na^+$
21. Which fatty acid salt is soluble in water solutions?
    a. $RCOO^-K^+$   b. $(RCOO^-)_2Mg^{2+}$   c. $(RCOO^-)_2Ca^{2+}$   d. none are soluble
22. Phosphatides are not:
    a. amine-containing lipids   b. glycerol-based lipids   c. sphingosine-based lipids
23. Which of the lipids does not incorporate glycerol as part of its structure?
    a. cephalins   b. lecithins   c. sphingomyelins
24. One distinguishes between a cephalin and a lecithin on the basis of:
    a. the amino alcohol unit incorporated in the molecule
    b. the presence or absence of a phosphate group
    c. the presence or absence of a sugar unit

## Chapter 20 - Lipids

25. The amino alcohol incorporated in the compound below is called:

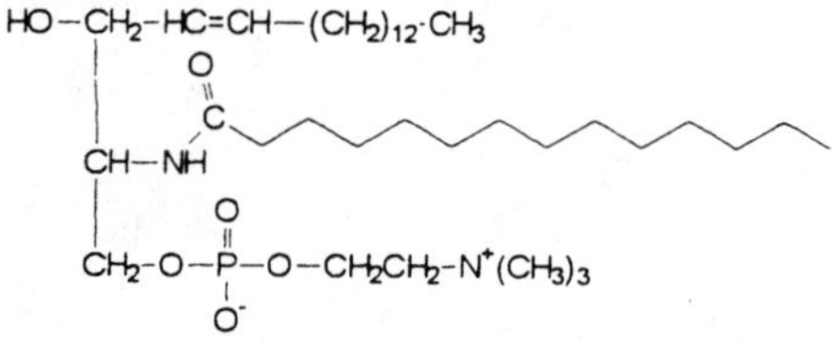

    HO–CH$_2$–HC=CH–(CH$_2$)$_{12}$·CH$_3$
    |
    CH–NH–C(=O)–(CH$_2$ chain)
    |
    CH$_2$–O–P(=O)(O$^-$)–O–CH$_2$CH$_2$–N$^+$(CH$_3$)$_3$

    a. lecithin    b. cephalin    c. cerebroside    d. sphingosine    e. cholesterol

26. Sphingomyelins are lipids based on sphingosine rather than glycerol. In all other respects, sphingomyelins resemble:
    a. glycosides    b. phosphatides    c. triglycerides

27. The process by which materials are moved across cell membranes with an energy input is called:
    a. passive transport    b. active transport    c. facilitated diffusion

28. Proteins that span the lipid bilayer of the cell membrane are called:
    a. long proteins    b. integral proteins    c. peripheral proteins    d. conjugated proteins

29. A phospholipid that is abundant in egg yolk is:
    a. lecithin    b. cephalin    c. cerebroside    d. sphingosine    e. cholesterol

30. The bile salts are not:
    a. fatty acids    b. emulsifying agents    c. steroids

31. This compound is:

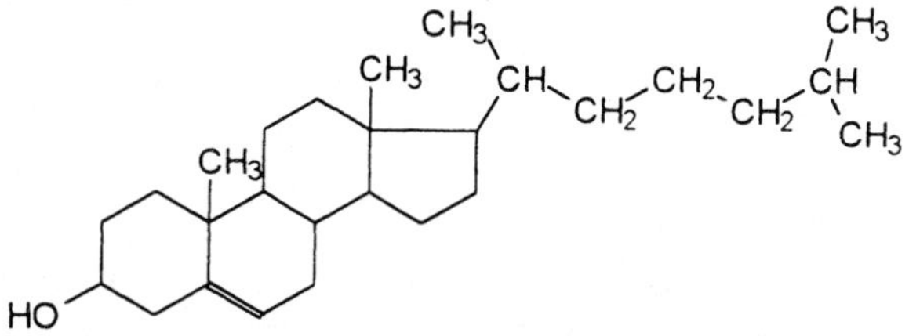

    a. cholesterol    b. prostaglandin    c. a steroid

32. Cholesterol is not:
    a. an alcohol    b. a lipid    c. saponifiable    d. a steroid

33. The prostaglandins are derivatives of:
    a. cholesterol    b. a saturated fatty acid    c. glycerol    d. arachidonic acid

34. In their physiological action, the prostaglandins resemble:
    a. carbohydrates    b. cholesterol    c. enzymes    d. hormones

35. Which special arrangement of polar lipids comes closest to the structure of cell membranes?
    a. bilayers    b. micelles    c. monolayers

Answer Questions 36 through 40 by referring to the following structure.

CH$_2$–O–C(=O)–CH$_2$CH$_2$CH$_2$CH$_2$CH$_2$CH$_2$CH$_2$CH$_2$CH$_2$CH$_2$CH$_2$CH$_2$CH$_2$CH$_2$CH$_2$CH$_2$CH$_3$
|
CH–O–C(=O)–CH$_2$CH$_2$CH$_2$CH$_2$CH$_2$CH$_2$CH$_2$CH=CHCH$_2$CH=CHCH$_2$CH=CHCH$_2$CH$_3$
|
CH$_2$–O–C(=O)–CH$_2$CH$_2$CH$_2$CH$_2$CH$_2$CH$_2$CH$_2$CH=CHCH$_2$CH=CHCH$_2$CH=CHCH$_2$CH$_3$

36. The compound is a:
    a. glyceride    b. phosphatide    c. sphingolipid    d. steroid

37. Hydrolysis of the compound would not yield:
    a. choline    b. glycerol    c. stearic acid

# Chapter 20 - Lipids

38. A high proportion of this type of lipid is found in:
    a. animal fats   b. cell membranes   c. vegetable oils
39. Compared to a similar but saturated fat, this compound will turn rancid:
    a. more rapidly   b. less rapidly
40. If a galactose unit were substituted for the third fatty acid unit, the compound would be a:
    a. cephalin   b. cerebroside   c. glycolipid

## ANSWERS

| | | | |
|---|---|---|---|
| 1. e | 11. a | 21. a | 31. a |
| 2. c | 12. d | 22. c | 32. c |
| 3. c | 13. b | 23. c | 33. d |
| 4. d | 14. a | 24. a | 34. d |
| 5. a | 15. c | 25. d | 35. a |
| 6. a | 16. d | 26. b | 36. a |
| 7. b | 17. c | 27. b | 37. a |
| 8. c | 18. b | 28. b | 38. c |
| 9. d | 19. c | 29. a | 39. a |
| 10. e | 20. b | 30. a | 40. c |

# E      Hormones

**KEY WORDS**

| | | | | |
|---|---|---|---|---|
| *hormone* | *pituitary* | *steroidal* | *ductless glands* | *endocrine* |
| *receptor sites* | *androgen* | *aldosterone* | *hypothalamus* | *cortisone* |
| *estrogen* | *progestins* | *inflammation* | *testosterone* | *estradiol* |
| *prostaglandins* | *mineralocorticoid* | *glucocorticoid* | *anabolic steroid* | *paracrine* |

**SUMMARY**

E.1 The Endocrine System
   A. *Hormones* are organic compounds that serve as chemical messengers. They are synthesized by the *ductless glands* of the *endocrine* system and are transported by the circulatory system to other parts of the body where they bring about marked physiological changes. Most hormones are either steroids, proteins, or peptide-like compounds.
   B. The messenger system
      1. A nerve impulse signals the *hypothalamus* to secrete releasing factors.
      2. The releasing factors stimulate the *pituitary* to secrete pituitary hormones.
      3. The pituitary hormone travels to some target endocrine gland and causes it to secrete its hormones.
      4. The released hormones bind at *receptor sites* on their target tissue, and the tissue responds with altered metabolism. This whole messenger system is usually regulated by some sort of feedback or servo mechanism.
   C. Refer to Table E.2 in the text for a list of human hormones and their physiological effects.

E.2 Adrenocortical Hormones
   A. *Aldosterone* is called a *mineralocorticoid* and is involved in regulating the exchange of $Na^+$, $K^+$, and $H^+$.
   B. *Cortisone* is a steroidal hormone involved in carbohydrate metabolism and is a *glucocorticoid*. These hormones increase glucose production and mobilize fatty acids. Cortisone and cortisol were once used widely in medicine to reduce *inflammation* and to treat arthritis, but today they have been largely replaced by a synthetic analog, prednisolone.

*Aldosterone*      *Cortisone*      *Prednisolone*

E.3 Sex Hormones
   A. The sex hormones are also *steroidal* hormones. They are responsible for stimulating and maintaining sexual characteristics.
      1. *Androgens* are male sex hormones, such as testosterone, that are secreted by the testes.

## Selected Topic E - Hormones

2. **Estrogens** are female sex hormones, such as estradiol, secreted by the ovaries. Progesterone prepares the uterus for pregnancy and prevents the further release of eggs from the ovaries.
B. Anabolic steroids are usually derivatives of testosterone. Dianabol will increase muscle bulk and strength, but there are serious health hazards with its use.

*Testosterone*   *Estradiol*   *Progesterone*

Progesterone serves as an effective birth control drug when injected. Oral birth control pills usually combine an estrogen with a *progestin* (a synthetic mimicker of the hormone progesterone). The estrogen regulates the menstrual cycle, while the progestin signals a state of false pregnancy so ovulation does not occur.

C. Conception and Contraceptives
   1. The menstrual cycle
      a. FSH (follicle-stimulating hormone) is released by the pituitary.
      b. FSH causes the follicle to secrete estrogen hormones that prepares the uterus for the fertilized eggs.
      c. As levels of estrogen increase, the levels of FSH decrease. LH (luteinizing hormone) is now released by the pituitary and brings about ovulation.
      d. A third hormone (LTH, or luteotrophic hormone) secreted by the pituitary influences the development of tissue called the corpus lumen.
      e. If the egg is not fertilized, the corpus lumen disintegrates and, with the blood vessels that rupture, forms the menstrual discharge.
   2. The "pill" is a mixture of two synthetic analogs of female hormones to deceive the body into thinking it is pregnant (false pregnancy).
   3. Mifepristone is a "morning after" pill that blocks the action of progesterone, which is essential for maintenance of pregnancy.
   4. Danazol is a male contraceptive, but most male contraception is by a surgical procedure to block sperm emission – the vasectomy.
D. Hormone replacement therapy
   1. Menopause usually occurs in the late forties. When ovulation no longer occurs, estrogen levels drop, and menstrual cycles cease.
   2. Side effects of menopause such as irritability, "hot flashes," anxiety, and fatigue can be overcome by daily doses of estrogens. Such therapy may also significantly reduce the risk of osteoporosis and coronary disease.

E.4 Prostaglandins
A. **Prostaglandins** are hormone-like substances synthesized from a 20–carbon, polyunsaturated fatty acid called arachidonic acid.

Prostaglandin $E_2$

### Selected Topic E - Hormones

B. Prostaglandins are among the most potent body regulators. They all induce smooth muscle contraction, lower blood pressure, and contribute to the inflammatory response. Aspirin obstructs the synthesis of prostaglandins from arachidonic acid by inhibiting the enzyme cyclooxygenase, which processes arachidonic acid.

C. One clinical use of prostaglandins has been to cause uterine contractions to induce abortion.

## DISCUSSION

Hormones are like vitamins in that they are relatively small organic compounds that are needed in small amounts. However, unlike the vitamins, hormones can be synthesized in the body by the endocrine glands. Also, rather than functioning as enzyme "helpers," or coenzymes, the hormones are "chemical messengers" that help regulate our body's metabolism. The hormones also vary considerably in structure, but those listed in Table E.2 are either proteins, amino acid derivatives, or steroids. Perhaps the most notable feature of the sex hormones discussed in this unit is the extraordinary similarity of compounds that elicit such distinctive responses in our bodies.

## SELF-TEST

1. Hormones function as:
    a. chemical messengers    b. coenzymes    c. provitamins
2. Hormones are synthesized by the:
    a. stomach    b. rod cells    c. cone cells    d. ductless glands of the endocrine system
3. Which gland is responsible for the production of releasing factors that trigger the pituitary gland?
    a. adrenal cortex    b. hypophysis    c. hypothalamus
4. Which of these hormones plays a major role in regulating the exchange of $K^+$?
    a. testosterone    b. estrogen    c. prostaglandin E2    d. progesterone    e. aldosterone
5. Which are male sex hormones?
    a. androgens    b. estrogens    c. progestins
6. Which component of the typical birth control pill is responsible for regulating the menstrual cycle?
    a. androgen    b. estrogen    c. progestin
7. Which compound is used in the synthesis of prostaglandins?
    a. cortisone    b. estrogen    c. arachidonic acid    d. aspirin
8. T / F  Cortisone is a vitamin that exhibits anti-inflammatory properties.
9. T / F  All hormones are steroids synthesized in trace amounts by the endocrine glands.
10. T / F  Hormones are secreted by endocrine glands located in the target organs.
11. T / F  Male sex hormones have been used to treat breast cancer in women.
12. T / F  The presence of estrogen in contraceptive pills is thought to be responsible for most of the undesirable side effects of these pills.

## ANSWERS

| | | | |
|---|---|---|---|
| 1. a | 4. e | 7. c | 10. F |
| 2. d | 5. a | 8. F | 11. T |
| 3. c | 6. b | 9. F | 12. T |

# 21 Proteins

## KEY WORDS

| | | | | |
|---|---|---|---|---|
| L-α-amino acid | globular | protein | prosthetic group | zwitterion |
| isoelectric pH | primary structure | salt bridge | "R" group | polypeptide |
| electrophoresis | secondary structure | disulfide | side chain | peptide bond |
| α-helix | tertiary structure | hydrophobic | N-terminal | C-terminal |
| β-pleated sheet | quarternary | net charge | denaturation | dipeptide |
| tripeptide | fibrous | silk | wool | collagen |
| residues | hydrogen-bonds | gelatins | dalton | diuretic |
| amino acid sequence | | | | |

## SUMMARY

21.1 General Properties of Amino Acids

A. **Proteins** are essential for all cells and function in body building and maintenance. Proteins are copolymers composed from about 20 different amino acids.

B. The common amino acids differ from one another only in the size and nature of the **"R" group** or amino acid **"side chain"** attached to the α-carbon. Note: Some of the amino acids could fit more than one of the groups listed below. See Table 21.1 in the text for structures and abbreviations.

$H_2N-C^\alpha H-COOH$
   $|$
   $R$

| "R" – type | Amino acids |
|---|---|
| Hydrogen | Gly |
| Hydrocarbon | Ala, Val, Leu, Ile, Pro |
| Alcohol-containing | Ser, Thr |
| Sulfur-containing | Met, Cys |
| Aromatic | Phe, Tyr, Trp |
| Basic | Lys, Arg, His |
| Acidic | Asp, Glu |
| Amide | Asn, Gln |

C. Amino acids could also be classified by polarity (nonpolar, polar-neutral, polar-acidic, polar-basic) or the size of the **R** group.

D. The functional variety of the "R" groups permits various types of side-chain interactions, such as hydrogen bonding, salt bridges, disulfide bond formation, and hydrophobic interactions.

E. Amino acids are solids at room temperature and form internal salts called **zwitterions**. This ionic species is also the dominant form in solution at neutral pH conditions.

$$H_3N^+-\overset{H}{\underset{R}{C}}*-\overset{O}{\underset{O^-}{C}}$$

(*The central carbon attached "α" to the carboxylate group is called the α-carbon or $C^\alpha$).

F. Configuration: Most of these monomers are **L–α–amino acids**, indicating that the carboxyl and amino groups are both on the same carbon and that they all have the same L–configuration about that carbon.

   1. Glycine is an exception, having two hydrogens on its α-carbon, so it does not have L- and D- forms.
   2. Proline is also unusual in that it is a secondary amine, whereas all of the others are primary amines.

# Chapter 21 - Proteins

21.2 Reactions of Amino Acids
   A. Acid–Base Properties
     1. Amino acids can act either as weak acids or weak bases and thus make good buffers in living systems.
     2. Charge behavior – all amino acids have at least two titratable groups. Some have additional titratable groups in their side chains. At low pH (<2) when these groups are all in their conjugate acid forms, all amino acids are positively charged because of the protonated amino group. At high pH (>12) all amino acids are negatively charged because of the presence of the carboxylate ion. The net charge on any given amino acid varies with pH from (+) at low pH to (–) at high pH.
     3. The *isoelectric* pH is that pH at which the *net charge* on the molecule is zero.

       Low pH (+ charge)         pH = pI (no net charge)         High pH (− charge)

     4. Amino acids and proteins can be separated on the basis of relative mobilities in an electric field. This process is called *electrophoresis*.
   B. Amino acids also undergo other reactions typical of carboxylic acids and amines. The most important such reaction is the polymerization reaction through peptide bond formation.

21.3 The Peptide Bond
   A. The *peptide bond* is a special name given to the amide linkage joining amino acid *residues*.

The nitrogen in the peptide bond is not basic because of delocalization of its unshared pair of electrons. This gives the C–N bond partial double-bond character and constrains the atoms of the peptide bond to be planar.
   B. *Dipeptides* have two amino acid residues linked by one peptide bond; *tripeptides* have three amino acid residues, tetrapeptides have four, etc.
   C. Peptides have an *N-terminal* (or free amino) end and a *C-terminal* (carboxyl) end and are named starting from the N-terminal end according to the sequence of amino acid residues present. Amino acid residues are named by dropping the "*-ine*" ending of the amino acid and adding "*-yl.*"

*Glycyl alanyl cysteine = Gly–Ala–Cys = G–A–C (one-letter codes)*

D. **Proteins** are linear polymers of hundreds of amino acids linked by peptide bonds; *polypeptides* are proteins.

## 21.4 The Sequence of Amino Acids
A. Each protein is a highly specific polymer usually with just one biological function. It is crucial that the amino acids of each protein be present in the proper order (***amino acid sequence*** or ***primary (1°) structure***) for the protein to function correctly. Often, a given protein, whether isolated from horse, human, or yeast, will have a very similar amino acid sequence.
B. By convention, amino acid sequences are read from the N-terminal end of the polypeptide. This is important because Gly–Ala–Asp (GAD) is just as different from Asp–Ala–Gly (DAG) as the word "dog" is from "god."
C. Living systems make (and need) from 4 thousand to 100 thousand different proteins from the same small set of amino acid monomers. Each polypeptide or ***protein*** can be likened to a different "word" made from the same set of 20 "letters" in this alphabet. Because proteins are so vital to living systems, the amino acids have been called "the alphabet of life." (Note: Protein "words" are very long, often 100 to 500 amino acids, or a.a.)

$H_3N^+$– ACGGDKLVIMTWAEVLMLHALLSTAGCLHKHKPSLIVHLVAPRDVALIMCS–$COO^-$
(1° structure of a small protein consisting of 51 amino acid residues using the one letter codes for a.a.)

D. We will see later that the sequence of amino acids is dictated by the DNA sequence. The sequences of several genomes have now been determined. The entire yeast genome has ~ 13,000,000 bases; the sequence of the human genome is expected to be complete early in the 21$^{st}$ century.

## 21.5 Peptide Hormones
A. Amino acids play important physiological roles in addition to being the building blocks for proteins. Many small peptides are hormones (oxytocin) or neurotransmitters (γ-aminobutyric acid).
B. Some peptides of physiological importance follow.
 1. Vasopressin and oxytocin are cyclic nonapeptide pituitary hormones. With only 2 out of 9 amino acids different, these two hormones produce some similar effects, yet very different effects. Vasopressin is an anti*diuretic* hormone that stimulates the constriction of blood vessels to increase blood pressure. Oxytocin stimulates uterine contractions and ejection of milk.

```
   Phe-Tyr-Cys                         Ile-Tyr-Cys
            \                                    \
             S                                    S
              S                                    S
             /                                    /
   Gln-Asn-Cys-Pro-Arg-Gly  Vasopressin  Gln-Asn-Cys-Pro-Leu-Gly  Oxytocin
```

 2. Bradykinin is a nonapeptide that also lowers blood pressure.
 3. Angiotensin II is an octapeptide (Asp–Arg–Val–Tyr–Ile–His–Pro–Phe) produced in the kidneys. It is a powerful vasoconstrictor important to the control of hypertension.

## 21.6 Classification of Proteins
A. Proteins are unique polymers in that each type of protein molecule has a definite composition and amino acid sequence (1° structure).
B. Proteins can be classified on the basis of solubility or function.
 1. ***Fibrous*** proteins are insoluble in water. They usually perform structural, connective, or protective functions. Examples are collagen, myosin, and elastin.
 2. ***Globular*** (~spherical) proteins are usually soluble in aqueous media. Examples are the albumins, globulins, and nearly all enzymes (Chapter 22).

# Chapter 21 - Proteins

C. For a protein molecule to function properly, it must also have the correct three-dimensional structure. The structures of proteins can be described in terms of 4 levels.
  1. *Primary structure* – linear sequence of amino acids linked by peptide bonds.
  2. *Secondary structure* – a local, fixed arrangement of the polypeptide chain usually stabilized by hydrogen bonds; *α-helix* and *β-sheet*.
  3. *Tertiary structure* – the three-dimensional structure of how a protein chain is folded in space.
  4. *Quaternary structure* – the arrangement of subunits in a multisubunit protein.

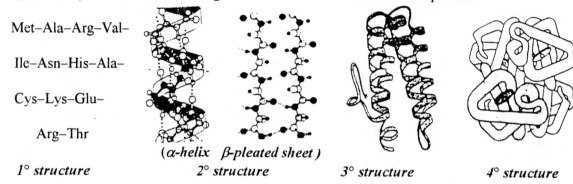

Met–Ala–Arg–Val–Ile–Asn–His–Ala–Cys–Lys–Glu–Arg–Thr

1° structure      (α-helix  β-pleated sheet)  3° structure      4° structure
                  2° structure

## 21.7 Secondary Structure of Proteins
  A. X-ray studies by Linus Pauling and others led to the discovery of the α-helix and β-pleated sheet. These secondary structures are stabilized by intrachain hydrogen-bonds.
  B. The physical properties of wool and silk reflect their secondary structures.
  1. *Wool* is a natural protein fiber. The structure of wool is that of a right-handed *α-helix*. Segments of this structure are also found in many globular proteins. There are 3.6 amino acid residues per turn of an α-helix. The helix coil is stabilized by intrasubunit hydrogen bonds from the carbonyl oxygen of one residue to the -N–H of another residue in the next turn. The helical structure permits wool to stretch like the coils of a spring.
  2. *Silk* is a protein fiber (fibrion) produced by silkworms. It is composed mostly of glycine (45%), alanine (30%), serine (12%), and few other amino acids (13%). Over much of the amino acid sequence, every other amino acid residue is a glycine (–Gly–Ser–Gly–Ala–Gly–Ala–Gly–, etc.). The structure of silk is found to be a *β-pleated sheet* arrangement, in which the polypeptide chains are nearly fully extended and hydrogen-bonded to one another. This gives silk its great strength, flexibility, and resistance to stretching.

## 21.8 Tertiary Structure of Proteins
  A. Many proteins have nearly spherical shapes, with nonpolar (hydrophobic) residues on the inside and polar residues on the surface ("oil-drop model").
  B. Hydrogen bonds are highly directional, giving rise to ordered secondary and tertiary structures.
  C. Disulfide bonds are strong, covalent bonds. Disulfide bonds are rare in intracellular proteins but more common in extracellular proteins (proteases, antibodies, etc.).
  D. Hydrophobic interactions, the process whereby nonpolar groups coalesce in water, are the result of the solvent properties of water wanting to hydrogen-bond to other water molecules at the exclusion of the nonpolar units, and not the result of the relatively weak attraction via dispersion forces of the nonpolar units.
  E. Hair is protein, and the disulfide linkages in human hair can be broken (reduced) and re-formed (oxidized) to produce a "permanent wave."

F. Intramolecular forces stabilizing the tertiary structure of proteins:
   1. *Hydrogen bonds*   2. *Salt bridges*   3. *Disulfide linkages*   4. *Hydrophobic interactions*

   (1)         (2)         (3)         (4)

21.9 Quaternary Structure of Proteins
   A. Quarternary structure describes the arrangement of subunits and thus is present only in oligomeric proteins. Quarternary structure is stabilized by the same set of interactions that maintain tertiary structure.
   B. Myoglobin/Hemoglobin
      1. Globular proteins have nearly spherical, three-dimensional structures, which makes them water soluble as colloidal dispersions.
      2. Myoglobin is a well-studied example of a globular protein. It is a heme protein that serves to store oxygen in muscle tissue. The heme group is called a *prosthetic group*. It has 153 amino acid residues that form eight helical segments, that pack together to produce a nearly spherical molecule. Hemoglobin consists of four globin chains ($\alpha_2\beta_2$), each with an iron-containing heme group. John Kendrew and Max Perutz shared the Nobel Prize in 1962 for their studies that led to the elucidation of the three-dimensional structure of myoglobin and hemoglobin.
   C. Collagen: the protein of connective tissues
      1. *Collagen* is the principal protein of connective tissue. As much as 60% of all mammalian protein is collagen, which is found in skin, bones, tendons, and teeth.
      2. Collagen is not readily digestible but can be converted to digestible *gelatins* by boiling in water.
      3. Collagen is rich in the amino acids glycine, proline, and hydroxyproline.
      4. Collagen consists of 3 protein chains wound in a triple helix, with glycines packed on the inside of the helix. The chains are held by hydrogen bonds, dispersion forces, and covalent cross-links.

21.10 Electrochemical Properties of Proteins
   A. Polypeptide chains usually carry a variety of positive and negative charges resulting from the ionization properties of certain amino acid side chains.
   B. The charge will vary with pH, but at pH 6 the following R groups contribute to the overall charge.
      1. (+) charge:   His, Lys, Arg ("basic" amino acids)
      2. (−) charge:   Glu, Asp ("acidic" amino acids)
   C. All proteins (and amino acids) carry a positive (+) charge at very low pH and a negative (−) charge at very high pH. Most proteins are least soluble at their isoelectric pH, where the net charge is zero. Bacteria produce lactic acid during the spoilage of milk, lowering the pH of milk from 6.6 to about 4.6, near the isoelectric pH of casein, which proceeds to separate from milk as white curds.

21.11 Denaturation of Proteins
   A. The tertiary structure of protein molecules is stabilized by a great many weak forces. It is possible to modify a protein molecule so it no longer folds or functions properly; this process is called *denaturation*.
   B. Denaturation can occur in many ways:
      1. Heat                2. Ultraviolet radiation        3. Changes in pH
      4. Organic solvents    5. Heavy metals ($Pb^{2+}$, $Hg^{2+}$)   6. Alkaloid reagents (tanning)

## Chapter 21 - Proteins

C. Sometimes, a denatured protein will refold to its correct (functional) tertiary structure when the denaturant is removed. This implies that the information needed to form the correct tertiary structure is implicit in the amino acid sequence.

D. Lead and mercury compounds are quite toxic. Lead compounds used to be commonly used as pigments in paints and in gasoline to improve its "octane" rating. Mercury is also quite toxic and was the cause of "hatter's disease" made famous by the *Mad Hatter* in *Alice's Adventures in Wonderland*.

## DISCUSSION

As in the preceding two chapters, Chapter 21 concentrates on structure. The proteins are more comparable in structure to the carbohydrates than to the lipids. Like the polysaccharides, they are polymers. Unlike most polysaccharides, the proteins are copolymers–incorporating more than twenty different monomers; and the monomers are amino acids, not sugars. It is this increased structural complexity that permits proteins to play so many different roles in the body. We'll point out now what will become obvious as you familiarize yourself with these compounds. Except for glycine, all the amino acids can be considered variations on the alanine structure; that is, each has a functional group attached at the side-chain methyl group of alanine. This fact should make them a bit easier to memorize. You should learn the structures of several representative amino acids as shown below.

| Nonpolar Side Chains | Polar Side Chains | Ionizable Side Chains | |
|---|---|---|---|
| | | Acidic | Basic |
| Glycine | Serine | Aspartic acid | Lysine |
| Alanine | Cysteine | 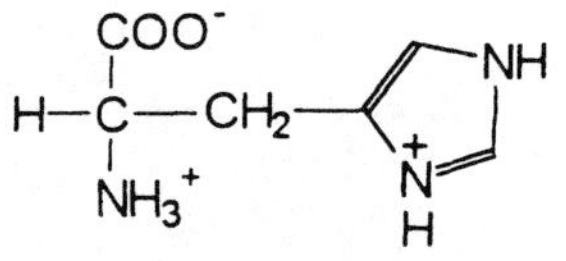 | Histidine |
| Phenylalanine | Cystine (**cystine** is the disulfide-linked form of two cysteines) | | |

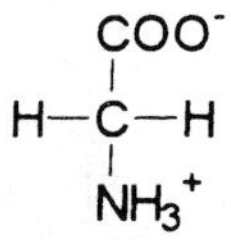

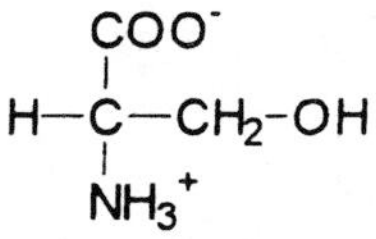

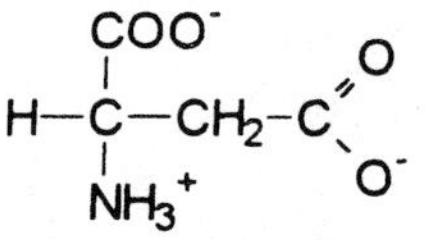

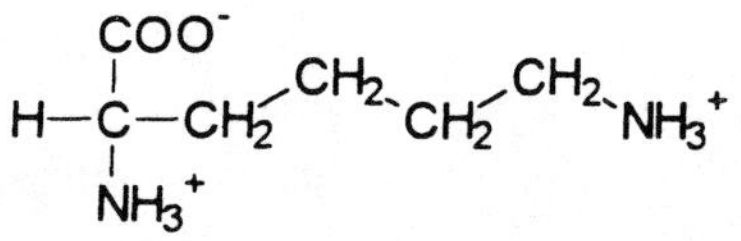

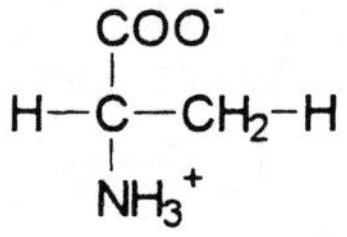

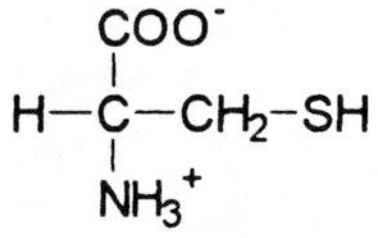

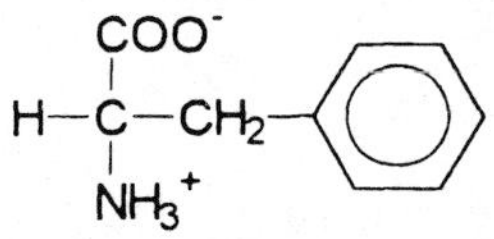

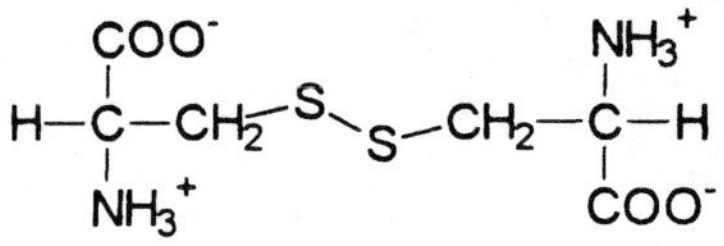

One aspect of the chemistry of amino acids and proteins that frequently causes confusion is the conjugate acid and conjugate base (c.a. and c.b.) forms of the compounds. In Section 21.2 in the text, the structures of a simple amino acid in acidic and basic solutions are given. Let's look at a slightly more complex situation in which the amino acid contains an acidic or basic side chain. Consider lysine and aspartic acid in a solution of low pH. Remember – low pH means acidic, and acidic means that lots of protons are around. If there are lots of protons available, every group on the amino acid that can carry a proton does so.

# Chapter 21 - Proteins

Thus, at low pH aspartic acid and lysine look like this.

| | | |
|---|---|---|
| **At low pH** ($< 2$) | $CH_2-COOH$<br>\|<br>$H_3N^+-CH-COOH$ | $(CH_2)_4-N^+H_3$<br>\|<br>$H_3N^+-CH-COOH$ |
| Every ionizable group has a proton. | aspartic acid (+1) | lysine (+2) |

At high pH the solution is basic and all available protons are plucked from the amino acids.

| | | |
|---|---|---|
| **At high pH** ($> 12$) | $CH_2COO^-$<br>\|<br>$H_2N-CH-COO^-$ | $(CH_2)_4NH_2$<br>\|<br>$H_2N-CH-COO^-$ |
| Every group has lost its proton. | aspartic acid (–2) | lysine (–1) |

As the pH of a solution is changed from high to low, one after another of the ionizable groups picks up a proton. The strongest bases (those with amino groups) react first, then the carboxylate groups ($-COO^-$) react. In cases in which two similar groups are present, each picks up a proton at a characteristic pH. Without knowing the individual $pK_a$ values, you would have no way of predicting which of the two similar groups reacts first, but we can show you the progressive change for our two model compounds, Asp and Lys.

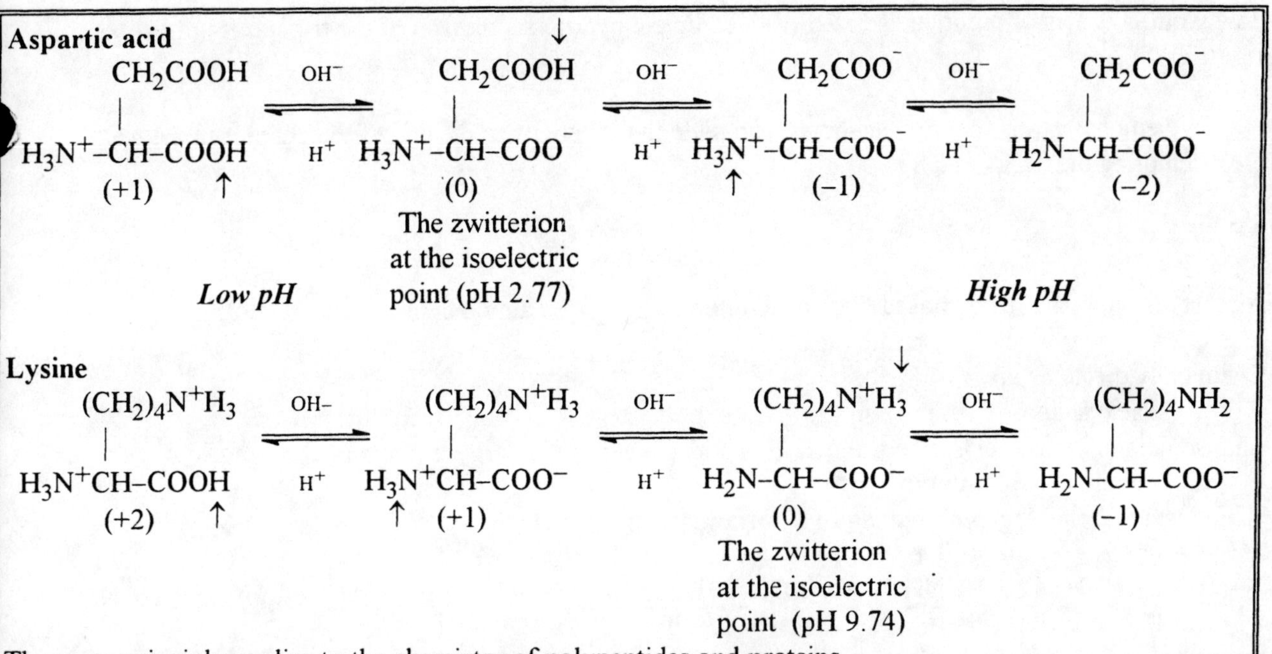

**Aspartic acid**

$CH_2COOH$ — $OH^-$ / $H^+$ ⇌ $CH_2COOH$ — $OH^-$ / $H^+$ ⇌ $CH_2COO^-$ — $OH^-$ / $H^+$ ⇌ $CH_2COO^-$
\|  \|  \|  \|
$H_3N^+-CH-COOH$  $H_3N^+-CH-COO^-$  $H_3N^+-CH-COO^-$  $H_2N-CH-COO^-$
(+1)  (0)  (–1)  (–2)

The zwitterion at the isoelectric point (pH 2.77)

*Low pH* ............................................. *High pH*

**Lysine**

$(CH_2)_4N^+H_3$ — $OH^-$ / $H^+$ ⇌ $(CH_2)_4N^+H_3$ — $OH^-$ / $H^+$ ⇌ $(CH_2)_4N^+H_3$ — $OH^-$ / $H^+$ ⇌ $(CH_2)_4NH_2$
\|  \|  \|  \|
$H_3N^+CH-COOH$  $H_3N^+CH-COO^-$  $H_2N-CH-COO^-$  $H_2N-CH-COO^-$
(+2)  (+1)  (0)  (–1)

The zwitterion at the isoelectric point (pH 9.74)

The same principle applies to the chemistry of polypeptides and proteins.

# Chapter 21 - Proteins

## PROBLEMS

The problems that follow are meant to supplement those in the text. They focus attention on the details of the molecular structure of simple peptides (and their constituent amino acids).

1. For the following structure, pick out the peptide bond, a disulfide bond, and an ionizable side chain.

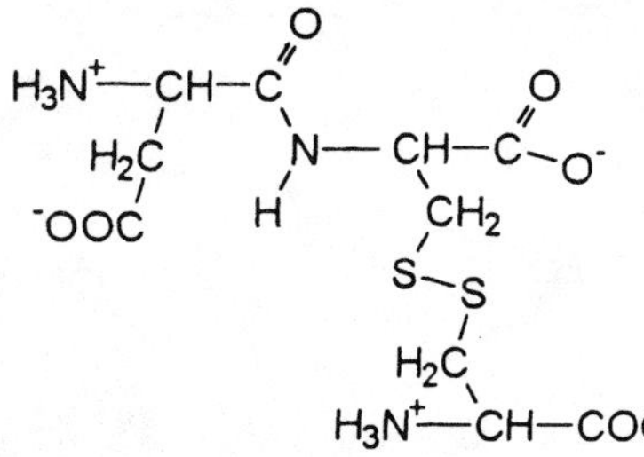

$$\begin{array}{c}
\text{H}_3\text{N}^+\text{—CH—C} \overset{\text{O}}{\underset{}{\diagdown}} \\
\text{H}_2\text{C} \quad \text{N—CH—C} \overset{\text{O}}{\underset{}{\diagdown}} \text{O}^- \\
\text{}^-\text{OOC} \quad \text{H} \quad \text{CH}_2 \\
\text{S—S} \\
\text{H}_2\text{C} \\
\text{H}_3\text{N}^+\text{—CH—COO}^-
\end{array}$$

2. Draw the complete structural formulas of two different dipeptides that incorporate serine and cysteine.
3. Draw the products of complete hydrolysis of:

$$\text{H}_3\text{N}^+\text{-CH-}\overset{\text{O}}{\underset{}{\text{C}}}\text{-NH-CH}_2\text{-}\overset{\text{O}}{\underset{}{\text{C}}}\text{-NH-CH-}\overset{\text{O}}{\underset{}{\text{C}}}\text{-O}^-$$
$$\quad \text{CH}_3 \qquad\qquad\qquad\qquad \text{CH}_2\text{OH}$$

4. Draw the abbreviated formulas (e.g., Gly–Ala–Ser) for all tripeptides that incorporate one unit each of glycine, alanine, and serine.
5. How would the sets of products isolated from complete hydrolysis of each of the tripeptides of Problem 4 differ?

We emphasize again that you should complete the questions at the end of the chapter in the text before attempting the **SELF-TEST**.

## SELF-TEST

1. Almost all proteins are composed from a set of about _____ amino acids.
    a. 4    b. 10    c. 20    d. 50    e. 100
2. Chemically, proteins are:
    a. nucleic acids    b. polyamides    c. polyesters    d. polysaccharides
3. The isomers of amino acids incorporated in peptides and proteins are members of the:
    a. D-family    b. L-family
4. Which is a reasonable representation of a portion of a protein chain?

    a. $-\text{CHCNHCCHNHCNH}-$  b. $-\text{CH}_2\text{CHNHCCH}_2\text{CHNHC}-$  c. $-\text{CHCNHCHCNHCHCNH}-$
       R    R                        R         R                        R    R    R

5. Which compound is not an alpha amino acid?

    a. $\text{HO-CH}_2\overset{\text{NH}_2}{\underset{}{\text{CH}}}\text{COOH}$   b. $\text{H}_2\text{N-}\overset{\text{COOH}}{\underset{}{\text{CH}}}\text{-CH}_3$   c. $\text{H}_2\text{N-}\overset{\text{CH}_3}{\underset{}{\text{CH}}}\text{-CH}_2\text{COOH}$   d. $\text{H-}\overset{\text{H}}{\underset{\text{COOH}}{\text{C}}}\text{-NH}_2$

6. $\text{H}_3\text{N}^+\text{-}\overset{\text{CH}_3}{\underset{}{\text{CH}}}\text{-CONH-CH}_2\text{CONH-}\overset{\text{CH}_2\text{OH}}{\underset{}{\text{CH}}}\text{-COO}^-$ is a(n):
    a. amino acid    b. tripeptide    c. dipeptide    d. polypeptide    e. protein

## Chapter 21 - Proteins

7. Which is not considered evidence of the zwitterionic structure of amino acids?
   a. They show greater solubility in water than in nonpolar solvents.
   b. They have high decomposition points.
   c. They can polymerize to form proteins.
8. Amino acid side chains do not include:
   a. hydrocarbon groups    b. ionized groups    c. polar groups
   d. a phenol group    e. phosphate esters
9. Cysteine is:
   a. $^+NH_3$           b. $^+NH_3$              c. $NH_3^+$                d. $SH\ NH_3^+$
      $CH_2COO^-$         $CH_3CHCOO^-$            $CH_2CH_2COO^-$             $CH_2-CHCOO^-$
10. Which amino acid contains a hydroxyl group?
    a. cysteine    b. leucine    c. lysine    d. serine
11. Which amino acid has an imidizolium side chain group?
    a. Trp    b. His    c. Arg    d. Lys    e. Ile
12. Which amino acid's "R" group can form H-bonds?
    a. Ile    b. Gly    c. Val    d. Ser    e. Leu
13. γ-Aminobutyric acid is:
    a. an essential amino acid
    b. the principal amino acid incorporated in the protein of silk
    c. a chemical neurotransmitter found in the brain
14. To indicate the order of a segment of peptide as Lys–Gly–Ala–Cys is to describe its _____ structure.
    a. primary    b. secondary    c. tertiary    d. quaternary
15. If we say that a protein contains an α-helix, we are describing its _____ structure.
    a. primary    b. secondary    c. tertiary    d. quaternary
16. Which amino acid side chain will participate primarily in hydrophobic interactions to maintain the tertiary structure of a protein?
    a. Asp    b. Lys    c. Trp    d. Ser    e. Glu
17. By describing the relative position of the four polypeptide chains of the hemoglobin molecule, we specify its structure as.
    a. primary    b. secondary    c. tertiary    d. quaternary
18. The structure of collagen can be described as a(n):
    a. alpha helix    b. double helix    c. triple helix    d. pleated sheet
19. Which term is not used in describing protein structure?
    a. alpha helix    b. double helix    c. pleated sheet    d. triple helix
20. Which amino acid has a phenol functional group in its side chain?
    a. serine    b. valine    c. glutamine    d. tyrosine    e. proline
21. Which of the following terms cannot be used to describe the structure of alanine at its isoelectric point?
    a. dipolar ion    b. electrically neutral    c. inner salt    d. zwitterion    e. peptide
22. In a strongly acidic solution, which form of Ala will predominate?
    
    a. $CH_3-\overset{NH_2}{CH}-COOH$    b. $CH_3-\overset{^+NH_3}{CH}-COO^-$    c. $CH_3-\overset{NH_2}{CH}-COO^-$    d. $CH_3-\overset{^+NH_3}{CH}-COOH$

23. In a solution of very high pH, aspartic acid molecules would exist as:
    
    a. $H_2N-\overset{CH_2COOH}{CH}-COOH$    b. $H_2N-\overset{CH_2COO^-}{CH}-COO^-$    c. $H_3N^+-\overset{CH_2COOH}{CH}-COO^-$
    
    d. $H_3N^+-\overset{CH_2COO^-}{CH}-COO^-$    e. $H_3N^+-\overset{CH_2COOH}{CH}-COOH$

# Chapter 21 - Proteins

24. Which amino acid would be expected to form a salt bridge with glutamic acid?
    a. Asp     b. Ser     c. Lys     d. Leu     e. Trp
25. Which amino acid will form disulfide bonds?
    a. Cys     b. Ser     c. His     d. Pro     e. Leu
26. Which amino acid side chain would normally be expected to be positively charged at pH 4?
    a. Gly     b. Asp     c. Leu     d. Pro     e. His
27. At the isoelectric point of proteins:
    a. the proteins are least soluble
    b. the proteins contain no charged groups
    c. the proteins have a large excess of positive charge
    d. the pH of the solution is always 7
28. Oxytocin and vasopressin are polypeptide:
    a. enzymes     b. hormones     c. structural material
29. Which is *not* a globular protein?
    a. albumin     b. collagen     c. myoglobin
30. Which compound does *not* contain heme as a prosthetic group?
    a. albumin     b. hemoglobin     c. myoglobin
31. Which of the following processes is least likely to have occurred during the denaturation of a protein?
    a. disruption of hydrogen bonds     b. hydrolysis of peptide bonds
    c. cleavage of disulfide bonds     d. disruption of salt bridges
32. Which is *not* a denaturing agent?
    a. heat     b. $CH_3CH_2OH$     c. $Hg^{2+}$     d. alkaloidal reagents     e. $H_2O$
33. Which type of protein is more easily denatured?
    a. fibrous     b. globular
34. T / F  An essential amino acid is one which must be incorporated in every protein.
35. T / F  The most abundant amino acid in silk protein is glycine.
36. T / F  Wool fibers are considerably more elastic than silk fibers because the secondary structure of wool protein is alpha helical.
37. T / F  The amino acid sequences of hemoglobins from humans and insects show no variation with species.
38. T / F  The peptide bonds in proteins are identical to the bonds that link monomer units in the nylon polymer.
39. T / F  In the Van Slyke analysis, proteins are treated with ninhydrin and produce a purple color.
40. T / F  The hydrogen bonds formed between different peptide linkages play a major role in establishing both the pleated sheet and the α-helix conformations in proteins.

# Chapter 21 - Proteins

## ANSWERS

**Problems**

1.

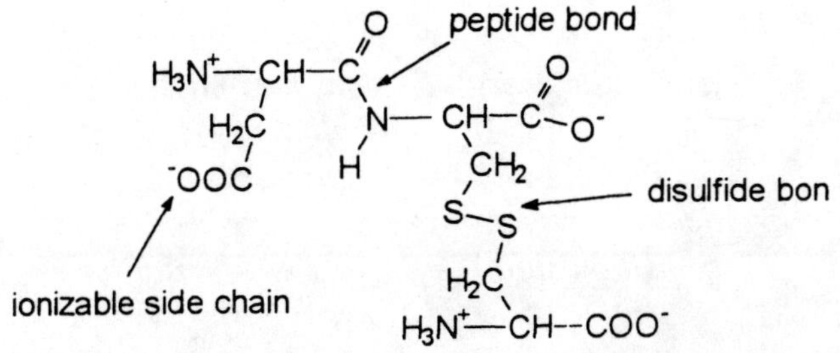

[Structure showing a dipeptide with labels: peptide bond, ionizable side chain, disulfide bond. The structure shows H$_3$N$^+$—CH—C(=O)—NH—CH—C(=O)—O$^-$ with side chains CH$_2$—COO$^-$ and CH$_2$—S—S—CH$_2$—CH(NH$_3^+$)—COO$^-$]

2.  
$\quad$ CH$_2$OH $\quad$ CH$_2$SH $\qquad\qquad$ CH$_2$SH $\quad$ CH$_2$OH  
H$_3$N$^+$–CH–C(=O)–NH–CH–COO$^-$ $\quad$ and $\quad$ H$_3$N$^+$–CH–C(=O)–NH–CH–COO$^-$

3. $\quad$ H$_3$N$^+$–CH–COO$^-$ $\qquad$ H$_3$N$^+$–CH$_2$–COO$^-$ $\qquad$ H$_3$N$^+$–CH–COO$^-$  
$\qquad\quad$ CH$_3$ $\qquad\qquad\qquad\qquad\qquad\qquad\qquad\qquad$ CH$_2$OH

4. Gly–Ala–Ser; $\quad$ Gly–Ser–Ala; $\quad$ Ala–Gly–Ser;  
   Ala–Ser–Gly; $\quad$ Ser–Gly–Ala; $\quad$ Ser–Ala–Gly

5. There would be no difference. Hydrolysis of each tripeptide would yield a mixture of glycine, alanine, and serine.

## Self-Test

| | | | |
|---|---|---|---|
| 1. c | 11. b | 21. e | 31. b |
| 2. b | 12. d | 22. d | 32. e |
| 3. b | 13. c | 23. b | 33. b |
| 4. c | 14. a | 24. c | 34. F |
| 5. c | 15. b | 25. a | 35. T |
| 6. b | 16. c | 26. e | 36. T |
| 7. c | 17. d | 27. a | 37. F |
| 8. e | 18. c | 28. b | 38. T |
| 9. d | 19. b | 29. b | 39. F |
| 10. d | 20. d | 30. a | 40. T |

# 22 Enzymes

**KEY WORDS**

*enzyme*  *activation energy*  *enzyme activity*  *chemotherapy*
*biocatalyst*  *lock-and-key*  *optimum pH*  *apoenzyme*
*enzyme assay*  *induced-fit*  *denatured*  *cofactor*
*primary structure*  *active site*  *allosteric*  *coenzyme*
*substrate*  *transition state*  *feedback inhibition*  *holoenzyme*
*absolute specificity*  *saturation*  *proenzyme*  *inhibitor*
*enzyme-substrate complex*  *competitive*  *noncompetitive*  *turnover number*
*stereospecific*  *antimetabolite*  *irreversible inhibitor*  *neurotransmitters*
*linkage specific*  *group specific*  *optimal temperature*  *ribozymes*
*amino acid sequence*  *effectors*  *activator*  *antibiotic*

**SUMMARY**

Structure-function: The amino acid sequence or primary structure of an enzyme reveals little about the mechanism of action of that enzyme. An enzyme functions or is "active" only when it has the proper tertiary structure. Knowledge of both primary and tertiary structures is needed to understand the mode of action of protein molecules. One of the most thoroughly studied classes of proteins are enzymes, the biocatalysts response for cellular reactions.

22.1 Classification and Naming of Enzymes
  A. **Enzymes** are proteins that serve as **biocatalysts**. There is usually a specific enzyme to catalyze each reaction in cellular metabolism. **Ribozymes** are RNA biocatalysts.
  B. The first enzymes discovered were given common names, that often reflected their source (pepsin) or substrate (maltase). A **substrate** is the reactant in an enzyme-catalyzed reaction. For example, the substrate cleaved by maltase is the disaccharide maltose.
  C. Enzymes are generally classified according to the type of reaction they catalyze. Most enzyme names end in "–ase."
    1. Oxidoreductases – redox reactions; dehydrogenases
    2. Transferases – transfer of groups; transaminases, kinases
    3. Hydrolases – hydrolysis reactions; proteases, lipases
    4. Lyases – removal of groups; decarboxylases
    5. Isomerases – isomeric conversions
    6. Ligases – formation of new bonds; synthetases

22.2 Characteristics of Enzymes
  A. Many enzymes require the help of small nonprotein molecules (**cofactors**) to be active with their substrates.
    1. *Apoenzyme* – protein part of the enzyme
    2. *Cofactor* – the nonprotein component needed for activity, usually a metal ion or a small organic molecule
    3. *Coenzyme* – an organic cofactor; many vitamins are related to coenzymes
    4. *Holoenzyme* – active enzyme-cofactor complex

B. *Proenzymes* (or zymogens) are proteins that can be converted to enzymes, usually by loss of a few amino acid residues. Examples are pepsinogen, trypsinogen, procarboxypeptidase. Many enzymes that are secreted for activities outside the cell (e.g., enzymes used in digestion) are synthesized as proenzymes to protect the cell before they are secreted.

22.3 Mode of Enzyme Action
  A. Enzymes act by providing an alternative pathway for a chemical reaction with a lower *activation energy* for the *transition state*. This involves formation of *enzyme-substrate complexes*, [E-S].

  E + S ⇌ [E-S] ⇌ E + P

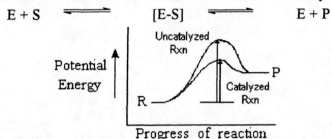

  B. Models of enzyme action: *"lock-and-key"* with *"induced fit"*
    1. The substrate binds to a cleft on the surface of the enzyme, called the active site. Part of the active site will contribute substrate specificity and part will be responsible for catalysis. The active site will normally consist of amino acid side chains that are close together in space or in the tertiary structure of the enzyme but may be far apart in the amino acid sequence or primary structure.
    2. The topography of the active site ("lock") is usually complementary to the *transition state* of the substrate ("key"), so a portion of the binding energy can be used to distort ("induced fit") the substrate toward the product, thus providing a pathway of lower *activation energy*.

  C. Enzymes, like all catalysts, do not alter the equilibrium constant for the reaction but merely speed up both forward and reverse reactions. The *turnover number* (the number of substrate molecules converted to product/minute/molecule) provides an indication of how rapidly the enzyme "turns over" substrate, and it varies from only a few to several million!

22.4 Specificity of Enzymes
  A. Most enzymes are highly specific catalysts, acting on only one or a few related substrates.
    1. A few enzymes have *absolute specificity* – they have activity with only one substrate (urease).
    2. Many enzymes are *stereospecific* – they have specificity for one stereoisometric substrate form (L-LDH).
    3. Enzymes with *group specificity* act on molecules that have the same functional group.
      a. Trypsin – splits peptide bonds that are located on the carboxyl side of Lys or Arg.
      b. Urease – catalyzes a single reaction, the hydrolysis of urea.
    4. *Linkage - specific* enzymes act on a particular type of bond (lipases).
  B. Enzyme specificity is very important in overall regulation of cellular metabolism.

22.5 Factors That Influence *Enzyme Activity*
  A. Concentration of substrate: The rates ($v_o$) of enzyme-catalyzed reactions initially increase with increasing substrate concentration or [S], but then reach a *"saturation"* condition at which no further increase in enzyme activity occurs.

## Chapter 22 - Enzymes

B. **Concentration of enzyme**: When the concentration of substrate is in great excess, the rate of an enzyme catalyzed reaction will increase as the enzyme concentration increases. This is the basis for clinical and biochemical *enzyme assays*.

C. **Temperature**: Most reaction rates are influenced by temperature, and enzyme reactions also initially increase with increasing temperature. However, enzymes are proteins with delicate three-dimensional structures that can be *denatured* (unfolded) by high temperatures. Many enzymes have an *optimal temperature* for maximal activity. This is often near 37 °C – body temperature.

D. **pH**: A change in pH can alter the charge distribution of an enzyme and render it inactive. Most enzymes have a *pH optimum* at which they are most active. This usually corresponds closely with the pH of the body in which the enzyme works.

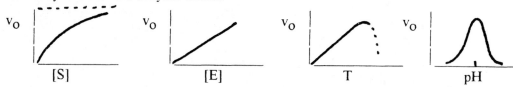

### 22.6 Enzyme Inhibition

A. *Inhibitors* are molecules that bind to enzymes and make them less active; thus small concentrations of an inhibitor or poison can have big effects by decreasing catalytic activity.

B. *Irreversible inhibition* (e.g., poisons) usually involves covalent modification of the enzyme, often at its active site. Many heavy metal ions like those of mercury and lead will react with the sulfhydryl groups on an enzyme, rendering it inactive. This is an example of noncompetitive, irreversible inhibition. With nerve poisons, nerve signals are propagated across the synapse by *neurotransmitters*, small molecules such as acetylcholine. Once the impulse has been relayed, it is important that the acetylcholine be hydrolyzed to acetate and choline so the receptors can ready themselves for the next impulse and not be in a continuous "on" state. Insecticides such as malathion are organic compounds of phosphorus that bind as competitive inhibitors to cholinesterase, thus making it unavailable to break down acetylcholine.

C. Reversible Inhibition
   1. *Competitive* inhibitors bear a close structural resemblance to the substrate, bind at the active site, and thus "compete" with the natural substrate for the enzyme's specificity pocket. Bacteria require *p*-aminobenzoic acid to synthesize THF (tetrahydrofolate). Sulfanilamide was one of the early sulfa drugs whose potency was due to its being able to serve as a competitive inhibitor of *p*-aminobenzoic acid and block the synthesis of THF (See F.6).
   2. *Noncompetitive* inhibitors bind at "other" sites on the enzyme and in binding alter the structure and activity of the enzyme.

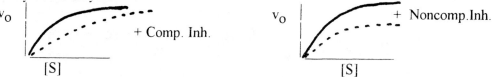

C. Enzyme Regulation and Allosterism (optional)
   1. In addition to their active sites, many enzymes possess regulatory or *allosteric* sites for binding *effectors* or modulators of their activity. The binding of the effector to the enzyme alters its structure to either increase (*activator*, positive effector) or decrease (negative effector) its activity. Such enzymes are called allosteric enzymes.

x = effector (activator)

Less active E     More active E

2. Allosteric enzymes are usually found near the beginning of a multistep sequence of reactions. The final product of the sequence is often a negative effector. Once its concentration builds to a sufficient level, the effector can bind to the allosteric enzyme and stop the whole sequence. This type of allosteric regulation is called "*feedback inhibition*."

```
     E₁       E₂      E₃     E₄      E₅
  A  →  B  →  C  →  D  →  E  →  F
  ↑
  stop
  ⎨_____⎬
```

$A \to F$; when the [F] is high, it can bind to enzyme $E_1$ and shut down its own synthesis by slowing down the synthesis of the precursor compound, B.

## 22.7 Chemotherapy

A. *Antimetabolites* possess structures closely related to the normal metabolite (substrate). Sulfa drugs resemble *p*-aminobenzoic acid and thus are readily incorporated by bacteria into a "false" form of folic acid, which serves as a competitive inhibitor of folic acid.

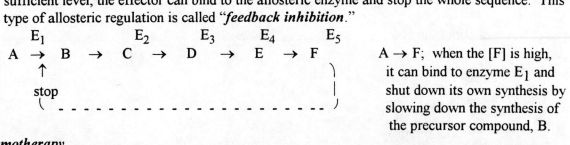

*p*-Aminobenzoic acid      Sulfanilamide

B. *Antibiotics* are compounds produced by one microorganism that are toxic to another organism.

1. Penicillin is an inhibitor of the enzyme transpeptidase.

[Structure of Penicillin G shown, with Cysteine and Valine portions labeled, and R = phenyl-CH₂–]

**Penicillin G**

2. Other antibiotics or synthetic analogs include aureomycin, streptomycin, tetracyclines, and chloramphenicol.

[Structure of Tetracycline shown]

Tetracycline

C. Enzymes are finding increased uses in industry (e.g., soaps, meat tenderizers) and in medicine. Clinical analysis of enzymes in body fluids or tissues is now a common diagnostic tool. Elevated levels of an enzyme (LDH) in the blood can point to a tissue that has been damaged (heart), acid phosphatase (prostate), alanine aminotransferase (liver), etc.

D. Advances in recombinant DNA technology will open a multitude of new applications for these efficient biocatalysts.

# Chapter 22 - Enzymes

## DISCUSSION

Enzymes are delicate protein structures with active sites for catalyzing chemical reactions. This enables cellular processes to occur under relatively mild conditions of pH and temperature. Many enzymes have sophisticated regulatory properties that allow their activity to be turned on or shut down. In Section 22.1 we presented the classification of enzymes by the type of reaction they catalyzed. The best way to illustrate what we mean by "type of reaction" is to write a representative equation. Remember that these examples are only illustrative. We could have written many others, some seemingly quite different.

Types of enzymes with examples

### I. Oxidases (oxidoreductases)

$$\underset{\text{Lactate}}{CH_3\text{-}CH(OH)\text{-}COO^-} + NAD^+ \underset{}{\overset{\text{lactate dehydrogenase}}{\rightleftarrows}} \underset{\text{Pyruvate}}{CH_3\text{-}CO\text{-}COO^-} + NADH + H^+$$

### II. Transferases

$$\underset{\text{Glucose}}{\text{Glucose}} + ATP \overset{\text{glucokinase}}{\rightleftarrows} \underset{\text{Glucose-6-phosphate}}{\text{Glucose-6-phosphate}} + ADP$$

### III. Hydrolases

$$\underset{\text{Acetylcholine}}{(CH_3)_3N^+\text{-}CH_2CH_2\text{-}O\text{-}CO\text{-}CH_3} + H_2O \overset{\text{acetylcholine-esterase}}{\rightleftarrows} \underset{\text{Choline}}{(CH_3)_3\overset{+}{N}\text{-}CH_2CH_2\text{-}OH} + \underset{\text{Acetate}}{^-O\text{-}CO\text{-}CH_3}$$

### IV. Lyases

$$\underset{\text{Pyruvate}}{CH_3\text{-}CO\text{-}COO^-} \overset{\text{pyruvate decarboxylase}}{\rightleftarrows} \underset{\text{Acetaldehyde}}{CH_3\text{-}CO\text{-}H} + CO_2$$

### V. Isomerases

$$\underset{\text{UDP-glucose}}{\text{UDP-glucose}} \overset{\text{UDP-glucose epimerase}}{\rightleftarrows} \underset{\text{UDP-galactose}}{\text{UDP-galactose}}$$

### VI. Ligases

$$CoA\text{-}SH + \underset{\text{Fatty acid}}{^-O\text{-}CO\text{-}R} + ATP \overset{\text{acyl-CoA synthetase}}{\rightleftarrows} \underset{\text{Thio ester of fatty acid}}{CoA\text{-}S\text{~}CO\text{-}R} + AMP + Pi$$

# Chapter 22 - Enzymes

## SUMMARY

Answer the problems at the end of the chapter in the text. Then complete the SELF-TEST.

## SELF-TEST

1. A compound that catalyzes a chemical reaction in a living organism is called a(n):
      a. carbohydrate     b. enzyme     c. lipid     d. vitamin
2. Enzymes belong to which class of organic compounds?
      a. carbohydrates     b. esters     c. hydrocarbons     d. lipids    e. proteins
3. The compound that has a reaction catalyzed by an enzyme is called a(n):
      a. activator     b. coenzyme     c. cofactor     d. substrate
4. Which type of enzyme is classified as an oxidoreductase?
      a. dehydrogenase     b. lipase     c. pepsin     d. peptidase     e. protease
5. Kinases belong to which class of enzymes?
      a. hydrolases     b. isomerases     c. ligases
      d. lyases     e. oxidases     f. transferases
6. To which of the classes listed in question 5 does a synthetase belong?
      a   b   c   d   e   f
7. Which type of enzyme listed in question 5 would catalyze the conversion of L-alanine to D-alanine?
      a   b   c   d   e   f
8. The substrate pyruvate is a 3-C keto-acid that can be reduced to lactate by lactate dehydrogenase (LDH). What amino acid R-group in the enzyme might be expected to interact with the $-COO^-$ group of pyruvate?
      a. Glu     b. Leu     c. Asp     d. Ieu     e. Arg
9. An apoenzyme is always a(n):
      a. protein     b. nonprotein organic molecule     c. inorganic ion
10. Which is not found as a cofactor of an enzyme?
       a. a protein     b. a vitamin     c. an inorganic ion
11. A <u>noncovalently</u> bonded vitamin derivative required for the functioning of an enzyme system is called a(n):
       a. activator     b. apoenzyme     c. coenzyme
12. When all of the active sites on enzyme molecules are saturated, an increase in _____ concentration will not increase the rate of the reaction.
       a. enzyme     b. substrate
13. An optimum pH of enzymes operating in the stomach is in the ___ range.
       a. acidic     b. neutral     c. basic
14. The optimum temperature for most enzymes operating in the human body is:
       a. 273 K     b. 37 °C     c. 98.6 °C     d. 120 °C
15. The allosteric site on a regulatory enzyme binds the:
       a. effector     b. proenzyme     c. substrate
16. Which is <u>not</u> a proenzyme?
       a. pepsinogen     b. prothrombin     c. trypsinogen     d. all are proenzymes
17. Pepsin is a protease that acts in the:
       a. mouth     b. stomach     c. intestines
18. What will happen to the rate of an enzyme catalyzed reaction already at very high [S] when more substrate is added?
       a. increase     b. decrease     c. no change

## Chapter 22 - Enzymes

19. What will happen to the rate of an enzyme catalyzed reaction at very high [S] when a competitive inhibitor is added?
    a. increase     b. decrease     c. no change
20. Which is not a poison that affects enzymes?
    a. $CN^-$     b. $Fe^{3+}$     c. $Pb^{2+}$     d. $AsO_4^{3-}$
21. Which poison does not act by tying up sulfhydryl groups of an enzyme?
    a. $CN^-$     b. $AsO_4^{3-}$     c. $Pb^{2+}$
22. EDTA is an effective antidote for acute poisoning by:
    a. $CN^-$     b. $AsO_4^{3-}$     c. $Pb^{2+}$
23. The combination of the apoenzyme and its cofactor produce an active enzyme or:
    a. vitazyme     b. holoenzyme     c. ribozyme     d. coenzyme
24. Which protein has the larger polypeptide chain, carboxypeptidase or procarboxypeptidase?
    a. carboxypeptidase     b. procarboxypeptidase
25. Activation of an enzyme can occur through:
    a. conversion of a proenzyme to the enzyme
    b. combination of an apoenzyme with a cofactor
    c. release of an inhibitor from the allosteric site of a regulatory enzyme
    d. all of the above
26. Which compound is responsible for transmitting an impulse across a nerve synapse?
    a. choline     b. acetylcholine     c. cholinesterase
27. Which of the compounds listed in Question 26 is the enzyme responsible for resetting the receptor to "off" after transmission of a signal across the synapse?
    a. choline     b. acetylcholine     c. cholinesterase
28. Phosphorus-containing compounds serve as:
    a. pesticides     b. nerve gases     c. intermediates in carbohydrate metabolism
    d. a and b     e. a, b, and c
29. When the [S] is low, what will happen to the rate of the enzyme catalyzed reaction when the [S] is doubled?
    a. no change     b. increase 10×     c. increase 2×     d. decrease 10×     e. decrease 2×
30. If one effector (Ef) activates one enzyme E-A, which in turn activates 100 molecules of enzyme E-B, each of which activates 100 molecules of enzyme E-C, and each molecule of enzyme E-C converts 1000 substrate molecules to product, what is the enhancement factor of the effector "Ef"?
    a. 100×     b. 10,000×     c. 100,000×     d. 1,000,000×     e. 10,000,000×
31. T / F  Enzymes are heat-stable catalysts.
32. T / F  Factors that affect the activity of an enzyme include temperature, pH, enzyme concentration, and substrate concentration.
33. T / F  Saliva contains the enzyme pepsin.
34. T / F  The active site of an enzyme is the point at which an effector attaches.
35. T / F  The carbohydrase lysozyme was the first enzyme to have its three-dimensional structure determined by scientists.
36. T / F  A stereospecific enzyme would catalyze only reactions involving both members of a mirror image pair of isomers.
37. T / F  If an enzyme exhibits absolute specificity, it catalyzes a single reaction.
38. T / F  The formation of an activated enzyme-substrate complex is postulated to explain the catalytic effect of an enzyme.
39. T / F  Addition of an enzyme changes the speed and the equilibrium position of a reaction.
40. T / F  The induced-fit theory states that the substrate must fit precisely the active site of the enzyme for the catalyst to be most effective.

# Chapter 22 - Enzymes

41. T / F  The amino acid side chains associated with the active site of an enzyme must be adjacent in the primary sequence of the protein.
42. T / F  The inhibitor of an allosteric enzyme is frequently the end product of the sequence of reactions controlled by the enzyme.
43. T / F  CPK and GOT are abbreviations for the enzymes used in detergent formulations and meat tenderizer.
44. T / F  The term "regulatory enzyme" is applied to any enzyme that catalyzes a chemical reaction.
45. T / F  Conversion of a proenzyme to the active enzyme frequently involves cleavage of a portion of the protein chain.

Matching: Match the enzymes listed in column A with their type of reactions illustrated in column B.

**Column A**

a. carbohydrase
b. dehydrogenase
c. isomerase
d. peptidase
e. transferase

**Column B**

___46. $\sim\underset{\|}{\overset{O}{C}}-NH-CH_2\sim\ +\ H_2O\ \rightleftharpoons\ \sim\underset{\|}{\overset{O}{C}}-OH\ +\ H_2N-CH_2\sim$

___47. $H_3^+N-\underset{H}{\overset{CH_3}{C}}-COO^-\ \rightleftharpoons\ ^-OOC-\underset{H}{\overset{CH_3}{C}}-N^+H_3$

___48. 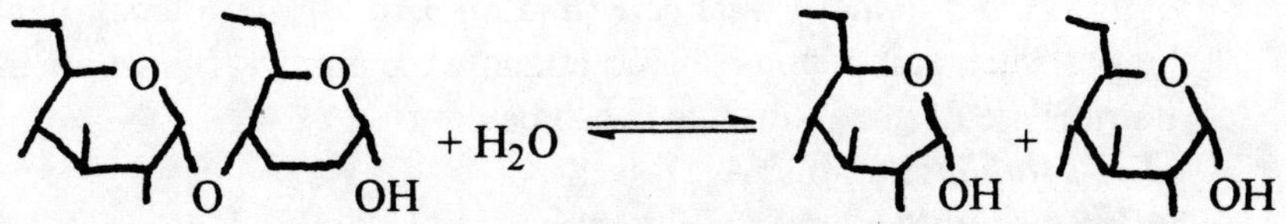 [disaccharide] + $H_2O\ \rightleftharpoons$ [monosaccharide] + [monosaccharide]

___49. $\underset{CH_3-CH_2}{OH}\ \rightleftharpoons\ CH_3-\overset{O}{\overset{\|}{C}}-H\ +\ 2H$

___50. $CH_3-\overset{O}{\overset{\|}{C}}-OH\ +\ ATP\ \rightleftharpoons\ CH_3-\overset{O}{\overset{\|}{C}}-O-\overset{O}{\overset{\|}{P}}-O^-\ +\ ADP$ (with $O^-$ below P)

## ANSWERS

| | | | | |
|---|---|---|---|---|
| 1. b | 11. c | 21. a | 31. F | 41. F |
| 2. e | 12. b | 22. c | 32. T | 42. T |
| 3. d | 13. a | 23. b | 33. F | 43. F |
| 4. a | 14. b | 24. b | 34. F | 44. F |
| 5. f | 15. a | 25. d | 35. T | 45. T |
| 6. c | 16. d | 26. b | 36. F | 46. d |
| 7. b | 17. b | 27. c | 37. T | 47. c |
| 8. e | 18. c | 28. e | 38. T | 48. a |
| 9. a | 19. c | 29. c | 39. F | 49. b |
| 10. a | 20. b | 30. e | 40. F | 50. e |

# F                                Vitamins

**KEY WORDS**

*vitamin*      *deficiency*      *provitamin*      *fat-soluble*      *water-soluble*
*thiamine*      *biotin*      *disease*      *calciferol*      *riboflavin*
*folic acid*      *scurvy*      *rickets*      *niacin*      *lipoic acid*
*beriberi*      *tocopherol*      *pantothenic acid*      *cyanocobalamin*      *pellagra*
*B complex*      *pyridoxine*      *ascorbic acid*      *retinol*      *antioxidant*
*prothrombin*      *phytochemical*

**SUMMARY**

F.1   What Are Vitamins?
     A. ***Vitamins*** are organic substances that our bodies need for good health but cannot synthesize; they must be included in one's diet.
     B. Some vitamins were discovered early because of vitamin-***deficiency diseases*** such as ***scurvy*** (vitamin C deficiency), ***beri beri*** (thiamine, or $B_1$, deficiency), and ***pellagra*** (niacin deficiency). The first such compounds characterized were amines, hence the name "vitamin."
     C. Vitamins are divided into two broad categories.
         1. ***Fat-soluble*** vitamins: A, D, E, K
         2. ***Water-soluble*** vitamins: B complex and C

F.2   Vitamin A
     A. Vitamin A is an unsaturated alcohol called ***retinol***. It was first isolated from fish oils but is also found in eggs and dairy products. Excess vitamin A is stored in the body and large excesses can be harmful. Carrots and certain other vegetables contain a carotenoid pigment, β-carotene, which is a ***provitamin*** that can be converted into vitamin A.

                                                                                                  Vitamin A

     B. Signs of vitamin A deficiency are night blindness and dried, or keratinized, mucous membranes.
     C. Vitamin A and the visual cycle
         1. Vitamin A is converted to 11-*cis*-retinal, which combines with opsin to form rhodopsin.
         2. When light strikes rhodopsin, 11-*cis*-retinal is converted to the trans isomer, triggering an electrical impulse (vision) and splitting rhodopsin to form opsin and the free aldehyde.
         3. The all-*trans* retinal is converted back to 11-*cis*-retinal and combined with opsin to complete the visual cycle (refer to Selected Topic D).

### F.3 Vitamin D

A. Vitamin D is actually several related compounds. Vitamin $D_2$, or ergo*calciferol*, is formed by the action of sunlight on the steroid ergosterol.

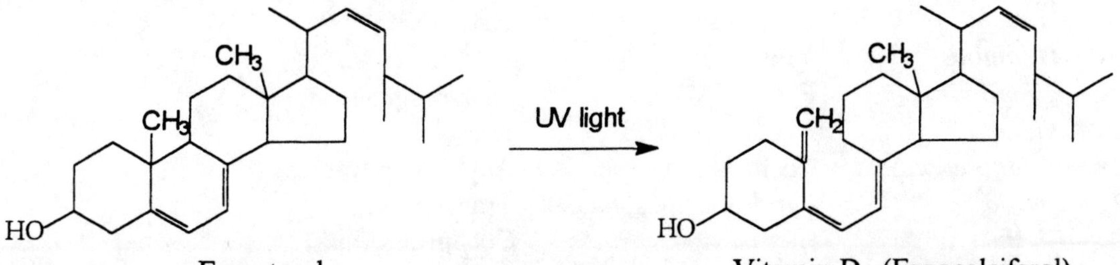

Ergosterol → UV light → Vitamin $D_2$ (Ergocalciferol)

B. Vitamin D promotes the uptake of calcium and phosphorus. A deficiency of vitamin D results in abnormal bone formation, a condition known as *rickets*.

C. Vitamin D is the "sunshine vitamin." Sunlight can convert 7-dehydro-cholesterol in the skin to vitamin D. It is also found in milk and fish oils. Large excesses of vitamin D are dangerous.

### F.4 Vitamin E

A. Vitamin E is a mixture of compounds called *α-tocopherols*. The tocopherols are phenols that are *antioxidants*. The loss of vitamin E's antioxidant effect is believed to be responsible for the symptoms of vitamin E deficiency. Vitamin E protects vitamin A, and vitamin E deficiency usually also leads to vitamin A deficiency.

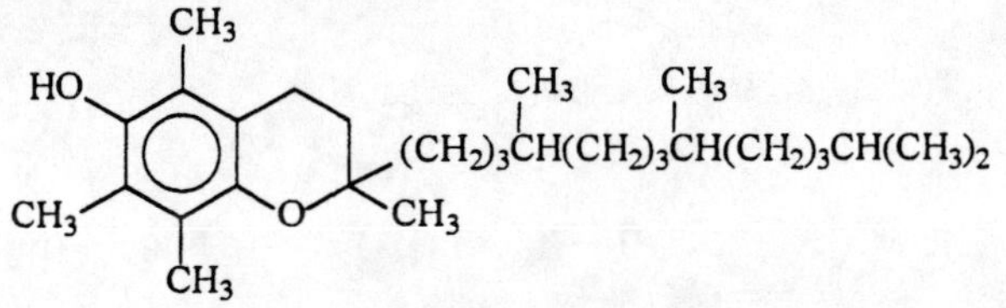

Vitamin E (α-Tocopherol)

B. Vitamin E is also stored in the body but is not as toxic in excess as are vitamins A and D. Good sources of vitamin E are wheat germ oil, vegetables, egg yolk, and meat.

### F.5 Vitamin K

A. There are many compounds with vitamin K activity. Vitamin K has a fused-ring structure related to naphthalene, with one ring also being a quinone. Often the hydrocarbon "tail" is similar to that found in vitamins A and E.

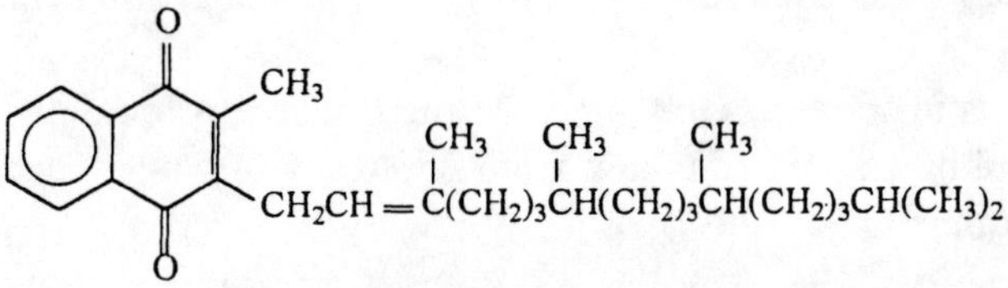

Vitamin $K_1$

B. Vitamin K is necessary for the function of *prothrombin*, an enzyme precursor involved in blood clotting. Symptoms of vitamin K deficiency are bleeding under the skin, which results in ugly "bruises" from minor blows.

C. Good sources of vitamin K are spinach and leafy green vegetables. Bacteria in the large intestine also produce the vitamin, and it can be absorbed from them.

## Selected Topic F - Vitamins

F.6 The B Complex
A. Vitamin *B complex* refers to a group of water-soluble vitamins. Many coenzymes are vitamin B derivatives. Water-soluble vitamins are not stored by the body. (Refer to Table F.1 in the text.)

B.
| Vitamin | Coenzyme | Reaction |
|---|---|---|
| $B_1$ – *Thiamine* | TPP | Decarboxylation |
| $B_2$ – *Riboflavin* | FMN, FAD | Dehydrogenation |
| $B_3$ – *Niacin* | NADH, NADPH | Redox reactions |
| $B_5$ – *Pantothenic acid* | Coenzyme A | Acyl group transfer |
| $B_6$ – *Pyridoxine* | Pyridoxal phosphate | Transamination |
| – *Biotin* | Biotin | $CO_2$ group transfer |
| – *Folic acid* | Tetrahydrofolate | 1 C transfer |
| $B_{12}$– *Cyanocobalamin* | dA cobalamin | Alkyl group transfer |
| – *Lipoic acid* | Lipoamide | Acyl group transfer |

C. Structures of some B vitamins

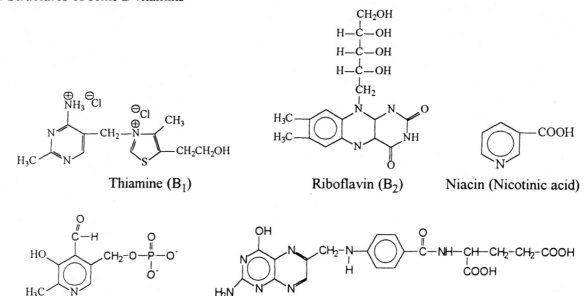

Thiamine ($B_1$)   Riboflavin ($B_2$)   Niacin (Nicotinic acid)

Pyridoxal-5'-phosphate   Folic acid

F.7 Vitamin C
  A. Vitamin C is *ascorbic acid*. It is required for hydroxylation reactions important to the synthesis of collagen.
  B. Vitamin C deficiency results in scurvy. Only 40 to 75 mg of ascorbic acid per day are necessary to prevent scurvy. Linus Pauling advocated taking from 250 to 15,000 mg of vitamin C per day. Citrus fruits are rich in vitamin C.

Ascorbic acid (Vitamin C)   $\xrightarrow{O_2}$   Dehydroascorbic acid (Oxidized vitamin C)

## Selected Topic F - Vitamins

F.8 Antioxidants and Disease Prevention
   A. *Antioxidants* (Vitamin E, C) prevent damage from free radicals, reactive molecules with unpaired electrons.
   B. *Phytochemicals* are nonnutrient, plant-derived compounds that have biological activity.

## DISCUSSION

This unit, like the preceding ones, deals with complex molecules. However, there is no structural feature common to all vitamins. Proteins are polyamides, and it is possible to describe in a general way the structural features common to thousands of different protein molecules. The same can be said about carbohydrates and even lipids; that is, they have similar structural features. But vitamins cannot be so easily categorized by structure. Because vitamins have such complex structures, their chemistries frequently intimidate students (who, because of exams, worry a great deal about being able to draw a structure like vitamin $B_{12}$). Therefore, let us first comment that there is not one chemist in a hundred (or more) who can draw the structure of vitamin $B_{12}$ from memory. So why did we bother to show the structures of all of these molecules? Because we want you to see that molecular architecture is more than a chemist's playground; it frequently determines the state of one's health. Consider the discussion of vitamin A in section F.2. This is a large, relatively nonpolar molecule. We've discussed polarity, hydrogen bonding, and solubility many times. Here you have a concrete example of the significance of this "chemistry." Vitamin A is soluble in nonpolar media like the fatty tissue of the body. It does not form hydrogen bonds with water in sufficient numbers to make it soluble in aqueous body fluids. That means that it is not rapidly excreted from the body with such fluids. Thus, individuals can build up a reserve and protect themselves against the effects of deprivation. It is also true that this same property of vitamin A permits one to overdose on the vitamin. In other words, you can store too much of it – the body doesn't automatically dump the excess over one's immediate needs.

In contrast, look at vitamin C (F.7). It's a relative of the carbohydrates–lots of hydroxyl groups, very polar, and very water–soluble. Now, perhaps you can understand some of the controversy surrounding Linus Pauling's recommendation to take massive doses of vitamin C. What difference does it make, say some scientists, whether you ingest 250 or 15,000 mg of vitamin C when evidence suggests that the aqueous body fluids wash out all but 200 mg? With these facts in mind, spend some time just looking at (not memorizing) some of the vitamin structures. Notice that vitamins A, D, E, and K, the fat-soluble vitamins, are all lipid-like. They have lots of carbon and hydrogen and very little else. There's an occasional oxygen, but mostly there are long, nonpolar chains of carbon.

Now look at the B complex vitamins (Figures F.3-F.8). All of them contain nitrogen. Even more important, however, is the fact that they all contain a relatively high proportion of groups that can interact through hydrogen bonding. The variation in structure among the B vitamins is great, but $B_2$ (riboflavin) is typical. Like most of the fat-soluble vitamins, it has a side chain. Now look closely at the riboflavin side chain. It carries 4 hydroxyl groups, and there are four nitrogens and two other oxygens in the molecule. Look at vitamin $B_{12}$ (cyanocobalamin). The laboratory synthesis of this complex molecule was regarded as one of the outstanding achievements of organic chemistry in this century. What should you know about the structure? Certainly that it contains cobalt, which is somewhat unusual, but primarily you should know that it is loaded with groups that confer water solubility – amides and other nitrogen-containing functions, hydroxyl groups, and a phosphate group.

You should use the problems at the end of the chapter in the text to organize for yourself some pertinent data about each of the vitamins. Go through all of the problems in the text before taking the **SELF-TEST**.

**Selected Topic F - Vitamins**

**SELF-TEST**

1. Vitamins are:
    a. amines required by an organism for good health
    b. organic compounds produced in trace amounts by the endocrine glands
    c. organic molecules that an organism requires in trace amounts but cannot synthesize for itself
    d. steroids that act as sex hormones
2. Which is <u>not</u> a fat-soluble vitamin?
    a. vitamin A    b. vitamin C    c. vitamin D
    d. vitamin E    e. vitamin K
3. Which is <u>not</u> a water-soluble vitamin?
    a. cholecalciferol    b. cyanocobalamin    c. niacin    d. vitamin $B_6$
4. Which is <u>not</u> a member of the B complex?
    a. biotin    b. thiamine    c. folic acid    d. niacin
    e. pantothenic acid    f. riboflavin    g. retinol
5. The plant pigment named β-carotene is a:
    a. coenzyme    b. contraceptive    c. hormone    d. provitamin    e. vitamin
6. A critical event in the chemistry of vision involves the conversion of:
    a. a cis isomer to a trans isomer
    b. a D isomer to an L isomer
    c. an ortho isomer to a para isomer
    d. ergosterol to calciferol
7. Which is <u>not</u> true of rhodopsin?
    a. It is a complex of a protein and a derivative of vitamin A.
    b. It is the visual pigment found in some receptor cells of the retina.
    c. It is converted to vitamin A by the absorption of light.
8. Which is <u>not</u> true of vitamin D?
    a. It is formed from steroidal precursors by the absorption of ultraviolet light.
    b. It is called the "sunshine vitamin."
    c. A deficiency of this vitamin results in abnormal bone formation.
    d. No harmful effects have been documented for overdoses of this vitamin.
9. Vitamin E is:
    a. an antioxidant
    b. frequently missing from the diet of vegetarians
    c. approved by medical authorities for the prevention of aging
10. The vitamin associated with blood clotting is vitamin:
    a. A    b. B complex    c. C    d. D    e. E    f. K
11. The B complex vitamins are frequently incorporated in:
    a. coenzymes    b. provitamins    c. contraceptives    d. rhodopsin
12. The vitamin thiamine is associated with the disease:
    a. scurvy    b. rickets    c. beriberi    d. baldness    e. senility
13. The B complex vitamins can be stored in almost unlimited quantities in the:
    a. adipose tissue    b. bone marrow    c. liver
    d. retina    e. They are not stored in significant amounts in the body.

**Selected Topic F - Vitamins**

14. Vitamin C activity is exhibited by:
    a. several pigments isolated from various colored plants
    b. steroid-like compounds found in the skin of various animals
    c. a carbohydrate-like compound found in citrus fruit
15. The coenzymes NADH and NADPH are derived from this vitamin:
    a. riboflavin    b. thiamine    c. ascorbic acid    d. niacin    e. Vitamin E

**Matching:** Match the names in column B with the vitamins in column A.

| Column A | Column B |
|---|---|
| 16. vitamin $B_1$ | a. ascorbic acid |
| 17. vitamin D | b. calciferol |
| 18. vitamin $B_2$ | c. cyanocobalamin |
| 19. vitamin $B_{12}$ | d. retinol |
| 20. vitamin E | e. riboflavin |
| 21. vitamin A | f. thiamine |
| 22. vitamin C | g. α-tocopherol |

Match the deficiency disease or symptom in column D with the relevant vitamin in column C.

| Column C | Column D |
|---|---|
| 23. ascorbic acid | a. scurvy |
| 24. cyanocobalamin | b. hemorrhage |
| 25. vitamin A | c. pellagra |
| 26. thiamine | d. pernicious anemia |
| 27. vitamin E | e. rickets |
| 28. vitamin K | f. beriberi |
| 29. vitamin D | g. sterility |
| 30. niacin | h. night blindness |

**ANSWERS**

| | | |
|---|---|---|
| 1. c | 11. a | 21. d |
| 2. b | 12. c | 22. a |
| 3. a | 13. e | 23. a |
| 4. g | 14. c | 24. d |
| 5. d | 15. d | 25. h |
| 6. a | 16. f | 26. f |
| 7. c | 17. b | 27. g |
| 8. d | 18. e | 28. b |
| 9. a | 19. c | 29. e |
| 10. f | 20. g | 30. c |

# 23 Nucleic Acids and Protein Synthesis

**KEY WORDS**

| | | | | |
|---|---|---|---|---|
| purine | nucleotide | replication | codons | pyrimidine |
| phosphodiester | semiconservative | translation | DNA, RNA | double helix |
| discontinuous | tRNA, rRNA, mRNA | deoxyribose | base pairing | transcription |
| anticodon | triplets | differentiation | genome | mutation |
| nucleoside | restriction enzymes | gene | polymerase | operator |
| exons | introns | mutagen | ribose | ribosome |
| cytosine | thymine | uracil | adenine | guanine |
| antiparallel | template | DNA ligase | universal | degenerate |
| plasmid | DNA fingerprinting | sugar-phosphate | genetic disease | substitution |
| insertion | initiation codon | termination codon | "A", "P" sites | intervening |
| deletion | recombinant DNA | RFLP | PCR | chromosome |
| clone | genetic code | genetic disease | nucleic acid | |

**SUMMARY**

23.1 Nucleotides

    A. Two kinds of *nucleic acid* polymers

        1. *DNAs* (*deoxyribonucleic* acids) are the genetic material generally found in the cell nucleus. DNAs consist of the bases A, G, T, or C linked to a *2'-deoxyribose* sugar and an inorganic phosphate.

        2. *RNAs* (*ribonucleic* acids) have many roles. RNAs consist of the bases A, G, U, or C linked to a *ribose* sugar and an inorganic phosphate.

    B. Nucleic acids are polymers of *nucleotides*. Each nucleotide consists of a *purine* or *pyrimidine* nitrogeneous base, a ribose sugar, and an inorganic phosphate.

        1. *Pyrimidines*

Cytosine (C)    Thymine (T)    Uracil (U)

        2. *Purines*

Adenine (A)    Guanine (G)

# Chapter 23 - Nucleic Acids and Protein Synthesis

C. *Nucleosides*: A *nucleoside* is a purine or pyrimidine base linked to the C1' position of a ribose sugar.
  1. Ribonucleoside = ribose + (A, G, C, or U)
  2. Deoxyribonucleoside = 2'-deoxyribose + (A, G, C, or T)

Adenosine (Nucleoside)

  3. Nomenclature:

| Base | Sugar | Nucleoside |
|---|---|---|
| Adenine | Ribose | Adenosine |
| Guanine | Ribose | Guanone |
| Uracil | Ribose | Uridine |
| Thymine | Deoxyribose | Deoxythymidine |
| Cytosine | Deoxyribose | Deoxycytidine |

D. *Nucleotides* (Sugar + Base + Phosphate): Nucleotides are C5' phosphoesters of nucleosides.
  1. Adenosine, the nucleoside, becomes adenosine monophosphate (AMP), the nucleotide.

Adenosine monophosphate or AMP ( nucleotide )

  2. Nomenclature
    i. Two systems
      a. drop "*–ine*" or "*osine*" ending and add "*–ylic acid*" ending: uridylic acid, adenylic acid
      b. as a nucleoside monophosphate: uridine monophosphate, UMP
    ii. Use "*deoxy*" or "*d*" for deoxyribose nucleotides AMP vs. dAMP
  3. Roles of nucleotides
    i. Monomers for building DNAs: dAMP, dGMP, dCMP, dTMP
    ii. Monomers for building RNAs: AMP, GMP, CMP, UMP
    iii. Energy exchange: ATP, ADP, GTP, GDP
    iv. Coenzymes: NADH, $FADH_2$
    v. Secondary messenger: c–AMP
    vi. Allosteric effectors of regulatory enzymes

# Chapter 23 - Nucleic Acids and Protein Synthesis

23.2 The Primary Structure of Nucleic Acids
   A. Nucleic acids are linear polymers of nucleotides. DNAs are synthesized from dNTPs, and RNAs are formed from NTPs (N = nucleoside). The nucleotides are connected by *3',5' phosphodiester* bonds. Each strand has one free 5' hydroxyl and one free 3' hydroxyl group.

   B. The primary structure of a nucleic acid is the sequence of the bases attached to the ribose. DNA sequences are always read in the 5' → 3' direction.
      For example:   5'–AGGTCTCAAGCTATAAGCCATCATC–3'
   C. Nucleic acid sequences are determined by using gel electrophoresis of radiolabeled nucleic acid fragments prepared with *restriction enzymes*. The human *genome* project seeks to determine the primary structure of human DNA estimated to consist of ~3.3 billion base pairs. In 1996 the sequence of the yeast genome of ~13 million base pairs was completed.

23.3 The Secondary Structure of Nucleic Acids
   A. Chargaff (1950) observed that the composition of most DNAs, where %A = %T and %G = %C, implied that these bases must be paired, A to T and G to C.
   B. *Double helix* of Watson and Crick (1953)
      1. Watson and Crick won the Nobel Prize for their DNA model of two *antiparallel* strands in a *double helix*, with the sugar-phosphate backbone on the outside and complementary *base pairs* on the inside. Single-stranded RNAs can also fold back on themselves to create short, double helical segments.
      2. Complementary *base pairing* of A = T and G ≡ C arises due to the complementary hydrogen bonds made between these purine-pyrimidine pairs.
      3. The double helix has ~10 bp/turn with a helical repeat of ~34Å or 3.4 nm.
      4. The antiparallel strands create "major" and "minor" grooves that can bind various proteins to form nucleoprotein complexes.
      5. Human DNA consists of about 6,600,000,000 nucleotides arranged on 23 chromosomes.
   C. The double helix of DNA permits us to understand the processes of replication and protein synthesis.

## Chapter 23 - Nucleic Acids and Protein Synthesis

23.4 Replication of DNA
   A. Genes are the basic units of heredity. A *gene* is that portion (a section or a series of sections) of DNA in a chromosome that codes for the synthesis of a particular protein. Each amino acid in the protein is encoded by a three-base sequence. Cells in the human body have 26 *chromosomes* with enough DNA to form over 5 billion base pairs. In theory, each cell in the body is capable of producing all of the different proteins that the body needs. In practice, different cells become specialized during *differentiation* and use only a small portion of their genetic content.
   B. DNA *Replication*: Synthesis of DNA from a DNA template (see Figure 23.9)
      1. The double helix of DNA is unwound by unwinding proteins to yield a single-stranded *template*.
      2. RNA and DNA *polymerase* enzymes then produce the growing complementary nucleic acid strands by using nucleotide triphosphates as a source of nucleotides and available energy.
      3. The process is *semiconservative* in that each double helix produced contains one new and one original strand of DNA.
      4. The synthesis proceeds in the 5' → 3' direction in a *discontinuous* fashion, i.e., short segments of DNA are synthesized as the DNA double helix is unwound. These are later linked together by a *DNA ligase* enzyme.
   C. "*DNA fingerprinting*" refers to the process of making cuts and then separating the resulting DNA fragments by size using gel electrophoresis. The sizes of the DNA fragments are unique to that individual and can be used as a "fingerprint" in criminal investigations or to screen for genetic diseases (*RFLPs* = Restriction Fragment Length Polymorphisms).
   D. *PCR* (Polymerase Chain Reaction) is a method developed in the 1980's that exploits the use of a heat-stable DNA polymerase and primers to allow the researcher to start with minute amounts of DNA and amplify it several million-fold in just a few hours. This has proven very useful in providing the amounts of DNA needed for DNA fingerprinting and gene isolation.

23.5 Transcription: Synthesis of RNA
   A. DNA directs the synthesis of RNAs during *transcription*. In this process a DNA molecule is partially unwound, then a limited portion of 1 of the 2 DNA strands is used to direct the synthesis of a complementary RNA molecule (A ⇒ U; G ⇒ C; T ⇒ A; C ⇒ G).
   B. RNAs of three basic types of single-stranded nucleic acid
      1. *Messenger RNA (mRNA)* contains the "*codons.*" Codons are 3-base *triplets* that code for the amino acid sequence of the protein to be synthesized (see Table 23.3 for the *genetic code*.)
      2. *Ribosomal RNA (rRNA)* is the most abundant type of RNA in the cell and represents a major component of the *ribosome,* which is the site of protein synthesis.
      3. *Transfer RNAs (tRNA)* are low, molecular-weight nucleic acid molecules, which can contain about 90 nucleotides and transport activated amino acids to the ribosome. The X-ray structures of several tRNAs have been determined. Most tRNAs are L-shaped, having the 3-base "*anticodon*" at one end and the attachment site for the amino acid at the other (3') end.

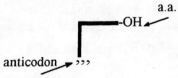

23.6 Protein Synthesis and the Genetic Code
   A. The *genetic code* shows how the codon triplets in mRNA specifies the particular amino acid.
   B. There are 4 N bases used in mRNA (A, U, G, C) or (4 × 4 × 4 = 64) possible triplets to specify the 20 common amino acids, e.g., UUU = phenylalanine, AUG = methionine (see Table 23.3).
      1. This code is essentially *universal*, and is used by all plant, animal, and bacterial cells.

Chapter 23 - Nucleic Acids and Protein Synthesis

  2. The code is *degenerate* in that amino acids are specified by multiple codons.
  3. AUG codes for methionine (or f-met; *N*-formylmethioine) and is also the *initiation codon*.
  4. Three of the 64 triplets do not code for amino acids and are used as *termination codons*.
  5. The $2^{nd}$ base is most important (A/G $\Rightarrow$ polar; C/U $\Rightarrow$ nonpolar).
C. *Translation* – the decoding of the mRNA molecule and its use in directing the synthesis of a protein molecule
  1. Protein biosynthesis takes place on ribosomes in the cytoplasm of the cell.
  2. The mRNA with its triplet *codons* travels from the nucleus to the cytoplasm.
  3. Each tRNA can transport only one kind of amino acid as specified by a 3-base sequence called its *anticodon*. The anticodon of a tRNA can form a complementary base pair with the corresponding codon in the mRNA bound to the ribosome.
  4. Protein biosynthesis is initiated as the mRNA, the first activated tRNA (f-met tRNA in bacteria), and the ribosomal subunits bind together to form a complex.
  5. The ribosome binds tRNAs to both the *"P"* (peptidyl) *site* and the *"A"* (aminoacyl) *site* with their anticodons paired with the codons on the mRNA. A peptide bond is formed when the "peptide" attached to the tRNA in the "P" site is transferred to make a new peptide bond to the free amino group of the amino acid attached to the other tRNA in the "A" site (see Figure 23.13).
  6. During a translocation step, the ribosome moves "one codon" along the mRNA.
  7. Another aminoacyl tRNA binds to the "A" site at the next codon, and another peptide bond forms.
  8. Steps 6 and 7 are repeated until protein biosynthesis is terminated by "stop" codons.
D. Genetic regulation
  1. A human cell contains about 100,000 genes. Many of the genes in plants and animals are segmented, with the parts that are *ex*pressed called *exons* and the parts that are not expressed, referred to as *intervening* sequences, or *introns*.
  2. We now know that there are *operator* genes, promotor genes, regulatory genes, and repressor molecules that help regulate the expression of proteins. The majority of DNA in a human cell is not used to code for amino acids.

23.7 Mutations and *Genetic Disease*
A. Any chemical or physical change that alters the sequence of bases in a DNA is termed a *mutation*, and the causative agent is called a *mutagen*. Mutants can involve *substitution*, *insertion*, or *deletion* of a base.
  1. UV light produces a thymine (T=T) dimer.
  2. Hydroxylamine ($NH_2OH$) deaminates cytosine so that it base pairs with adenine (C'=A) instead of guanine (C≡G).
  3. Nitrous acid ($HNO_2$) can convert cytosine to uracil.
B. Over 1200 *genetic diseases* in humans are caused by gene mutations. Many such diseases have been traced to a problem with one enzyme or one protein (see Table 23.4). The human genome project is helping to identify the locations of genes associated with many genetic diseases.
  1. PKU (phenylketonuria) – phenylalanine hydroxylase
  2. Sickle cell anemia – hemoglobin
  3. Albinism – tyrosinase
  4. Galactosemia – galactose 1-P-uridyl transferase
  5. Tay-Sachs disease – hexosaminidase A

# Chapter 23 - Nucleic Acids and Protein Synthesis

23.8 Genetic Engineering: Biotechnology
  A. *Recombinant DNA* refers to the splicing together of DNA from different species. It is possible, using recombinant DNA technology, to isolate a gene, and insert it into a *plasmid*, or circular piece of DNA, and then into a bacterium in such a way that the bacterial cell will produce multiple copies of the *cloned* gene and also the corresponding protein of the new gene (see Fig. 23.19).
   1. *Restriction enzymes* (endonucleases) are used to cut out the gene of interest and open up the plasmid. There are over 100 restriction enzymes available; each one is specific for cutting at a particular DNA sequence. The endonuclease *Eco*RI cuts double-stranded DNA at the sequence –GAATTC–.
   2. The sticky ends of the foreign DNA are complementary to those of the nicked plasmid.
   3. A DNA ligase seals the foreign DNA segment into the plasmid.
   4. The modified plasmid can be inserted into treated *E. coli* cells.
  B. Human insulin and other proteins can now be produced by bacteria. This technique holds great promise for producing large quantities of otherwise very rare proteins such as interferon and growth hormone.
  C. Genetic screening (DNA fingerprinting) often involves a distinctive pattern from RFLPs. Small amounts of DNA can be amplified by the use of polymerase chain reaction (*PCR*) methodology.
  D. The human genome project is a cooperative international effort to determine the base sequence of all 23 pairs of chromosomes. It is estimated that there are about 3.3 billion bp in the human genome.

## DISCUSSION

In the **SELF-TEST** we expect you to be able to recognize the distinguishing features of nucleic acids, nucleotides, and nucleosides. We'll also expect you to recognize which type of compound is being discussed from its name. Although you will not be asked to draw the complete structure of the various bases, you should recognize a purine (two fused heterocyclic rings) and a pyrimidine (one heterocyclic ring) when you see one. You should also know which bases are purines (adenine and guanine) and which are pyrimidines (cytosine, thymine, and uracil).

To give you a warm-up before the **SELF-TEST**, try the following problems, which review other points covered in the chapter.

**Problems:** In each labeled drawing, there is an **error**. You are being asked to **spot the error**.

1. A typical nucleoside

# Chapter 23 - Nucleic Acids and Protein Synthesis

2. A nucleotide obtained from the hydrolysis of DNA

3. A typical base pair in a nucleic acid

4. Beginning of DNA replication

```
A C G
: :: :::
T G C

T G C
: :: :::
A C G
```

5. DNA serving as a template for the formation of a mRNA molecule

```
A C C U G U

U G G A C A
:: ::: ::: :: ::: ::
A C C U G U
```

6. Formation of a protein molecule at the ribosome

# Chapter 23 - Nucleic Acids and Protein Synthesis

**SELF-TEST**

1. An individual unit of heredity is called a:
    a. nucleotide    b. nucleoside    c. gene    d. protein
2. DNA segments that interrupt a gene and are <u>not</u> involved in directing polypeptide synthesis are called:
    a. genettes    b. gamma genes    c. exons    d. introns
3. Human DNA consists of approximately how many base pairs?
    a. 13 million    b. 4 million    c. 200,000    d. 1 billion    e. 3 billion
4. If a nucleic acid is completely hydrolyzed, which type of compound is <u>not</u> one of the products?
    a. a purine    b. a pyrimidine    c. phosphoric acid
    d. an amino acid    e. a sugar
5. Which set of bases does <u>not</u> make up a base pair usually found in nucleic acids?
    a. adenine-thymine    b. cytosine-guanine    c. uracil-thymine    d. adenine-uracil
6. The t-RNA molecules incorporate _____ sugars.
    a. ribose    b. deoxyribose
7. Which group is <u>not</u> part of the backbone of a strand of nucleic acid?
    a. sugar unit    b. base unit    c. phosphoric acid unit
8. Base pairing is accomplished through the formation of:
    a. hydrogen bonds    b. phosphate linkages    c. hemiacetal linkages
9. The nucleus of a human body cell has how many chromosomes?
    a. 2    b. 12    c. 23    d. 36    e. 46
10. Which base is <u>not</u> normally found in RNA?
    a. adenine    b. cytosine    c. guanine    d. thymine    e. uracil
11. Which molecule is a nucleoside?
    a. cytosine    b. cytidine    c. cytidine monophosphate
    d. deoxycytidine monophosphate
12. Which contains the codon?
    a. DNA    b. mRNA    c. tRNA    d. the protein molecule
13. Which molecule carries the anticodon?
    a. mRNA    b. tRNA    c. the ribosome    d. the protein molecule
14. There are about how many base pairs per turn of the double helix?
    a. 2    b. 6    c. 10    d. 16    e. 20
15. If the triplet were 5'–U–G–C–3', the anticodon (5' → 3') would be:
    a. ACG    b. GCA    c. AUG    d. AGC
16. Where does gene replication take place?
    a. in a cell nucleus    b. at a ribosome complex    c. at the cell membrane
17. Where does the transcription of information from DNA to mRNA take place?
    a. in a cell nucleus    b. at a ribosome complex    c. at the cell membrane
18. After replication, each daughter DNA molecule:
    a. contains only purines or pyrimidines, but not both
    b. contains one strand of the parent molecule
    c. is the mirror-image isomer of the other daughter molecule
19. Which process occurs at the ribosome complex?
    a. replication of DNA    b. transcription of mRNA    c. translation to protein
20. When active protein synthesis is taking place in the cell, which material is <u>not</u> required at the ribosomes?
    a. DNA    b. mRNA    c. tRNA    d. growing protein chain

# Chapter 23 - Nucleic Acids and Protein Synthesis

21. The compound illustrated here is:

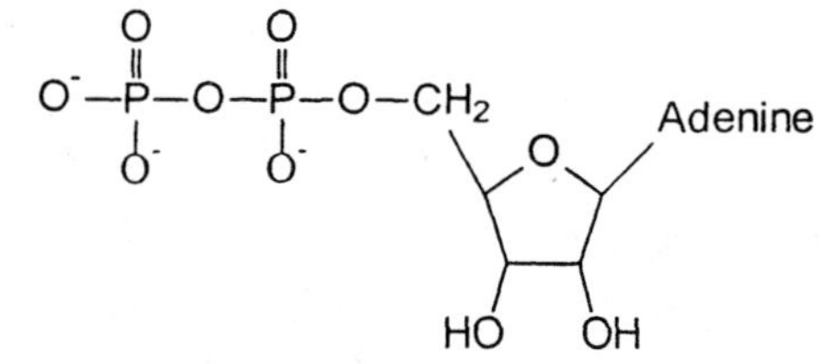

    a. AMP   b. c–AMP   c. ADP   d. ATP

22. The strands of DNA run in which direction?
    a. parallel   b. antiparallel

23. About how many bases (minimum) are needed to code for a 240 amino acid protein?
    a. 80   b. 480   c. 720   d. 240   e. 120

Answer Questions 24 through 29 by referring to the following structure.

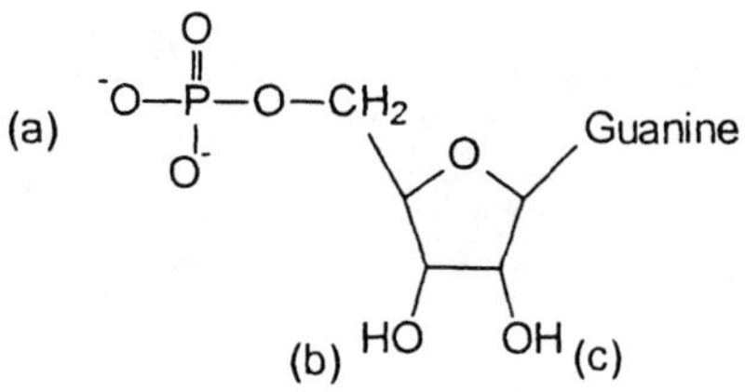

24. The compound is a:
    a. nucleoside   b. nucleotide   c. nucleic acid

25. The compound is:
    a. guanine   b. guanine monophosphate   c. deoxyguanine
    d. guanosine monophosphate   e. deoxyguanosine monophosphate

26. The compound incorporates a:
    a. purine   b. pyrimidine

27. The compound could be incorporated in:
    a. DNA   b. RNA

28. Three of the –OH groups have been labeled a, b, and c. Which of these would not be used in formation of the nucleic acid polymer?
    a   b   c

29. If the compound were incorporated in a nucleic acid, which base would not appear in the same polymer?
    a. adenine   b. cytosine   c. guanine   d. thymine   e. uracil

30. T / F  The molecular weights of nucleic acids are generally greater than those of proteins.
31. T / F  In nucleoproteins the basic side chains of the protein form salt bridges with the base pairs of the nucleic acids.
32. T / F  Nucleotides are formed in the hydrolysis of nucleosides.
33. T / F  Adenylic acid is identical to adenosine monophosphate.
34. T / F  It is the presence of the ribose unit in the nucleic acid RNA that makes the compound an acid.
35. T / F  The pairing of a purine with a pyrimidine permits the strands of a double helix to maintain a constant spacing.
36. T / F  It is impossible for base pairing to occur in single-stranded RNA.
37. T / F  Transfer RNA contains both the anticodon and the amino acid called for by the codon.
38. T / F  The codons that do not call for a specific amino acid signal the termination of protein synthesis.
39. T / F  Thymine and uracil are both purines.
40. T / F  Some codons call for more than one kind of amino acid.

# Chapter 23 - Nucleic Acids and Protein Synthesis

# ANSWERS

## Problems

1. A nucleoside would not have a phosphate group attached to the sugar ring.
2. The sugar unit is ribose. DNA would yield only deoxyribose.
3. Both bases are purines. A typical base pair would include a purine and a pyrimidine.
4. The bases on the complementary strands of the DNA molecule are not complementary. T was paired with T, G with G, etc. In DNA a strand carrying T and G and C would be matched with one carrying A and C and G.
5. Among the bases attached to the double-stranded DNA molecule is uracil. This base is found only in RNA.
6. The tRNAs are pairing with doublets rather than with the correct triplets.

## Self-Test

| | | | |
|---|---|---|---|
| 1. c | 11. b | 21. c | 31. F |
| 2. d | 12. b | 22. b | 32. F |
| 3. e | 13. b | 23. c | 33. T |
| 4. d | 14. c | 24. b | 34. F |
| 5. c | 15. b | 25. d | 35. T |
| 6. a | 16. a | 26. a | 36. F |
| 7. b | 17. a | 27. b | 37. T |
| 8. a | 18. b | 28. c | 38. T |
| 9. e | 19. c | 29. d | 39. F |
| 10. d | 20. a | 30. T | 40. F |

# G  Viruses

**KEY TERMS**

| | | | | |
|---|---|---|---|---|
| *virus* | *protein coat* | *RNA virus* | *DNA virus* | *retrovirus* |
| *AZT* | *carcinogen* | *benign tumor* | *malignant* | *oncogene* |
| *protease inhibitors* | *epidemiological* | *antimetabolites* | *benzpyrene* | *cisplatin* |
| *reverse transcriptase* | *Ames test* | *5-fluorouracil* | *methotrexate* | *BHT* |
| *tumor-suppressor* | | | | |

**SUMMARY**

G.1 The Nature of Viruses
  A. *Viruses* are infectious agents composed of a nucleic acid core and a ***protein coat***. Infectious diseases of viral origin include the common cold, polio, rabies, hepatitis, chicken pox, measles, mumps, herpes, and AIDS.
  B. Viruses can be subdivided into two classes, depending on the type of nucleic acid present.
   1. ***DNA Viruses***: Viral DNA is replicated in the host cell and directs the production of coat protein(s) to assemble new viruses.
   2. ***RNA Viruses***: Most RNA viruses act in a manner similiar to DNA viruses, but some RNA viruses called ***retroviruses*** synthesize DNA by a process that is the opposite of transcription, using an enzyme called ***reverse transcriptase***. The human immunodeficiency virus (HIV) that causes AIDS is an example of a ***retrovirus***.

G.2 Antiviral Drugs
  A. Because few reactions are unique to viruses, it has been difficult to develop drugs unique for treating viruses, thus vaccines have been widely used to control viruses.
  B. Antiviral compounds can act in a variety of ways.
   1. Block cell membrane receptors
   2. Prevent viral particles from releasing nucleic acid
   3. Block replication of the viral nucleic acid
  C. Acyclovir (Zovirax®) was one of the earliest antiviral agents. Acyclovir is used to control genital herpes, chicken pox, shingles, and cold sores. Note that the structure of acyclovir closely resembles that of deoxyguanosine. It is a substrate for the viral thymidine kinase, that helps convert it to the triphosphate, which in turn can block chain elongation at the DNA polymerase.

Acyclovir

G.3 Human Immunodeficiency Virus (HIV) and AIDS
  A. HIV is the virus that causes AIDS. The virus uses its glycoproteins to attach and enter T-cells, where it replicates and destroys the cell. Without T-cells, the AIDS victim is susceptible to pneumonia or other infectious diseases.

B. **AZT** (azidothymidine or Retrovir®), ddI (2',3'-dideoxyinosine), and ddC(2',3'-dideoxycytidine) are nucleoside analog drugs approved by the FDA to treat AIDS. Nucleosides missing the 3'-OH group prevent further replication once they are incorporated into the growing polynucleotide.

AZT / azidothymidine        DDI / 2',3'-dideoxyinosine

C. A second class of anti-AIDS drugs inhibit the specific viral-induced HIV protease. Examples of drugs that operate as *protease inhibitors* are saquinavir (Invirase®), ritonavir (Norvir®) and nelfinavir (Viracept®).
D. A third class of anti-AIDS drugs is the nonnucleoside reverse transcriptase inhibitors or noncompetitive inhibitors of reverse transcriptase, such as nevirapine (Viramune®), and abacavir (Ziagen®).
E. The most effective treatments for AIDS has been to use a "cocktail" of a protease inhibitor in combination with one or more of the reverse transcriptase inhibitors.

G.4 Cancer and Carcinogens
   A. *Carcinogens* are cancer-causing substances that promote the growth of tumors.
      1. *Benign* tumors grow slowly and do not invade neighboring tissues.
      2. *Malignant* tumors are "cancers" that invade and destroy neighboring tissue.
   B. Most cancers (~80–90%) are caused by exposure to environmental factors (e.g., smoking). However, many of the carcinogens that we ingest are produced naturally in the environment.
   C. Testing for carcinogens
      1. Bacterial screening: the *Ames test* is a simple and relatively cheap screening test for mutagens, but it can also pick up carcinogens that are produced by metabolism.
      2. Animal tests: Very large doses are tested on laboratory animals.
      3. *Epidemiological* methods: These involve the use of statistical analysis of affected population, and the probable common cause, e.g., cigarette smoking and lung cancer.
         D. Chemical carcinogens: Aromatic hydrocarbons such as 3,4-*benzpyrene* formed from the incomplete combustion of organic materials (cigarette smoking, barbecuing, etc.) are chemical carcinogens. Other examples are β-naphthylamine and benzidine (dye industry), dimetylnitrosoamine, and vinyl chloride.

G.5 Genetic Basis of Cancer
   A. The genes that are implicated in the process of turning normal cells into cancerous ones are called *oncogenes* and usually are genes that are involved in regulating cell growth and division.
   B. Tumor-suppressor genes code for proteins that act as signals to inhibit cell growth. These can be inactivated when mutated.

### G.6 Chemicals Against Cancer

A. Many of the compounds used in cancer chemotherapy are designed to inhibit DNA synthesis. Because cancer cells are undergoing rapid growth, they are generally affected to a greater extent than normal cells.

B. Examples of *Antimetabolites*
  1. *Cisplatin* – a *cis* isomer of a platinum complex that binds to and blocks DNA replication.
  2. *5-Fluorouracil* – incorporates into DNA and slows the rate of replication.
  3. 6-Mercaptopurine – substitutes for adenine in a nucleotide.

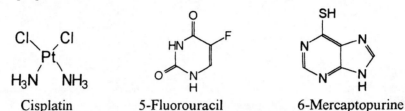

    Cisplatin      5-Fluorouracil      6-Mercaptopurine

  4. *Methotrexate* – an analog of folic acid that prevents the enzyme from using good folic acid.

Methotrexate

Folic Acid

C. Actinomycin (molds), alkaloids (plants), sex hormones, the food preservative butylated hydroxytoluene (*BHT*), vitamins A and C, broccoli, and many other substances have been correlated with some positive value in protecting against various forms of cancer.

## SELF-TEST

1. Which of the following diseases is *not* caused by a virus?
      a. mumps    b. polio    c. AIDS    d. gonorrhea    e. measles    f. warts
2. AIDS is an example of a _____ virus.
      a. RNA    b. DNA    c. tRNA    d. mRNA    e. retrovirus
3. AZT, an anti-AIDS drug, is structurally similar to:
      a. an amino acid    b. purine    c. nucleoside    d. nucleotide
4. A cancer is often synonymous with:
      a. pollutants    b. carcinogen    c. benign tumor    d. malignant tumor
5. Genes that trigger the formation of cancerous cells are called:
      a. regulatory genes    b. oncogenes    c. Z genes    d. X genes
6. Which compound among the following is *not* an anticancer drug?
      a. methotrexate    b. 5-fluorouracil    c. cisplatin    d. cytidine

7. 3,4 benzpyrene is a known:
    a. carcinogen    b. vitamin    c. amino acid    d. neurotransmitter    e. cofactor

**Matching:**

8. ___ Ames test
9. ___ actinomycin
10. ___ BHT
11. ___ antimetabolite
12. ___ AIDS

a. food additive, preservative
b. inhibits protein synthesis
c. methotrexate
d. bacterial screening for mutagens
e. retrovirus
f. carcinogen, barbecue
g. oncogenes

## ANSWERS

| | | | | | |
|---|---|---|---|---|---|
| 1. d | 3. c | 5. b | 7. a | 9. b | 11. c |
| 2. a, e | 4. d | 6. d | 8. d | 10. a | 12. e |

# 24 Metabolism and Energy

**KEY WORDS**

| | | | | |
|---|---|---|---|---|
| anabolism | respiration | exergonic | endergonic | metabolism |
| catabolism | coupling | free energy change | intermediates | metabolites |
| digestion | α-amylase | passive transport | dextrins | active transport |
| acetyl-CoA | catabolic | thioesters | cytochrome | chemiosmotic |
| pyruvate | respiratory chain | oxaloacetate | citrate | coenzyme A |
| mitochondria | NADH, FADH$_2$ | glycogenolysis | oxygen debt | citric acid cycle |
| anaerobic | electron carriers | electron transport | photosynthesis | Krebs cycle |
| aerobic | anabolic | energy-rich | lipoproteins | ADP, ATP |
| bile | creatine phosphate | metabolic pathway | myosin | actomyosin |
| actin | oxidative | lactic acid | lipases | TCA cycle |
| lipoproteins | phosphorylation | gastric juice | saliva | |

**SUMMARY**

Life requires a continuous input of energy. Green plants use *photosynthesis* to capture the sun's energy. Animals rely on the stored chemical energy found in carbohydrates, fats, and proteins. The coordinated chemical reactions that sustain life are called *metabolism*.

  1. *Catabolism* – reactions involved in breaking down foodstuffs
  2. *Anabolism* – biosynthetic reactions

Most foodstuffs represent reduced forms of carbon, which are ultimately oxidized to carbon dioxide and water. *Respiration* refers to all metabolic processes whereby gaseous oxygen is used to oxidize organic foodstuffs to $CO_2$, $H_2O$, and energy. Energy is released when bonds are formed, and energy is required to break bonds.

  1. *Exergonic* reactions release energy; this is typical of catabolic pathways.
  2. *Endergonic* reactions require energy input; this is typical of anabolic pathways.

24.1 ATP: Universal Energy Currency

  A. *ATP* (adenosine triphosphate) and *ADP* (adenosine diphosphate) release large amounts of Gibbs *free energy* upon hydrolysis. Such compounds are called "**energy-rich**" compounds, and the pyrophosphate bond a "high-energy" bond ($\Delta G°' \sim -7500$ cal/mol or $-7.5$ kcal/mol).

  B. ATP is a "middleweight" among energy-rich compounds, low enough to be produced when coupled with certain *catabolic* reactions but high enough to drive most *anabolic* reactions. ATP is often called the "energy currency" of the cell.

C. Coupled reactions
1. Nearly all anabolic reactions within a living cell are *endergonic*, or nonspontaneous. However, these reactions can be carried out by "*coupling*" the unfavorable reaction to another reaction that is very favorable. The $\Delta G$ for the coupled reaction is the sum of the individual $\Delta G$ values.
2. Energy coupling of reactions requires that the two reactions share a common intermediate.

| | | | |
|---|---|---|---|
| Glucose + fructose | → | sucrose + $H_2O$ | $\Delta G^{o'} = +7.0$ (unfavorable) |
| $H_2O$ + ATP | → | ADP + $P_i$ | $\Delta G^{o'} = -7.5$ (favorable) |
| Glucose + fructose + ATP | → | sucrose + ADP + $P_i$ | $\Delta G^{o'} = -0.5$ (also favorable) |

D. The "burning" of foodstuffs to carbon dioxide via metabolic pathways proceeds by a series of small steps that produce many chemical *intermediates*, called *metabolites*. Some of the reactions involving these metabolic intermediates are sufficiently exergonic that they can be coupled to drive the formation of ATP.

| | | | |
|---|---|---|---|
| Phosphoenol pyruvate (PEP) | → | pyruvate + $P_i$ | $\Delta G^{o'} = -12.8$ kcal/m (favorable) |
| $P_i$ + ADP | → | ATP | $\Delta G^{o'} = +7.5$ kcal/m (not favorable) |
| PEP + ADP | → | pyruvate + ADP | $\Delta G^{o'} = -5.3$ kcal/m (favorable) |

24.2 Digestion and Absorption of Major Nutrients
  A. *Catabolism* can be broken down into three stages (see Figure 24.5).
    1. Stage I – digestion and hydrolysis of biopolymers (polysaccharides → simple sugars).
    2. Stage II – conversion of monomers to acetyl-CoA or a Krebs (TCA) cycle intermediate.
    3. Stage III – oxidation of acetyl-CoA (TCA cycle) and production of ATP (oxidative phosphorylation).
  B. *Digestion* is the "hydrolytic process whereby food molecules are broken down into simpler chemical units that can be absorbed by the body." Our bodies contain a "tunnel" called the digestive tract (or alimentary canal) that begins at the mouth and ends at the anus. Foodstuffs that are digested during this passage can be absorbed and used by the body.
  B. Foodstuffs are acted on both mechanically and biochemically – mechanically, as in chewing to break down large food particles into smaller particles and, biochemically, as the various digestive enzymes operate to hydrolyze the biopolymers into smaller units that can be absorbed along the digestive tract. Carbohydrates, proteins, and lipids require different types of enzymes to break down their polymers. The biochemistry changes as we move along the digestive tract.
  C. The digestion of carbohydrates.
    1. Mouth: Food is lubricated with mucin; *saliva* contains *a-amylase* (ptyalin) that begins cleavage of glycosidic bonds.
    2. Stomach: A small amount of acid hydrolysis, but little carbohydrate digestion, takes place here.
    3. Small intestine: A second amylase converts starch and *dextrins* to maltose. Disaccharidases convert maltose and other disaccharides to primarily glucose, fructose, and galactose.
  D. Digestion of proteins: Intact proteins cannot be absorbed across intestinal membranes. Digestion of proteins into amino acids by proteases takes place primarily in the stomach and the small intestine.
    1. Stomach: *Gastric juice* contains HCl acid (pH ~1) and the proenzyme pepsinogen, which is converted to its active form as the enzyme pepsin.
        Pepsin         – cleaves after Trp, Tyr, Phe, Met, Leu

# Chapter 24 - Metabolism and Energy

    2. Small Instestine: Pancreatic juice delivers the proenzymes trypsinogen, chymotrypsinogen, and procarboxypeptidase. Intestinal juice also contains the exopeptidase (aminopepitdase) and di- and tripeptidases.

        Trypsin           – cleaves after Lys, Arg
        Chymotrypsin    – cleaves after Phe, Trp, Tyr
        Carboxypeptidase – cleaves off C-terminus

  E. Digestion of lipids: Fats are emulsified by the *bile* salts released by the gallbladder into the upper intestine, digested by *lipases*, absorbed into the lymph system, and finally enter the bloodstream. Being water insoluble, fats must be transported by *lipoproteins* in the blood.

  F. Absorption of most digested food takes place in the small intestine through the *villi*. Following absorption, most food molecules are carried via the bloodstream to the liver.

    1. Some substances (fatty acids and monoglycerides) can be absorbed by *"passive transport"* via simple diffusion or osmosis.

    2. Monosaccharides and amino acids cross the small intestinal wall (*villi*) by an energy-requiring process. Such processes are termed *"active transport."*

24.3 Overview of Stage II of Catabolism

  A. Metabolism is the set of coordinated chemical reactions that sustains life. A *metabolic pathway* is a subset of these reactions that describes the biochemical conversion of a given reactant to its desired end product (*one highway on the road map*).

  B. The details of stage II for different nutrients involve different metabolic pathways that will be presented later (carbohydrates, Chapter 25; lipids, Chapter 26; and proteins, Chapter 27). However, catabolism tends to converge in that sugars, fatty acids, and many amino acids all are broken down to produce a common end product – acetyl-coenzyme A or acetyl-CoA.

The structure of acetyl-coenzyme A *(Acetyl-CoA)*

  C. Acetyl-CoA is at a hub of metabolism, it can be oxidized to release energy to produce ATP, and it also is the starting material for the biosynthesis of fatty acids, phospholipids, cholesterol, etc.

Acetyl~CoA → Fatty Acids
Acetyl~CoA → Acetoacetyl~CoA → Steroids
Acetoacetyl~CoA → Cholesterol
Acetoacetyl~CoA → Ketone Bodies
Acetyl~CoA → Krebs Cycle

# Chapter 24 - Metabolism and Energy

## 24.4 The Krebs Cycle
A. *Acetyl-CoA* is the thioester of *coenzyme A* with acetic acid. *Thioesters* are "high-energy" compounds. The fate of acetyl-SCoA is described by the *Krebs cycle*, also known as the *TCA cycle* or *citric acid cycle*. (Refer to Figure 24.18)
B. The TCA cycle (Krebs cycle or citric acid cycle)
  1. The steps of the Krebs cycle take place in the *mitochondria*. The reactions involve condensation, dehydration, hydration, oxidation, decarboxylation, and hydrolysis.
  2. *Oxaloacetate (OAA)* initially condenses with acetyl-SCoA to form the 6-carbon *citrate*. The citric acid will undergo several oxidative steps to regenerate the OAA just consumed, thus completing the cycle.
  3. Regulation of the pathway occurs primarily at Steps 1 and 3. Citrate synthetase (Step 1) is inhibited by ATP and NADH. At Step 3, isocitrate dehydrogenase (ICDH) catalyzes an oxidative decarboxylation, producing $CO_2$ and NADH. ICDH is inhibited by high energy conditions such as high levels of ATP and NADH, but also is turned on by high levels of ADP. Step 4 is another oxidative decarboxylation, producing the second molecule of $CO_2$.
  4. Steps 5–8 regenerate 4-carbon OAA from 4-carbon succinyl-CoA, also producing GTP, $FADH_2$, and NADH.
C. During one turn of the cycle, the net effect is that a 2-carbon acetyl group is lost and 2 molecules of $CO_2$, 3 of NADH, 1 $FADH_2$, and 1 GTP are produced. The reduced coenzymes serve as *electron carriers* and will be regenerated when they donate their electrons to the *electron transport* chain.

## 24.5 Cellular Respiration
A. Oxidation is the removal of electrons. When foodstuffs are oxidized, something must be reduced. The most common *electron carriers* of these reducing equivalents are the two coenzymes NADH and $FADH_2$.
  1. $NAD^+ + H^+ + 2 e^- \rightarrow$ *NADH* (nicotinamide adenine dinucleotide)
  2. $FAD + 2 H \rightarrow$ *$FADH_2$* (flavin adenine dinucleotide)
B. The reactions of the Krebs cycle, electron transport, and oxidative phosphorylation take place within organelles called *mitochondria*. Mitochondria are called the "power plants" of the cell.
C. The passage of electrons from NADH and $FADH_2$ to oxygen to produce water does not take place directly, but occurs stepwise as the electrons are passed through a series of electron carriers (including heme-containing proteins called *cytochromes*) that make up the "*respiratory chain*." Each intermediate in the respiratory chain is first reduced by the addition of electrons and then re-oxidized as it passes those electrons on to the next carrier. These *electron carriers* are found associated with the inner mitochondrial membrane and can be divided into different types:
  1. Two electron carriers: NADH, $FADH_2$, and $CoQH_2$ dependent proteins.
  2. One electron carriers: various cytochromes that contain a heme-like group with iron that can flip oxidation states from +3 to +2
D. The *electron transport* chain is made up of four membrane associated complexes.
  1. NADH $\rightarrow$ Complex I $\rightarrow$ Complex III $\rightarrow$ Complex IV $\rightarrow$ oxygen
  2. $FADH_2$ $\rightarrow$ Complex II $\rightarrow$ Complex III $\rightarrow$ Complex IV $\rightarrow$ oxygen
  3. $CoQH_2$ helps move the electrons from Complexes I and II to Complex III.
  4. Cyt c is a membrane-associated protein that carries electrons from Complex III $\rightarrow$ IV.
  5. Only Complex IV, containing cytochrome oxidase, has the ability to transfer electrons to molecular $O_2$ to produce water.
E. Hydrogen cyanide and its salts release cyanide that can bind to and tie up the heme groups, thus inhibiting electron transfer and causing cell respiration to cease.
  1. Cyanide was used in the mass suicide at Jonestown, Guyana.

# Chapter 24 - Metabolism and Energy

  2. Cyanide is also found in nature in amygdalin and other cyanogenic glycosides. Amygdalin has been marketed as Laetrile® as a possible cancer treatment.
 F. Oxidative Phosphorylation
  1. *Catabolic* pathways utilize $NAD^+$ and FAD to oxidize the foodstuffs in our diets to carbon dioxide, thus producing large amounts of the reduced coenzymes NADH and $FADH_2$. These coenzymes must be reoxidized to $NAD^+$ and FAD to keep the catabolic pathways going.
  2. The hydrogens ($H^+ + e^-$) from NADH and $FADH_2$ are passed on to oxygen to form water. The energy released during this exergonic process is coupled to the synthesis of ATP by a process referred to as "*oxidative phosphorylation.*" It is possible to uncouple these processes and have electron transport without ATP synthesis. This is strong evidence supporting the "*chemiosmotic*" hypothesis for ATP synthesis.
   a. Oxidation of each NADH in the mitochondria by oxygen is accompanied by the production of 2.5 molecules of ATP.
   b. Oxidation of each $FADH_2$ by oxygen leads to the production of 1.5 ATP.
   c. Energetics summary:   1 NADH    = 2.5 ATP;
                            1 $FADH_2$  = 1.5 ATP;
     thus 1 acetyl-CoA = 3 NADH / 1 GTP / 1 $FADH_2$ = 10 ATP

  3. The efficiency of a metabolic pathway can be estimated by comparing the free energy value of all of the ATP molecules synthesized with the free energy released during the oxidation of the foodstuff. Different foods yield different amounts of energy per gram [fats (~9 kcal/g), carbohydrates and proteins (~4 kcal/g)] because the carbon atoms in fats, for example, are in a more highly reduced state than the carbon atoms in carbohydrates or proteins. However, all catabolic pathways are ~35–40% efficient at storing the energy released as newly synthesized ATP; the rest is lost as heat, which helps keep our bodies warm.

24.6 Muscle Power
 A. Exercise can make muscles stronger and do more work with less strain.
 A. Muscle contains the proteins *actin* and *myosin* in a loose complex called *actomyosin*. When ATP is added to isolated actomyosin, the muscle fibers contract, which implies that ATP is the energy source for muscle contraction.
 B. Muscle fibers are divided into two categories:
  1. "Fast-twitch" – Type IIB – white muscle; low in mitochondria and myoglobin; high in enzymes for *glycogenolysis*; designed for short bursts of vigorous work; sprinters have lots of white muscle.
  2. "Slow-twitch" – Type I – red muscle; high in mitochondria and myoglobin to supply the oxygen for *aerobic* respiration; designed for sustained, moderate levels of physical activity; long distance joggers have lots of red muscle; endurance training can increase the number of mitochondria in muscle fibers.
 C. *Oxygen debt* occurs during strenuous exercise when energy demands can exceed the ability to supply oxygen to muscle tissue. Muscle metabolism shifts to *anaerobic* processes that convert pyruvate to lactate in order to regenerate the $NAD^+$ needed to keep glycolysis operable. This *lactic acid* buildup leads to a pH drop and deactivation of muscle enzymes, described as "muscle fatigue," and can cause lactic acid acidosis. The overworked muscles are incurring an oxygen debt that needs to be repaid after the strenuous exercise is over.
 E. *Creatine phosphate* is a "high-energy" compound that serves as the storage form of energy in muscles. Muscle cells at rest contain four to six times as much creatine phosphate as ATP.
     Creatine phosphate + ADP   →    ATP + creatine

# Chapter 24 - Metabolism and Energy

## DISCUSSION

This chapter introduces the subject of metabolism and energy balance in living systems. Metabolism refers to the whole series of reactions needed to sustain life. We learned that our bodies are able to convert the foods we eat into useful forms of chemical energy. The rules that govern such energy conversions are found in the study of thermodynamics. We need to concern ourselves with only two aspects of thermodynamics: (1) the relationship between free energy changes and "spontaneity," and (2) the coupling of a very exergonic reaction with an endergonic reaction so the net coupled reaction is still exergonic.

What determines whether the reaction A → B is more favored than the reverse reaction, B → A? There are two factors that determine the "spontaneity," or favored direction, of a chemical reaction. The first factor is the tendency to minimize energy (rocks roll down a hill, not up). The second factor is the tendency to maximize entropy, which is related to probability or what is often called "randomness." For example, we postulate that gas molecules will disperse to fill a large container, even when no energy changes take place, because it is simply more probable that the gas molecules would disperse throughout the container rather than grouped together or ordered in one small part of the container. The Gibbs free energy change for a reaction measures the net effect of both of these factors.

$$\Delta G = \Delta H - T\Delta S$$

(where $\Delta G$ = **free energy** change, $\Delta H$ = **enthalpy** (energy) change, $\Delta S$ = **entropy** change)

It is not necessary for you to have a complete understanding of these quantities to appreciate the usefulness of the Gibbs free energy change, $\Delta G$. The relationships between spontaneity and $\Delta G$ are summarized below:

| $\Delta G$ | "spontaneity" | $K_{eq}$ | |
|---|---|---|---|
| + (endergonic) | nonspontaneous | <1 | (unfavorable) |
| 0 | system at equilibrium | 1 | |
| − (exergonic) | spontaneous | >1 | (favorable) |

We have stated that catabolic pathways (the ones that degrade foodstuffs) are generally considered to be exergonic, that is, energetically favorable with a negative $\Delta G$. For instance, recall that for the combustion of glucose: $C_6H_{12}O_6 + 6\ O_2 \rightarrow 6\ CO_2 + 6\ H_2O \qquad \Delta G^{o'} = -686$ kcal/mol

This is a very energetically favorable reaction. If we burn carbohydrates in air, we obtain $CO_2$, $H_2O$, and a lot of energy released as heat. Living cells metabolize ("burn") sugar slowly, employing several dozen metabolic steps in order to conserve some of the free energy released. The free energy cannot be deposited in a bank for later withdrawal; rather, it must be stored in the form of "high-energy" bond formation. This is accomplished by coupling the exergonic step in the catabolic pathway with another reaction that requires an input of energy to proceed, such as the synthesis of ATP from ADP. Reaction coupling requires that both the exergonic and endergonic reactions must share a common chemical intermediate. For example, the formation of sucrose actually proceeds through an intermediate, glucose-1-phosphate.

| | | |
|---|---|---|
| Glucose + ATP | → | Glucose-1-phosphate + ADP |
| Glucose-1-phosphate + fructose | → | sucrose + $P_i$ |
| Glucose + fructose + ATP | → | sucrose + ADP + $P_i$ |

# Chapter 24 - Metabolism and Energy

The free energy changes for the hydrolysis of some common phosphates (see also Table 24.1).

| Compound | | Hydrolysis Products | | $\Delta G°'$ (kcal/mol) |
|---|---|---|---|---|
| Phosphoenol pyruvate (PEP) | → | pyruvate | + $P_i$ | –12.8 |
| Creatine phosphate | → | creatine | + $P_i$ | –10.3 |
| ATP | → | ADP | + $P_i$ | – 7.5 |
| ADP | → | AMP | + $P_i$ | – 7.5 |
| Glucose-6-phosphate | → | glucose | + $P_i$ | – 3.8 |
| AMP | → | adenosine | + $P_i$ | – 3.4 |

The reactions listed above ATP in this table could be used to drive the synthesize ATP, and the hydrolysis of ATP can in turn be coupled to drive the synthesis of the phosphate esters listed below ATP in the table.

We've referred to metabolism, metabolites, and metabolic products frequently in past chapters. In this chapter we are taking an extended look at metabolic processes in human beings. The single most noticeable feature of metabolic reactions is that transformations, which can be summarized in one equation, usually proceed by mechanisms that involve many steps. The very complexity of these processes makes life possible, but it also makes studying the processes difficult.

How about the reaction patterns of the Krebs cycle? The Krebs cycle is designed to oxidize acetate to two molecules of carbon dioxide. A chemist can do this in the laboratory by burning acetic acid, that is, by carrying out the combustion of acetic acid. Cells are far more subtle. The Krebs cycle (Figure 24.18) starts by taking the acetic acid (activated by attachment to coenzyme A) and bonding it to one of the intermediates of the cycle (oxaloacetic acid). The resulting product is "manipulated" to produce an organic molecule that is especially suited to being decarboxylated. By "manipulated" we mean that water is removed, then replaced in a different position, and then hydrogen is removed. The product does just what it is supposed to do – it decarboxylates. The product from that reaction decarboxylates again, producing the two carbon dioxide molecules. Now all that remains is to manipulate the product succinate a bit to regenerate the starting molecule, OAA. To accomplish this, hydrogen is removed (fumarate), water is added (malate), and more hydrogen is removed (oxaloacetate). And there's oxaloacetic acid again. What we want to see is the reasonableness of the process. If a cell cannot set fire to acetic acid to achieve its ends, then it simply builds molecules (enzymes) that use multiple steps to accomplish the same results.

Oxidation is recurring in the Krebs cycle. It occurs at four different points in the cycle (Steps 3, 4, 6, and 8). The oxidizing agents required for these steps are regenerated by the electron transport system, which operates in conjunction with oxidative phosphorylation. It is here that ATP is actually synthesized in reactions coupled to the transport of electrons to oxygen to form water (oxidative phosphorylation).

The other topic in Chapter 24 that often causes confusion is the role of electron carriers, such as NADH and $FADH_2$, in the oxidation of foodstuffs to $CO_2$ and $H_2O$. As the fats, carbohydrates, and proteins in our diet are oxidized to $CO_2$, something else must be reduced (NADH and $FADH_2$). Water is produced when these reduced coenzymes are reoxidized and their electrons flow through the electron *transport* system (*ETS*) to reduce oxygen to water. The term *"oxidative phosphorylation"* refers to the coupled synthesis of ATP to the flow of electrons through the ETS. However, there are many intermediate electron carriers along the way. Most of these "reducing equivalents" are initially transferred to the coenzymes $NAD^+$ and FAD to produce NADH or $FADH_2$. These molecules then function as carriers of two electrons to deliver the electrons released during the oxidation of foodstuffs to $CO_2$ in the various

# Chapter 24 - Metabolism and Energy

pathways to the ETS found in the mitochondria. As electrons pass along the ETS, the energy released is indirectly coupled to the synthesis of ATP by a process known as oxidative phosphorylation. These processes are summarized below.

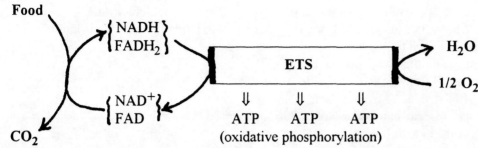

The rest of the material in Chapter 24 is reviewed in the **SELF-TEST**.

## SELF-TEST

1. The molecule described as the "energy currency" of metabolism is:
    a. glucose   b. stearic acid   c. cAMP   d. Vitamin A   e. ATP
2. Digestion of carbohydrates begins with saliva in the mouth and the action of the enzyme:
    a. insulin   b. α-amylase   c. kinase   d. mutase
3. The synthesis of fats for energy storage is an example of:
    a. catabolism   b. anabolism
4. Converting $CO_2$ to glucose is an example of a(n) _____ reaction.
    a. exergonic   b. metabolite   c. endergonic
5. The chemically significant feature of nucleotides like ATP and GTP is the _____ linkage.
    a. glycosidic   b. amide   c. pyrophosphate   d. peptide
6. The free energy of hydrolysis, $\Delta G°'$, for ATP is about:
    a. 12.5 kcal/mol   b. 7.3 kcal/mol   c. 3.5 kcal/mol
    d. –7.3 kcal/mol   e. –3.5 kcal/mol
7. The $K_{eq}$ for the reaction A → B is 10. Under conditions such that the concentration of A is ten times that of B, $\Delta G$ would be expected to be:
    a. negative   b. zero   c. positive
    (Refer to the list of $\Delta G°'$ values given in the **DISCUSSION** above to answer Questions 8–11.)
8. Which of the following compounds would <u>not</u> be considered a "high energy" compound?
    a. PEP   b. ATP   c. glucose-6-phosphate   d. ADP
9. The free energy change for the synthesis of glucose-6-phosphate from glucose and ATP is:
    a. –11.1   b. +11.1   c. –7.3   d. +4.0   e. –4.0
10. Calculate $\Delta G°'$ (kcal/mol) for the synthesis of ATP from ADP and creatine phosphate.
    a. –5.5   b. –3.2   c. +3.2   d. –17.8   e. +17.8
11. What is the free energy change (kcal) required to convert one mole of AMP to ATP via ADP?
    a. 13.8   b. –13.8   c. –2.2   d. +2.2   e. 16.0
13. Which is <u>not</u> another name for the Krebs cycle?
    a. citric acid cycle   b. Cori cycle   c. tricarboxylic acid cycle
14. The reactions of the Krebs cycle take place in the _____ of the cell.
    a. mitochondria   b. cytoplasm   c. nucleus   d. ribosomes
15. Which is <u>not</u> one of the intermediates in the Krebs cycle?
    a. citric acid   b. isocitric acid   c. fumaric acid   d. lactic acid   e. succinic acid

# Chapter 24 - Metabolism and Energy

16. Two molecules undergo decarboxylation in the Krebs cycle. Which does not?
    a. α-ketoglutaric acid   b. oxaloacetic acid   c. oxalosuccinic acid
17. Which molecule combines with acetyl-CoA to form citrate in the TCA cycle?
    a. OAA   b. pyruvate   c. fumarate   d. lactate   e. succinate
18. What process does not occur in electron transport coupled with oxidative phosphorylation?
    a. conversion of oxygen to water
    b. synthesis of the oxidizing agents $NAD^+$ and FAD
    c. synthesis of ATP from ADP
    d. oxidation of lactic acid to pyruvic acid
    e. All of the above occur in oxidative phosphorylation.
19. Which is not true of the cytochromes?
    a. They are iron-containing proteins.
    b. They participate in the series of reactions called the respiratory chain.
    c. Their action is strongly inhibited by carbon dioxide.
    d. All of the above are true of the cytochromes.
20. Which of the following would be described as a one electron carrier?
    a. NADH   b. cytochromes   c. $FADH_2$
21. The reduced form of nicotinamide adenine dinucleotide is represented as:
    a. $FADH_2$   b. FADH   c. $NAD^+$   d. $NADH_2$   e. NADH
22. The oxidized form of flavin adenine dinucleotide is given as:
    a. $FADH_2$   b. $FAD^+$   c. FAD   d. $NAD^+$   e. NADH
23. The synthesis of ATP coupled with the passage of electrons to molecular $O_2$ to form water is called:
    a. respiratory chain   b. electron transport chain   c. Krebs cycle
    d. oxidative phosphorylation
24. Which of the following metabolic processes does not occur in the mitochondria?
    a. saliva digestion   b. Krebs cycle   c. electron transport chain
    d. oxidative phosphorylation
25. The final electron acceptor in respiration is:
    a. $H_2O$   b. FAD   c. $NAD^+$   d. $O_2$   e. cytochromes
26. An exergonic reaction has a $K_{eq}$ that is:
    a. >1   b. 0   c. <1
27. The efficiency of most metabolic pathways is about:
    a. 5%   b. 15%   c. 35%   d. 75%   e. 100%
28. Cyanide is a lethal poison because it interferes with which of the following macromolecules?
    a. NADH   b. actomyosin   c. cytochromes   d. creatine kinase
29. Each NADH that is reoxidized through the ETS produces the equivalent of _____ ATP.
    a. 1   b. 2   c. 3   d. 5   e. 8
30. Each acetyl unit converted to carbon dioxide and water produces the equivalent of _____ ATP.
    a. 1   b. 4   c. 8   d. 12   e. 20
31. Actomyosin is:
    a. the enzyme that catalyzes the transfer of phosphate from creatine phosphate to ADP
    b. the protein complex that constitutes the contractile tissue of muscle
    c. the drug used to counter the effects of an accumulated oxygen debt
32. T / F  The iron in cytochromes may be in the +2 or +3 oxidation state.
33. T / F  The ΔG for a two-step process is equal to the product of the ΔGs for the individual steps.
34. T / F  Carbon dioxide represents a highly oxidized, high-energy form of carbon.
35. T / F  Creatine phosphate is stored in muscle tissue to help regenerate ATP for muscle contraction.
36. T / F  High levels of Type I muscle fibers are appropriate for aerobic oxidation.

# Chapter 24 - Metabolism and Energy

37. T / F  Cytochromes are proteins involved in electron transport.
38. T / F  Cyanide compounds act as poisons by inhibiting cytochrome oxidase ..

**ANSWERS**

| | | | | |
|---|---|---|---|---|
| 1. b | 9. b | 17. e | 25. c | 33. F |
| 2. a | 10. a | 18. c | 26. b | 34. F |
| 3. d | 11. d | 19. d | 27. c | 35. T |
| 4. a | 12. c | 20. d | 28. b | 36. T |
| 5. b | 13. b | 21. a | 29. c | 37. T |
| 6. b | 14. e | 22. d | 30. d | 38. T |
| 7. b | 15. c | 23. a | 31. b | |
| 8. d | 16. d | 24. c | 32. T | |

# 25 Carbohydrate Metabolism

## KEY WORDS

| | | | | |
|---|---|---|---|---|
| *hypoglycemia* | *fermentation* | *cytoplasm* | *hyperglycemia* | *glycogen* |
| *anaerobic* | *second messenger* | *insulin* | *hypoglycemia* | *glucagon* |
| *glycogenolysis* | *glucose tolerance* | *phosphorylase* | *diabetes mellitus* | *Cori cycle* |
| *epinephrine* | *adenylate cyclase* | *acetyl-CoA* | *blood sugar level* | *aerobic* |
| *pyruvate* | *gluconeogenesis* | *thioester* | *galactosemia* | *lactate* |
| *glycogenesis* | *phosphofructokinase* | *glycolysis* | *ATP, ADP, cAMP* | *cortisone* |
| *substrate-level phosphorylation* | *pyruvate dehydrogenase complex (PDC)* | *Krebs cycle* | *renal threshold* | *adrenaline* |

## SUMMARY

25.1 Glycolysis

  A. ***Glycolysis*** = "splitting of sugar." In glycolysis a 6-carbon glucose is converted to two 3-carbon pyruvates. Glycolysis is often referred to as the Embden–Meyerhof pathway.

  B. Summary of key features of ***glycolysis*** (refer to Figure 25.1). All steps take place in the *cytoplasm* of the cell. The splitting of glucose into two pyruvates requires 10 steps.

  C. Phase I – Priming phase (steps 1–5) – Glucose is converted into two molecules of glyceraldehyde-3-P with the investment of two molecules of ATP using two kinases, two isomerases, and aldolase.

   1. Glucose is phosphorylated to glucose-6-phosphate by hexokinase and ATP

   2. An isomerase converts glucose-6-P into fructose-6-P. Note: All common monosaccharides can ultimately be converted to fructose-6-phosphate. Thus the next step is a key regulatory point in the pathway.

   3. ***Phosphofructokinase*** converts fructose-6-P into fructose 1,6-bisphosphate. This key allosteric enzyme is inhibited by high-energy conditions such as high levels of ATP and citrate, and it is stimulated by low-energy conditions such as high levels of ADP. Note: Infants with *galactosemia* cannot tolerate milk because they lack the enzyme necessary to convert galactose (produced from the hydrolysis of milk sugar lactose) into glucose.

   4. Fructose 1,6-bisphosphate (six carbons) is split into two 3-carbon trioses, dihydroxyacetone phosphate (DHAP) and glyceraldehyde-3-P by the enzyme aldolase. This reaction is reversible, and the reverse reaction is an example of an aldol condensation.

   5. Another isomerase converts DHAP into a second molecule of glyceraldehyde-3-P (G-3-P).

  D. Phase II - production of ATP and NADH

   6. Step 6 is the ***substrate-level phosphorylation*** of an aldehyde (G-3-P) to the phosphate ester of a carboxylic acid. The energy released during the oxidation is partially conserved in the new phosphate ester bond that is produced in 1,3-bisphosglycerate. One molecule of NAD+ is consumed, and one molecule of NADH produced for each G-3-P at this step, or two NADHs for the two G-3-Ps.

   7. Step 7: Two ATPs are produced as the phosphate ester bonds are hydrolyzed in each 1,3-bisphosglycerate → 3-phosphoglycerate

   8. Step 8: A mutase converts 3-phosphoglycerate → 2-phosphoglycerate.

   9. Step 9: Enolase converts 2-phosphoglycerate → phosphoenolpyruvate (PEP) in a dehydration reaction.

   10. Step 10: Two more ATPs are produced as the two PEPs are converted to two pyruvates.

  E. Net reaction: Glucose → 2 pyruvate + 2 ATP + 2 NADH.

# Chapter 25 - Carbohydrate Metabolism

25.2 Metabolism of Pyruvate
  A. Pyruvate is a major metabolic intermediate. Its fate depends on the conditions of the cell. It is converted to acetyl-CoA under aerobic conditions; under anaerobic conditions it is converted into lactate (e.g., in muscle cells) or ethanol (e.g., in yeast).
  B. In *aerobic* oxidation of glucose, it is converted to *pyruvate*, which is then converted to *acetyl-CoA*, a high-energy *thioester*.
   1. Using five vitamin-related cofactors (TPP, lipoic acid, FAD, CoA, and NADH), this conversion is carried out by multiple copies of three different enzyme activities in one multienzyme complex, the *pyruvate dehydrogenase complex* (PDC).
   2. In addition to the acetyl-CoA, this step produces one molecule each of $CO_2$ and NADH
   3. The PDC is allosterically regulated by the energy level of the cell. E1 in the complex is turned off when a phosphate group is added by a kinase (regulation by covalent modification).
  C. Under *anaerobic* conditions, pyruvate is reduced to *lactate* to regenerate the $NAD^+$ needed in Step 6 of glycolysis to keep the glycolysis pathway going.
   1. High lactate levels result in muscle "fatigue." Most (70–80%) of the lactate produced in muscle diffuses out and is transported by the blood back to the liver to enter the Krebs cycle or to be remade into new glucose (gluconeogenesis).
   2. The *Cori cycle* describes the storage and utilization of glucose to maintain blood sugar levels.

```
        ┌---- (muscle cells) ---┐                    ┌----- (Liver cells) -----┐
          Glucose   →   Lactate    ← (Blood) →         Lactate   →   Glucose
        └---- (glycolysis) ----┘                    └--(gluconeogenesis)--┘
```

  D. Yeast and some other microorganisms regenerate $NAD^+$ by converting pyruvate → ethanol + $CO_2$. This process is known as *fermentation*. Pyruvate is decarboxylated to acetaldehyde, and the $NAD^+$ required for the pathway is regenerated when acetaldehyde is reduced to ethanol by alcohol dehydrogenase.
  E. Alcohol metabolism – Limited amounts of alcohol can be "detoxified" by the liver. However, this will not protect the fetus of an alcoholic mother, and excessive drinking can cause FAS (fetal alcohol syndrome) and mental retardation.
   1. More than 200,000 people die each year of alcoholism. Alcohol-impaired driving is the leading cause of death for those under 25 years of age.
   2. Chronic alcoholism can be treated with the drug disulfiram (Antabuse), which blocks the conversion of acetaldehyde to acetate. The increased levels of acetaldehyde bring on general discomfort, with nausea, vomiting, and blurred vision. However, acetaldehyde is chemically very similar to formaldehyde, and such treatment should be administered only by a physician.

$$CH_3CH_2 \diagdown \quad \overset{S}{\underset{\|}{\phantom{C}}} \quad \overset{S}{\underset{\|}{\phantom{C}}} \quad \diagup CH_2CH_3$$
$$\phantom{CH_3CH_2} N-C-S-S-C-N$$
$$CH_3CH_2 \diagup \phantom{N-C-S-S-C-N} \diagdown CH_2CH_3$$

Disulfiram (Antabuse)

25.3 ATP Yield From Glycolysis
  A. The energy balance sheet for anaerobic metabolism
   1. (Blood glucose)        Glucose → 2 lactates + 2 ATP
   2. (Fermentation in yeast)   Glucose → 2 ethanols + 2 ATP + 2 $CO_2$
  B. Recall that the complete oxidation of 1 mol of glucose should release 686 kcal of free energy and that 1 mol of ATP conserves about 7.5 kcal; thus anaerobic metabolism taps only ~ 2 to 3 % of the total energy available from the combustion of the glucose molecule. This is the big advantage of going to aerobic metabolism.

# Chapter 25 - Carbohydrate Metabolism

C. The conversion of glucose to $CO_2$ and water involves many oxidative steps and involves several pathways in different locations of the cell. These processes are summarized below.

| | | | |
|---|---|---|---|
| **Glycolysis** | Glucose | → 2 pyruvates + 2 NADH + 2 ATP | = 5–7* ATP |
| **PDC** | 2 pyruvates | → 2 acetyl-CoA + 2 NADH + 2 $CO_2$ | = 5 ATP |
| **Krebs cycle** | 2 acetyl-CoA | → 4 $CO_2$ + 6 NADH + 2 $FADH_2$ + 2 GTP | = 20 ATP |
| | Glucose | → 6 $CO_2$ + 2 ATP + 10 NADH + 2 $FADH_2$ + 2 GTP | = 30–32 ATP |

*Variation is due to the fact that NADH in the cytoplasm may be transported to the ETS in the mitochondria for reoxidation as either NADH (~2.5 ATP equiv.) or $FADH_2$ (~1.5 ATP equiv.).

D. The efficiency of a metabolic pathway can be estimated by comparing the free energy value of all of the ATP molecules synthesized with the free energy released during the oxidation of the foodstuff.

$$C_6H_{12}O_6 + 6\ O_2 + 32\ ADP + 32\ P_i \rightarrow 6\ CO_2 + 32\ ATP + 38\ H_2O$$

For the oxidation of glucose, $\Delta G°' = -686$ kcal/mol in released free energy vs. $-240$ kcal is conserved as new ATP (32 x $-7.5$ kcal/mole ATP) or 35% efficient.

E. Different foods yield different amounts of energy per gram [fats (~9 kcal/g); carbohydrates and proteins (~4 kcal/g)] because the carbon atoms in fats, for example, are in a more highly reduced state than the carbon atoms in carbohydrates or proteins. However, all catabolic pathways are ~35% efficient at storing the energy released as newly synthesized ATP; the rest is lost as heat, which helps keep our bodies warm. When we exercise and metabolism generates excess heat, sweat is used to help dissipate some of the extra heat energy.

25.4 Gluconeogenesis

A. Catabolic and anabolic pathways may seem to be similar to or simply the reverse of one another, but they are usually separate and involve irreversible steps, so that each pathway can be controlled or regulated.

B. The synthesis of glucose from lactate (or ethanol) is not the simple reverse of glycolysis (or fermentation). Steps 1, 3, and 10 in glycolysis are not reversible and thus require other enzymes to be invoked to synthesize new glucose or *gluconeogenesis* (see Figure 25.4).

C. The conversion of pyruvate back to PEP involves several new steps, as shown by the dashed arrows below. This process consumes the equivalent of 2 ATP per pyruvate converted into PEP.

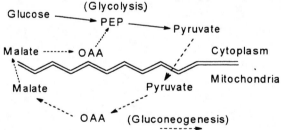

1. Pyruvate carboxylase catalyzes the carboxylation of 3-carbon pyruvate into 4-carbon oxaloacetate (OAA). This is a biotin-dependent enzyme and consumes ATP to provide energy to make the C–C bond.
2. OAA made in the mitochondria makes its way out into the cytoplasm where a GTP-dependent PEP carboxykinase converts OAA into PEP.

D. Steps 3–8 of gluconeogenesis are the reverse of steps 4–9 of glycolysis.

E. Steps 9 and 11 of gluconeogenesis are catalyzed by phosphatases rather than by using the reverse kinase reactions seen in steps 3 and 1 of glycolysis. Note that the removal of the phosphate is energetically favorable, producing inorganic phosphate, but the ATP invested is not returned. These phosphatases (like the kinases) are key complementary, regulatory points in the pathway.

F. The last enzyme, glucose-6-phosphatase, is produced only in liver and kidney cells. Thus gluconeogenesis takes place primarily in liver and kidney tissue.

# Chapter 25 - Carbohydrate Metabolism

G. Summary:   2 pyruvates + 2 GTP + 4 ATP → Glucose + 2 GDP + 4 ADP

25.5 Glycogen Metabolism
 A. Glycogen is the storage form of glucose, composed of chains of D-glucose joined with an α-(1–4) linkage with frequent branches joined by an α-(1–6) linkage. Its structure is similar to that of amylopectin in starch, but more highly branched. Glycogen is stored in the liver (~100 g) and muscle (~350 g) tissues. This does not represent very much stored energy. Additional carbohydrate in the diet is converted to fat, a more efficient form of fuel storage, and stored in adipose tissue. Muscle glycogen is broken down to provide glucose (energy) for active muscles, while liver glycogen is used as a reserve to help maintain proper blood glucose levels.
 B. ATP turnover is very high. There is insufficient ATP in resting muscle to sustain exercise for even a few seconds. A typical human will hydrolyze and resynthesize his own body weight in ATP each day. Creatine kinase catalyzes the reaction of creatine phosphate (stored in muscle tissue) with ADP to regenerate needed ATP until glycolysis can meet the demands. However, even the stored creatine phosphate can provide energy for only about 20 seconds of strenuous muscle activity. For longer periods of work, the body must rely on the metabolism of blood glucose or muscle glycogen.
 C. Glycogen degradation or breakdown (*glycogenolysis*) vs. synthesis (*glycogenesis*) (see Figure 25.5)
  1. *Glycogen phosphorylase* catalyzes the splitting off of a glucose unit as glucose-1-phosphate, which can be converted into glucose-6-phosphate in the glycolytic pathway, thus eliminating the first priming step in glycolysis.
  2. *Glycogen synthetase* catalyzes the addition of "activated" UDP-glucose units to elongate the growing glycogen chain by one glucose unit. UDP-glucose is formed from glucose and UTP.
     Glucose-1-phosphate + UTP → UDP-glucose + $PP_i$.
   The inorganic pyrophosphate formed breaks down into phosphate; thus the cost of activating one glucose is the equivalent of two ATP "high-energy" phosphate bonds.
  3. Glycogen phosphorylase and glycogen synthetase are highly regulated enzyme activities, involving kinases and phosphatases that are hormonally controlled. The regulation is complementary, so synthesis is slowed down when glycogen breakdown is activated and *vice-versa*.

25.6 Regulation of Carbohydrate Metabolism
 A. Blood glucose levels:
  1. After sugars enter the bloodstream, they are carried to the liver where they are phosphorylated. The liver helps regulate the *blood sugar level*.
   a. Normal: 80–100 mg of glucose per 100 mL of blood (or ~5–6 g, or 1 tsp., of total blood glucose in the body).
   b. *Hypoglycemia* – low blood sugar levels; excess insulin.
   c. *Hyperglycemia* – high blood sugar levels; lack of insulin, starvation.
  2. The brain uses ~120g of glucose per day compared to ~ 200g for the remainder of the body when at rest. Thus glucose must be supplied to the blood between meals.
  3. The liver helps maintain glucose levels by releasing glucose to the blood or absorbing excess glucose and converting it into glycogen or fat.
  4. The kidneys will excrete glucose into the urine when the blood glucose levels exceed the "*renal threshold*" of ~ 160 mg/100 mL blood. A *glucose tolerance test* is used to diagnose *diabetes mellitus*, a major cause of hypoglycemia.
 B. Glycolysis and gluconeogenesis are regulated pathways. As noted above, the interconversion of fructose-6-P to fructose-1,6-bisP is catalyzed by a kinase (glycolysis) or a phosphatase (gluconeogenesis). The kinase is inhibited by high-energy conditions such as high levels of ATP or

# Chapter 25 - Carbohydrate Metabolism

citrate and activated by low-energy conditions such as high levels of AMP. The phosphatase is inhibited by low energy conditons.

C. Hormonal regulation of blood sugar level (insulin / epinephrine / glucagon)
  1. Blood glucose is regulated primarily by hormones that control the synthesis or breakdown of glycogen in the liver.
  2. *Insulin*, a protein hormone secreted by the pancreas, promotes anabolic processes and lowers blood sugar levels by increasing the uptake and utilization of glucose, enhancing glycogenesis and suppressing gluconeogenesis, synthesis of new glucose.
  3. The hormones *glucagon* and *epinephrine* act to increase blood sugar levels by increasing the rate of glycogen breakdown into glucose. They are "primary messengers" that bind to their respective receptors, activating the enzyme *adenyl cyclase* to convert *ATP* → *cAMP*. The cAMP serves as a "*second messenger*" inside the cell that initiates a cascade of events that leads to the breakdown of glycogen to form glucose-1-phosphate.

$$ATP \longrightarrow PP_i + cAMP$$

  a. *Glucagon* is a polypeptide (29-mer) hormone that is also produced in the pancreas.
  b. *Epinephrine* (*adrenaline*) is the "fight-or-flight" hormone produced by the medulla of the adrenal glands.
  4. *Cortisone* and cortisol from the adrenal cortex are steroidal hormones that stimulate the synthesis of glucose from amino acids.
  5. The disease *diabetes mellitus* can result from faulty insulin production, its release, or lack of sufficient receptors.
    a. Type I, or insulin-dependent, diabetes is caused by a lack of insulin and is treated with daily insulin injections. Lack of insulin can lead to severe acidosis and diabetic coma.
    b. Type II (non-insulin-dependent) diabetes is more common, occurring later in life and resulting from beta cells not secreting enough insulin. Oral drugs can be taken to stimulate insulin release or sensitivity.

Chlorpropamide (Diabinese)

## DISCUSSION

In the aerobic oxidation of glucose, glucose is first converted to pyruvate. The oxidation of pyruvate to $CO_2$ involves the pyruvate dehydrogenase complex and the Krebs cycle, which produce several molecules of the reduced coenzymes NADH and $FADH_2$ (Chapter 24).

Let's take another look at the glycolysis or the Embden–Meyerhof pathway and compare it to the Krebs cycle, noting overall patterns that might make each of these metabolic pathways easier to comprehend. First, remember that both of these pathways are designed to produce energy, that is, to yield ATP molecules. Second, there is an obvious difference between the two series of reactions – one is cyclic

# Chapter 25 - Carbohydrate Metabolism

(the Krebs cycle) and the other is not (the Embden–Meyerhof pathway). The latter starts with glucose and ends with lactic acid. The former starts and ends with oxaloacetate.

Now let's focus on the Embden–Meyerhof pathway (Figure 25.1). Phosphate groups are added to sugar molecules, which split in two and pick up more phosphate until, finally, a high-energy phosphate (1,3-diphosphoglyceric acid) is formed. The beauty of this compound lies in its ability to transfer a phosphate group to ADP. That is its function. After the transfer, the remaining compound rearranges a bit to become another high-energy phosphate, phosphoenol pyruvate or PEP. PEP is also able to transfer phosphate to ADP. That leaves pyruvate, which can be reduced to lactate or fed into the Krebs cycle. In the Embden–Meyerhof pathway, a sugar derivative is oxidized at Step 6 and $NAD^+$ is reduced to NADH, but in fermentation the pyruvate is reduced and NADH is oxidized to yield $NAD^+$. Thus, there is **no net oxidation** or reduction in fermentation.

Below is a summary of the Embden–Meyerhof pathway, showing only glucose and its products. See if the comments give you a sense of the direction of the reactions, a sense of a grand design in which everything is done for a purpose. Also, you might notice again how the sacrifice of two ATP molecules during the early "priming" stage prepares the way for the synthesis later of four ATP molecules. (These systems reflect an old rule of business – you must sometimes spend money to make money.)

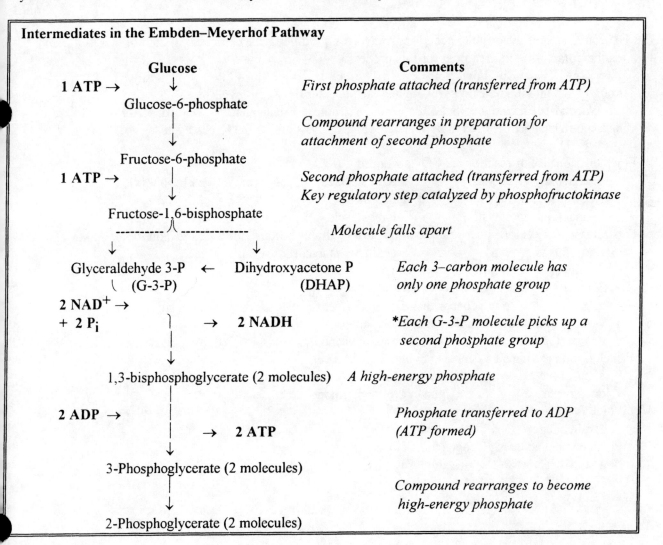

# Chapter 25 - Carbohydrate Metabolism

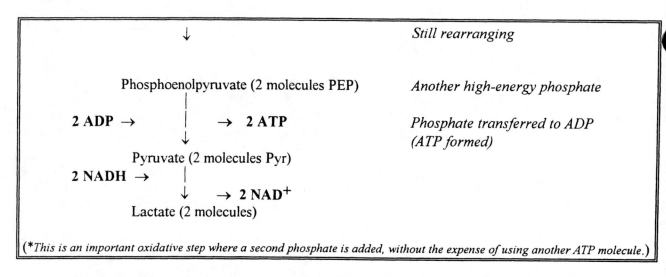

(*This is an important oxidative step where a second phosphate is added, without the expense of using another ATP molecule.)

Chapter 25 also considers the regulation of blood glucose levels. These topics are reviewed here in the **SELF-TEST** and in the end-of-chapter problems.

## SELF-TEST

1. The conversion of pyruvate to acetyl-CoA involves a multienzyme complex. Which of these vitamin related-cofactors is <u>not</u> required for this process?
    a. NADH    b. lipoic acid    c. FAD    d. ATP    e. thymine PP
2. Through what intermediate do pyruvate and lactate enter the Krebs cycle?
    a. acetyl-coenzyme A    b. ADP    c. fructose-6-phosphate    d. $NAD^+$
3. Which oxidizing agent is used in both anaerobic glycolysis and in the Krebs cycle?
    a. FAD    b. $NAD^+$    c. $O_2$
4. In which sequence of reactions is $CO_2$ produced?
    a. anaerobic glycolysis    b. Krebs cycle    c. oxidative phosphorylation
5. In which sequence of reactions is ATP <u>not</u> formed?
    a. anaerobic glycolysis    b. Krebs cycle    c. oxidative phosphorylation
6. Fats release more calories per gram than do carbohydrates because the average oxidation number of carbon in fats is _____ than it is in carbohydrates.
    a. higher    b. lower
7. The Embden–Meyerhof pathway is also known as:
    a. glycolysis    b. glyconeogenesis    c. Krebs cycle    d. oxidative phosphorylation
8. An overdose of insulin produces a condition called:
    a. galactosemia    b. hyperglycemia    c. hypoglycemia
9. Which hormone triggers a decrease in blood sugar levels?
    a. cortisone    b. glucagon    c. insulin
    d. epinephrine    e. human growth hormone
10. Which is <u>not</u> a pancreatic hormone?
    a. adrenaline    b. glucagon    c. insulin
11. Which compound does glycogen most closely resemble?
    a. amylopectin    b. amylose    c. cellulose    d. glucose

## Chapter 25 - Carbohydrate Metabolism

12. The Cori cycle describes:
    a. the relationship between glycogenesis and glycogenolysis
    b. the conversion of acetic acid to carbon dioxide
    c. the interconversion of monosaccharides
13. Which compound serves as the intermediate through which all of the monosaccharides enter the Embden–Meyerhof pathway?
    a. galactose     b. fructose-6-phosphate     c. glucose-1-phosphate     d. pyruvate
14. Normal blood glucose levels are about:
    a. 10 g/L     b. 100 g/L     c. 100 mg/100 mL     d. 240 mg/L     e. 100 mg/mL
15. The molecule described as the "second messenger" is:
    a. ATP     b. adrenaline     c. glycogen     d. cAMP     e. glucagon
16. What is the end product of aerobic glycolysis?
    a. acetate     b. ethyl alcohol     c. glycogen     d. lactate     e. pyruvate
17. Which high-energy phosphate serves as the <u>direct</u> source of energy in muscle contraction?
    a. ATP     b. creatine phosphate     c. PEP
18. The key regulatory enzyme of the glycolytic pathway is the allosteric enzyme:
    a. glycokinase     b. aldolase     c. LDH
    d. phosphofructokinase     e. glucomutase
19. In fermentation, pyruvate is converted to _____ and then to ethanol.
    a. acetaldehyde     b. lactate     c. acetate     d. fructose
20. For every mole of glucose degraded to pyruvate, _____ moles of ATP and _____ moles of NADH are produced.
    a. 2, 2     b. 2, 1     c. 2, 0     d. 5, 0     e. 4, 1
21. The biggest difference between glycolysis and gluconeogenesis is the step(s) that re-form:
    a. fructose-6-P     b. pyruvate     c. PEP     d. glucose-6-P     e. glyceraldehyde-3-P
22. The only oxidative step in glycolysis occurs with the oxidation of:
    a. fructose-6-P     b. pyruvate     c. PEP     d. glucose-6-P     e. glyceraldehyde-3-P
23. T / F   Glucose is stored in the body in the form of glycogen.
24. T / F   The glucose tolerance test is used to diagnose galactosemia.
25. T / F   There are no redox reactions in the Embden–Meyerhof pathway.
26. T / F   In periods of normal (not strenuous) activity, energy is supplied to muscles through aerobic pathways.
27. T / F   While the Embden–Meyerhof pathway supplies the ATP required for muscular activity, the body incurs an oxygen debt.

## ANSWERS

| | | | | | |
|---|---|---|---|---|---|
| 1. d | 6. a | 11. a | 16. e | 21. c | 26. T |
| 2. a | 7. a | 12. a | 17. a | 22. e | 27. T |
| 3. b | 8. c | 13. b | 18. d | 23. T | |
| 4. b | 9. c | 14. c | 19. a | 24. F | |
| 5. b | 10. a | 15. d | 20. a | 25. F | |

# 26 Lipid Metabolism

**KEY WORDS**

| | | | |
|---|---|---|---|
| *adipocyte* | *fatty acid spiral* | *ketone bodies* | *adipose tissue* |
| *ketosis* | *obesity* | *β-oxidation* | *malonyl~SCoA* |
| *acidosis* | *lipid storage disease* | *acetyl–CoA* | *acetone* |
| *DHAP* | *fatty acid synthetase* | *triglycerides* | *cholesterol* |
| *lipases* | *essential fatty acid* | *body mass index* | *set–point theory* |

**SUMMARY**

The American Heart Association recommends that no more than 30% of the total calories in the diet be provided by lipids. Fatty acids in triacylglycerides can be saturated or unsaturated. The diet should contain 4–6 g of the "essential," polyunsaturated fatty acids like linoleic and linolenic acids.

26.1 Storage and Mobilization of Fats
  A. Fats are excellent fuels. Energy is supplied by the oxidation of carbohydrates and lipids. Lipids are more highly reduced than carbohydrates (or proteins) and thus release more energy when oxidized to $CO_2$ and $H_2O$ (~9.5 kcal/g vs. 4.2 kcal/g). Lipids can also be stored more efficiently with less water of hydration; thus lipids are the body's primary energy reserve.
    1. The average human carries a 20–30 day energy reserve as fat versus a 1–2 day reserve as glycogen.
    2. A camel's hump is mostly fat (adipose tissue).
  B. Fat is stored throughout the body in fat cells, or *adipocytes*, in a special kind of connective tissue called *adipose tissue*. Fats serve as a protective cushion around organs, help insulate against temperature changes, and are the body's most efficient form of energy storage. Phospholipids are important components of cell membranes.
  C. *Triglycerides* are stored as fat droplets in adipose tissue. Excess dietary carbohydrates are first used to replenish liver and muscle glycogen, and any remaining excess is then converted to fats for efficient storage.
  D. "Brown fat" contains specialized mitochondria that oxidize fat to produce heat instead of energy. This is important for hibernating animals and newborn infants.
  E. Mobilization of fats is regulated by hormones in a manner similar to that discussed for glycogen in Chapter 25. Epinephrine and glucagon, activated by low blood sugar levels indicative of an energy poor-state, bind to receptor proteins on adipose tissue.
    1. Hormonal signal → receptor binding → activation of adenylate cyclase → cAMP
    2. cAMP activates a kinase that turns on *lipases* and promotes release of fatty acids from adipose tissue.
    3. Lipases hydrolyze triglycerides, releasing free fatty acids and glycerol.
        triglycerides → free fatty acids + glycerol
  F. Glycerol metabolism
    1. Glycerol is transported to the liver, where it is converted in 2 steps to dihydroxyacetone phosphate *(DHAP)*.   Glycerol + ATP → Glycerol-3-phosphate + ADP
        Glycerol-3-phosphate + $NAD^+$ → DHAP + NADH

    2. DHAP follows the glycolytic pathway (DHAP → Pyr → PDC → TCA cycle).

# Chapter 26 - Lipid Metabolism

3. Summary: Glycerol to $CO_2$ (= 17.5 mol ATP)

    Glycerol → Glycerol-3-P → DHAP → Pyruvate → Acetyl-CoA → $CO_2$
    (−1 ATP)   (+1 NADH = 2 ATP)  (+1 NADH +2 ATP = 4 ATP)  (+1 NADH = 2.5 ATP)  (+10 ATP)

## 26.2 Fatty Acid Oxidation

A. Activation of the fatty acid chain. The free fatty acids react with ATP and then coenzyme A to form a fatty acyl–CoA and AMP.

    i) Fatty acid + ATP → fatty acyl adenylate + PP (→ 2 $P_i$)
    ii) Fatty acyl adenylate + HSCoA → fatty acyl~CoA + AMP

  1. These processes take place in the cytoplasm.
  2. The fatty acyl–CoA cannot cross the mitochonrial membrane. This requires a small carrier molecule, carnitine.

B. The *fatty acid spiral* (see Figure 26.4) or *β-oxidation* takes place within the mitochondria. The pathway repeats four steps in a spiral fashion: oxidation/hydration/oxidation/cleavage.

  1. Oxidation as dehydrogenation to produce a double bond and $FADH_2$
  2. Hydration of the double bond to produce a β-hydroxy fatty acyl–CoA
  3. β-oxidation of the hydroxyl group to form a β-keto ester of CoA and 1 NADH
  4. Attack at the β carbonyl C to cleave off a two carbon *acetyl~CoA* unit and to form a new shortened molecule of fatty acyl–CoA. If the resulting fatty acyl group has more than four carbons, go back to step 1. During the final pass through the cycle, the cleavage of the four carbon β-keto ester produces two molecules of acetyl–CoA (see summary of catabolic pathways). Note: Acetyl–CoA was also formed from glucose metabolism. It is important to note that excess carbohydrate is converted to fat, but that humans cannot use fat to synthesize carbohydrates.

## 26.3 ATP Yield From Fatty Acid Oxidation

A. The oxidation of fats produces approximately 9.5 kcal/g, twice as much per gram as the oxidation of carbohydrate or protein.

B. A 16–carbon palmitate fatty acyl–CoA requires 7 turns of the *fatty acid spiral* to produce:

    8 Acetyl–CoA    x 10    =    80 ATP
    7 $FADH_2$        x 1.5    =    10.5 ATP
    7 NADH        x 2.5    =    <u>17.5 ATP</u>
                                             108 ATP
    Activation costs (ATP → AMP)    <u>−2</u>
                                             106 ATP  Net yield

C. Consider the ATP yield for a triglyceride such as glycerol tripalmitate.
    1 glycerol + 3 palmitates = ~17.5 ATP + 3 x 106 ATP = ~ 335.5 ATP !!!!

## 26.4 Ketosis

A. The *ketone bodies* are acetoacetic acid, β-hydroxybutyric acid, and *acetone*. Acetoacetic acid is synthesized in the liver from 2 molecules of acetyl–CoA, and the other two ketone bodies are produced from acetoacetic acid.

B. Low levels of ketone bodies are normal components of the blood and are used as energy sources.

C. High levels of ketone bodies (or *ketosis*) can occur during starvation or illnesses such as diabetes mellitus.

D. Two of the ketone bodies are acids. Uncontrolled ketosis can lead to *acidosis* and death. Short oxygen supply and dehydration are symptoms of acidosis. Acidic blood cannot transport oxygen very well, and the kidneys eliminate lots of fluids, trying to remove the excess acids.

# Chapter 26 - Lipid Metabolism

26.5 Fatty Acid Synthesis
   A. The biosynthesis of fatty acids accomplishes the reverse of the oxidative steps employed during β-oxidation. However, fatty acid synthesis is catalyzed by a *fatty acid synthetase* complex in the cytoplasm.
   B. Malonyl–CoA and fatty acid synthesis
      1. The carbon source for fatty acid synthesis is acetyl–CoA, but only one molecule of acetyl–CoA is used directly during the synthesis. The others are activated by a biotin-dependent carboxylation reaction that forms *malonyl–CoA*.
      2. Malonyl–CoA condenses with the enzyme-bound acetyl–E to form acetoacetyl–E plus carbon dioxide. The carbon dioxide serves only to activate the alpha carbon position for the condensation reaction.
      3. The β-keto acetoacetyl–E is then reduced to the β-hydroxy compound, dehydrated to give the double bond, and hydrogenated to produce a saturated hydrocarbon chain.
      4. The growing fatty acyl–E is now ready for another cycle by undergoing condensation with another molecule of malonyl–CoA.
      5. Each repetition of the cycle adds two carbons to the length of the chain.

26.6 Obesity, Exercise, and Diets
   A. If you eat more food than you need, you gain weight. Obesity is an increasing problem in the developed nations. *Obesity* is the condition that exists when excessive fat is stored in adipose tissue and the individual becomes overweight (20% > ideal weight). Foods (carbohydrate, fat, or protein) eaten in excess of immediate energy requirements are stored as fat. This puts an extra load on the heart and is a major factor in diseases of the heart and circulatory system.
   B. Crash diets are often deficient in essential nutrients and can be harmful. Reducing the caloric intake by just 100 cal/day can result in the loss of a pound of fat (3500 cal) in 35 days. A person who is moderately active needs ~15 kcal/lb (e.g., 160 lb = 2400 kcal)
   C. Burning Calories with exercise:

   | **Activity:** | Walking | Bicycling | Basketball | Swimming | Mowing |
   |---|---|---|---|---|---|
   | **Kcal/hr :** | 420 | 420–600 | 360–660 | 360–750 | 450 |

   D. "*Set–Point theory*" postulates that the hypothalamus senses when the level of fatty acids in the blood drops below some "set" level, and that triggers the hunger response. Exercise may help lower this "set-point."
   E. The *body mass index* (BMI) is between 20–25 for most adults, BMI over 40 signals extreme obesity.

$$\text{BMI} = 700 \times \frac{\text{Body wt. (lb)}}{(\text{Height, in.})^2}$$

**Note: Optional Material** – Some courses will include the biosynthesis of other lipids presented below.

26.7 Biosynthesis of Fats
   A. Fats are synthesized from fatty acids and glycerol. The acetyl~CoA needed to synthesize fatty acids can come from the degradation of carbohydrates, proteins, or lipids. The same is true for glycerol, so an excess of any of the major foodstuffs can be converted to fat.
   B. Synthesis of triglycerides
      1. Two molecules of activated fatty acyl–CoA react with glycerol phosphate to produce phosphatidic acid.
      2. A phosphatase removes the phosphate to produce a diglyceride.

# Chapter 26 - Lipid Metabolism

3. The diglyceride reacts with another molecule of fatty acyl–CoA to complete the synthesis of the triglyceride.

26.8 Biosynthesis of Phospholipids
   A. Phospholipids are synthesized from a diglyceride and an activated form of the organophosphate group.
   B. In the biosynthesis of phosphoethanolamine, a cytidine nucleotide acts as the activator and carrier. CDP–ethanolamine reacts with the diacylglycerol to form the phosphatidylethanolamine.
   C. The phosphoethanolamine can be methylated to form lecithin, or the ethanolamine can be exchanged with serine to form cephalin.

26.9 Cholesterol and Some Other Steroids
   A. *Cholesterol* and other steroids are nonsaponifiable lipids.
   B. Cholesterol is synthesized from acetyl–CoA. Other steroids can be synthesized from cholesterol.

26.10 Lipid Storage Diseases
   A. Glycolipids are important components of brain and nerve tissue. Abnormal accumulation of these compounds can produce mental disorders and neurological problems.
   B. *Lipid-storage diseases* are genetic diseases caused by defective enzymes. Tay-Sachs disease, Gauchers disease, and Niemann-Pick disease are examples.

# DISCUSSION

We emphasized in Chapter 26 that fat metabolism cannot be divorced from the metabolism of carbohydrates (also see the summary figure on page 267). What may not have been quite so obvious is the similarity between the chemistry of the fatty acid cycle and that of the Krebs cycle. Since metabolic pathways may strike students as endless collections of unrelated reactions, we emphasize patterns and relationships whenever we can.

Let's recall some significant features of the Krebs cycle. During the first half of the Krebs cycle, a molecule that will release two carbon dioxide units is synthesized. In the latter part of the cycle, the remaining molecule is converted back to the original starting material. Here's a summary of these later reactions.

Krebs cycle (steps 6, 7, and 8, p. 686 of the text)

$$HOOC-CH_2-CH_2-C(=O)OH \xrightarrow{FAD \to FADH_2} HOOC-CH=CH-C(=O)OH \xrightarrow{H_2O} HOOC-CH(OH)-CH_2-C(=O)OH \xrightarrow{NAD^+ \to NADH + H^+} HOOC-C(=O)-CH_2-C(=O)OH$$

Now look at the analogous reactions from the fatty acid cycle.

Fatty acid cycle (excluding final step)

$$R-CH_2-CH_2-C(=O)SCoA \xrightarrow{FAD \to FADH_2} R-CH=CH-C(=O)SCoA \xrightarrow{H_2O} R-CH(OH)-CH_2-C(=O)SCoA \xrightarrow{NAD^+ \to NADH + H^+} R-C(=O)-CH_2-C(=O)SCoA$$

# Chapter 26 - Lipid Metabolism

Except that the fatty acid reacts in the form of its thioester, the reaction sequence is the same. Two saturated carbons are unsaturated, then the components of water are added, and finally the hydroxyl group is oxidized to a carbonyl group. In both cases, it is the carbonyl group that reacts in the next step. In the Krebs cycle, the carbonyl group adds acetyl coenzyme A. In the fatty acid cycle, the carbonyl group releases acetyl coenzyme A.

$$\text{Krebs Cycle: } HOOC-\overset{O}{\underset{\|}{C}}-CH_2-\overset{O}{\underset{\|}{C}}-OH + CH_3-\overset{O}{\underset{\|}{C}}-SCoA + H_2O \longrightarrow HOOC-\overset{OH}{\underset{|}{C}}(CH_2-\overset{O}{\underset{\|}{C}}-OH)-CH_2-\overset{O}{\underset{\|}{C}}-OH + CoASH$$

$$\text{Fatty Acid Cycle: } R-\overset{O}{\underset{\|}{C}}-CH_2-\overset{O}{\underset{\|}{C}}-SCoA + CoA\sim SH \longrightarrow R-\overset{O}{\underset{\|}{C}}-SCoA + CH_3-\overset{O}{\underset{\|}{C}}-SCoA$$

Just as the Krebs cycle yields reduced species ($FADH_2$ and NADH) for the respiratory chain, so does the fatty acid cycle. Thus both of these processes generate ATP indirectly through oxidative phosphorylation. (Remember that the fatty acid cycle not only supplies $FADH_2$ and NADH formed in every turn of the cycle, but it also produces acetyl-coenzyme A. Acetyl-CoA then feeds into the Krebs cycle where it generates additional $FADH_2$ and NADH.)

The biosynthesis of fatty acids occurs through a near reversal of the fatty acid cycle. By "near" we mean that, except for the initial step in each turn of the reverse cycle, the sequence of reactions is essentially the same.

$$\overset{O}{\underset{\|}{C}}-CH_2-\overset{O}{\underset{\|}{C}} \longrightarrow \overset{HO}{\underset{|}{C}}H-CH_2-\overset{O}{\underset{\|}{C}} \longrightarrow -CH=CH_2-\overset{O}{\underset{\|}{C}} \longrightarrow -CH_2-CH_2-\overset{O}{\underset{\|}{C}}$$

Start with carbonyl                                                               End with saturated carbon

However, there are many differences in the way these reaction steps are carried out. Beta oxidation takes place in the mitochondria, while fatty acid biosynthesis occurs in the cytoplasm where the growing fatty acid molecule reacts as an enzyme complex (rather than as a coenzyme A ester). Also, whereas each two–carbon would be added in the form of acetyl-coenzyme A in a direct reversal, it is malonyl~CoA that adds (with the loss of carbon dioxide) in the near reversal. In the near reversal, the reducing agent is NADPH; whereas in a direct reversal, NADH and $FADH_2$ would be used. Finally, remember that phosphate (glycolysis) or coenzyme A, when attached to a molecule, serves to activate that molecule. In this sense, these groups play similar roles.

The rest of the material in Chapter 26 is reviewed in the problems at the end of the chapter and in the **SELF-TEST**.

# Chapter 26 - Lipid Metabolism

## SELF-TEST

1. The oxidation of 1 g of fat to carbon dioxide yields about:
    a. one calorie     b. four kilocalories     c. nine calories     d. nine kilocalories
2. The average percent body fat for an adult female is about _____ %.
    a. 5     b. 10     c. 15     d. 20     e. 25
3. Which is not true of fats?
    a. They supply more energy per gram than carbohydrates.
    b. They can be stored with less water than carbohydrates.
    c. They can be mobilized more quickly than carbohydrates.
    d. All of the above statements are true.
4. The lymphatic system transports fats:
    a. from the small intestine to the blood     b. from the small intestine to the cells
    c. from the blood to the cells
5. Normal fasting level for total cholesterol in blood plasma is about:
    a. 200 g/L     b. 10–20 mg/100 mL     c. 120 g/L     d. 120–250 mg/100 mL     e. 2 mg/100 mL
6. How is fat stored in the adipose tissue?
    a. as droplets of fat within cells     b. as bilayers forming cell membranes
    c. complexed with protein molecules
7. Which energy reserve is used first?
    a. depot fat     b. fat stored in the liver     c. glycogen
8. The CoA-thioesters of fatty acids need the help of _____ to cross the mitochondrial membrane.
    a. NADH     b. ATP     c. glucose     d. insulin     e. carnitine
9. Fatty acids that enter the fatty acid cycle are first activated by their conversion to:
    a. coenzyme-A thioesters     b. enzyme complexes     c. phosphate esters
10. Which is the product of β-oxidation?
    a. R–CH=CH–C(=O)–SCoA     b. R–CH(OH)–CH$_2$–C(=O)–SCoA     c. R–C(=O)–CH$_2$–C(=O)–SCoA
11. Which oxidizing agent is employed in the fatty acid cycle?
    a. FADH$_2$     b. NAD$^+$     c. NADP$^+$     d. O$_2$
12. What is the end product obtained from the fatty acid in the fatty acid cycle?
    a. acetyl-coenzyme A     b. dihydroxyacetone phosphate     c. malonyl-coenzyme A
13. If 8 FADH$_2$ and 8 NADH molecules enter the respiratory chain, how many ATP molecules can be produced?
    a. 11     b. 22     c. 27     d. 32     e. 40
14. Which category of foodstuff cannot supply acetyl–coenzyme A for the biosynthesis of fatty acids?
    a. carbohydrates     b. fats     c. proteins     d. All 3 types supply acetyl–CoA.
15. Which is not a path entered by acetyl-coenzyme A?
    a. synthesis of steroids     b. Krebs cycle     c. fatty acid synthesis
    d. formation of ketone bodies     e. glycogen synthesis
    f. Acetyl-coenzyme A follows all of the above routes.
16. Before fat biosynthesis begins, glycerol must be activated in the form of:
    a. its phosphate ester     b. its coenzyme A thioester
    c. UDP–glycerol     d. CDP–glycerol
17. For fat biosynthesis, fatty acids are activated in the form of:
    a. their phosphate esters     b. their coenzyme A thioesters
    c. UDP–acid     d. CDP–acid
18. Fast walking burns about _____ kcal/hr.
    a. 100     b. 200     c. 400     d. 800     e. 1000

# Chapter 26 - Lipid Metabolism

19. In phospholipid biosynthesis the cytidine nucleotide does which of the following?
    a. catalyzes the conversion of ethanolamine to choline
    b. acts as a carrier molecule or activator
    c. picks up a phosphate unit from ATP
20. The ketone bodies do not include:
    a. acetoacetic acid   b. acetone   c. β-hydroxybutyric acid   d. pyruvic acid
21. High concentrations of ketone bodies in the blood:
    a. cause diabetes mellitus      b. are a symptom of starvation
    c. increase the pH of the blood   d. All of the above are correct.
22. How many NADHs are produced during the β-oxidation of an 18–carbon fatty acid?
    a. 20       b. 10       c. 9       d. 8       e. 5
23. The initial activation of free fatty acids occurs in the _____ of the cell.
    a. mitochondria      b. cytoplasm      c. nucleus
24. The starting material for cholesterol biosynthesis is:
    a. glucose      b. oxaloacetate      c. pyruvate      d. acetate      e. lactate
25. In a person suffering from diabetes mellitus, which of the following does not lead to an increase in the production of ketone bodies?
    a. conversion of body tissues to fat metabolism
    b. reliance on gluconeogenesis for glucose required by cells
    c. breakdown of insulin to fatty acids
    d. All of the above lead to increased production of ketone bodies.
26. T / F  Fats are transported in the blood primarily as lipoproteins.
27. T / F  Blood lipid levels are not affected by normal body processes and remain relatively constant under most conditions.
28. T / F  The glycerol obtained from the hydrolysis of fats is fed into the Embden-Meyerhof pathway as dihydroxyacetone phosphate.
29. T / F  Condensation of malonyl-coenzyme A with a growing fatty acid chain produces fatty acids with an odd number of carbon atoms.
30. T / F  In phospholipid biosynthesis in humans, the cytidine nucleotide usually activates the diglyceride molecule.
31. T / F  Some ketone bodies are excreted in the urine of healthy individuals.
32. T / F  "Air hunger" accompanies acidosis because the ability of the blood to transport oxygen decreases with decreasing pH.
33. T / F  Most obesity is the result of glandular malfunction.
34. T / F  The lipid-storage diseases, such as Niemann-Pick disease, produce arteriosclerosis.
35. T / F  The oxidative step in β-oxidation is the formation of the β-alcohol.

**ANSWERS**

| | | | |
|---|---|---|---|
| 1. d | 11. b | 21. b | 31. T |
| 2. e | 12. a | 22. d | 32. T |
| 3. c | 13. e | 23. b | 33. F |
| 4. a | 14. d | 24. d | 34. F |
| 5. d | 15. e | 25. c | 35. F |
| 6. a | 16. a | 26. T | |
| 7. c | 17. b | 27. F | |
| 8. e | 18. c | 28. T | |
| 9. a | 19. b | 29. F | |
| 10. c | 20. d | 30. F | |

# 27 Protein Metabolism

## KEY WORDS

amino acid pool
catabolism
anabolism
nitrogen balance
carbamyl phosphate
turn over
urea cycle

essential amino acid.
nonessential amino acid
transamination
oxidative deamination
glutamine
ammonia
antihistamines

decarboxylases
gluconeogenesis
glutamate
urea cycle
arginine
glucogenic
α-ketoglutarate

ornithine
uric acid
gout
kwashiorkor
starvation
ketogenic

## SUMMARY

27.1 Overview of Amino Acid Metabolism
  A. Amino acids are *actively transported* across the intestinal wall and carried to the liver. Ingested protein satisfies two needs.
    1. Replaces nitrogen (N) eliminated as urea.
    2. Supplies essential amino acids that we are unable to synthesize.
  B. *Nitrogen balance* = $N_{in}$ minus $N_{out}$. Unlike fats and carbohydrates, we do not have a storage macromolecule for amino acids; rather there is a circulating *amino acid pool*.
    1. Positive nitrogen balance occcurs during growth and pregnancy.
    2. Negative nitrogen balance results from fasting, starvation, fever, and diet lacking essential amino acids.
  C. Protein molecules are constantly being broken down *(catabolism)* and resynthesized *(anabolism)* in the body. Liver proteins *turn over* every few days, while collagen molecules last for a few years.
  D. *Essential* amino acids are those that cannot be synthesized and need to be supplied in the diet (Lys, His, Met, Arg, Thr, Leu, Val, Ile, Phe, Trp). We need well–balanced (complete) protein that can supply all the needed essential amino acids, and not just protein (incomplete protein) in the diet. Plant protein is often "incomplete protein," lacking one or more of the essential amino acids in sufficient amounts. Eating a combination of complementary vegetables can provide an adequate diet.

27.2 Catabolism of Amino Acids
  A. Proteins we eat are hydrolyzed to amino acids, which become part of the amino acid pool. These amino acids can be used to synthesize new proteins. The nitrogen can be removed and the carbon skeleton metabolized for energy production or stored as glycogen and fat.
  B. Catabolism of amino acids – removal and excretion of nitrogen
    1. Transamination: The amino group removed is transferred to an □-keto acid (commonly this is □-ketoglutarate) to produce the corresponding amino acid (e.g., glutamate) and the new □-keto acid. Transaminases require pyridoxal phosphate (vit. B6) as a cofactor. The human body contains about 100 g of free amino acids, with about half of that coming from Glu and Gln.

    Alanine     + *α-ketoglutarate* → Pyruvate      + *Glutamate*
    Aspartate   + *α-ketoglutarate* → Oxaloacetate  + *Glutamate*
    etc.

# Chapter 27 - Protein Metabolism

    2. ***Oxidative deamination***: The glutamic acid formed during the transamination reactions can be converted back to α-ketoglutarate by oxidative deamination. The reaction occurs primarily in liver mitochondria. The $NH_4^+$ can be converted to urea for excretion.
        ***Glutamate***     [O] →    ***α-ketoglutarate*** + ammonia
    3. Carbon skeletons: After the nitrogen is removed from amino acids via transamination, the ketoacid carbon skeleton that remains is catabolized to a TCA cycle intermediate (see Fig. 27.3). This conversion is unique for each amino acid, involving from one to many steps.
  C. ***Gluconeogenesis*** is the synthesis of new glucose from glycerol or from amino acid carbon skeletons. This process requires six high-energy phosphates per glucose molecule synthesized. Although expensive, this process helps keep the brain fueled with glucose. Amino acids that can form intermediates of carbohydrate metabolism to synthesize glucose are referred to as ***glucogenic*** amino acids (most amino acids are glucogenic). Amimo acids that give rise to acetyl-CoA or acetylacetyl-CoA are called ***ketogenic*** amino acids (Leu).

27.3 Storage and Excretion of Nitrogen
  A. Carbohydrates can be stored in the body as glycogen, and the fats as triglycerides in the fat depots. Proteins are not stored, but a limited supply of amino acids circulates in the bloodstream and is called the ***amino acid pool.***
  B. ***Glutamine*** is a high nitrogen compound present in many tissues and in the blood for the temporary storage of N. Ammonia can also be removed from tissues by glutamine synthetase to form Gln:
        Glutamate   + $NH_4^+$ + ATP   →     Glutamine   +   ADP   +   water
  C. N in Gln can be donated and serve as the N source in the synthesis of purines, pyrimidines, etc.
  D. The excess ammonium ion is converted to urea and excreted. Levels of $NH_3$ greater than 5 mg/100 mL of blood are toxic to humans. Ammonia levels are normally kept at 1–3 µg/100 mL blood. Excess N must be excreted.
    1. Vertebrates excrete excess N as ***urea***.
    2. Birds and reptiles excrete excess N as ***uric acid***.
    3. Marine organisms (fish) excrete excess N as free ***ammonia***.
  E. Over 80% of the N from protein catabolism is excreted by the kidneys as urea. Urea is made primarily in the liver. The ***urea cycle*** (also deduced by Krebs) describes how N from amino acid ***catabolism*** is converted to urea. The kidneys filter about 100 L of blood each day, excreting approximately 30 g of urea in about 1.5 L of urine.
    1. Excess nitrogen is transferred to α-ketoglutarate to produce glutamate or to oxaloacetate to produce aspartate.
    2. $NH_3$ is produced by the oxidative deamination of α-ketoglutarate.
    3. $NH_3$ is activated by ATP and $CO_2$ to produce ***carbamyl phosphate***.
    4. Carbamyl phosphate condenses with ***ornithine*** to form citrulline.
    5. Citrulline condenses with aspartate to form arginosuccinate, which splits off fumarate, leaving arginine.
    6. ***Arginine*** is hydrolyzed to ***urea*** and ***ornithine***.
    7. Summary:     $HCO_3^-$ + $NH_4^+$ + 3ATP + aspartate + 2 $H_2O$ →
                          urea   +   2 ADP   +   2 $P_i$   +   AMP   +   $PP_i$   +   fumarate
  F. Nucleoprotein metabolism
    1. Nucleic acids are hydrolyzed during digestion to nucleotides and then to nucleosides. The nucleosides are absorbed and then split by nucleosidases to form ribose sugars and the purine and pyrimidine bases.

# Chapter 27 - Protein Metabolism

2. Purines (adenine, guanine) are metabolized to uric acid.
   a. Birds and reptiles excrete excess N as *uric acid* to conserve water.
   b. *Gout* is a metabolic disorder caused by the deposition of *uric acid* salts in cartilage.
3. Pyrimidines are metabolized to carbon dioxide, water, and urea.
4. The purine and pyrimidine bases can be synthesized from amino acids and other metabolites, and thus are not essential in the diet.

G. Genetic diseases of amino acid metabolism
   1. PKU or phenylketonuria – failure to convert phenylalanine to tyrosine.
   2. Albinism – failure to convert tyrosine into melanin pigments
   3. Maple syrup urine disease – failure to degrade branched-chain amino acids

## 27.4 Synthesis of Nonessential Amino Acids

A. Higher plants and many microorganisms are capable of synthesizing all their amino acids from $CO_2$, $H_2O$, and inorganic salts.

B. Animals can synthesize only about half of the amino acids. These are referred to as the *nonessential* amino acids.
   1. Ala:   from transamination of pyruvate
   2. Glu:   from transamination of α-ketoglutarate
   3. Asp:   from transamination of oxaloacetate
   4. Ser:   made from phosphoglycerate
   5. Gly:   made from serine
   6. Cys:   made from methionine
   7. Gln:   made from glutamate
   8. Asn:   made from aspartate
   9. Pro:   made from glutamate
   10. Tyr:  from hydroxylation of phenylalanine

## 27.5 Formation of Amino Acid Derivatives

A. Amino acids act as precursors of many important molecules.
   1. Gly / Asp / Gln are used in the synthesis of purines and pyrimidines.
   2. Gly and succinyl-CoA are used to synthesize heme.

B. *Decarboxylases* remove $CO_2$ to convert primary amino acids into primary amines. Many important body regulators and neurotransmitters are produced in this way. Like transaminases, most decarboxylases require pyridoxal-5'-phosphate (vit. $B_6$) as a cofactor.
   1. Histidine → $CO_2$ + histamine (inflammatory response, allegies, etc.)
   2. Tyrosine → $CO_2$ + tyramine (similar to norepinephrine)
   3. 5-HydroxyTrp → $CO_2$ + serotonin (brain neurotransmitter)
   4. 3,4-DihydroxyPhe → $CO_2$ + dopamine (neurotransmitter, precursor to norepinephrine)
       (Deficiency of dopamine is a primary cause of Parkinson's disease.)
   5. Glutamate → $CO_2$ + γ-aminobutyrate (GABA) (inhibits dopamine receptors)
   6. Ornithine → $CO_2$ + putrescine (one of polyamines, essential for all cells)

C. Many drugs are targeted to interact with amino acid derivatives. *Antihistamines* are very common drugs to fight allergies and ulcers.

Histamine              Cimetidine (Tagamet)

# Chapter 27 - Protein Metabolism

27.6 Relationships Among the Metabolic Pathways
  A. The metabolic pathways are interconnected (see summary on next page and Figure 27.6).
  B. *Catabolism* of all three types of major foodstuffs tends to converge to acetyl~SCoA and the Krebs cycle.
  C. *Anabolism* tends to diverge. Starting with a few key metabolites, the building blocks of biochemistry are synthesized, and from them millions of different biomolecules (e.g., proteins) can be formed.
  D. The chemistry of starvation
    1. A body totally deprived of food soon uses up its glycogen reserves and needs to convert to fat metabolism, thus bringing on ketosis. During starvation or uncontrolled diabetes mellitus, acetyl~SCoA concentration is high and oxaloacetate ketosis is followed by **acidosis**.
    2. During *starvation* the body will also break down its own proteins to try to meet its metabolic needs and to provide glucose to the brain via gluconeogenesis.
    3. *Kwashiorkor* is a protein-deficiency disease that produces emaciation, bloatedness, mental apathy, and eventual death.

# Chapter 27 - Protein Metabolism

## SUMMARY OF CATABOLIC PATHWAYS

# Chapter 27 - Protein Metabolism

## DISCUSSION

The summary on the previous page shows the interrelatedness of many aspects of the various metabolic pathways that were presented in Chapters 24 – 27. Although it appears to be a very complicated figure, take time to see how all of the separate pathways that were presented in the past four chapters are tied together in this one figure. Note that only 1 ATP is required to prime the glucose carbon skeleton for glycolysis when starting with glycogen instead of glucose, and remember that the second stage of glycolysis takes place twice for each glucose that enters the pathway. An (*) denotes a key enzyme that helps regulate the flux through the pathway. Note that the activation of fatty acids also takes place in the cytoplasm of the cell, but that β-oxidation like the TCA cycle occurs inside the mitochondria. Also shown, but less obvious, are examples of how amino acids such as Ala, Asp, Gln, and Glu are metabolized through the combination of transamination reactions to remove the nitrogen and oxidative deamination to produce ammonia for later removal as urea. The convergent nature of catabolism is evident as the metabolism of carbohydrates, fats, and proteins (amino acids) all converge to just a few common intermediates such as pyruvate, oxaloacetate, α-ketoglutarate, and acetyl~SCoA.

The study of the metabolism and biosynthesis of amino acids is both easier and more difficult than the study of comparable processes involving carbohydrates and fats. It is easier because most of the emphasis can be placed on two key reversible reactions, both of which involve the fate of the amine group--transamination and oxidative deamination.

The fact that amino acid metabolism neatly blends into carbohydrate metabolism also makes the former a little easier to understand. Many of the ketoacids that result when amine groups are removed from amino acids are already quite familiar to you as intermediates of the glycolytic pathway or the citric acid cycle. What makes this topic more difficult than the analogous material covered in Chapters 24 through 26 is the fact that the variety of amino acids is considerably greater than the variety of monosaccharides or fatty acids. All monosaccharides isomerize to common intermediates and then use the same metabolic pathways. Different length fatty acids cycle through the same spiral pathways (e.g., β-oxidation), varying only the number of turns required. But there is no common intermediate or common metabolic pathway for the complete degradation of amino acids. As Figure 27.6 indicates, they do all ultimately end up in the citric acid cycle. However, the same figure also indicates that their points of entry are quite varied. We've gotten around this difficulty by simply acknowledging it and not bothering to specify all the details. One metabolic pathway that we do not cover in detail is the urea cycle, which describes the fate of most of the nitrogen removed from amino acids. The amino acid arginine is cleaved by arginase into urea and another amino acid, ornithine. Ornithine is converted back into arginine by the steps of the urea cycle; thus ornithine serves a role similar to OAA in the other Krebs cycle.

Arginine + H$_2$O → Urea + Ornithine

# Chapter 27 - Protein Metabolism

The following diagram summarizes the source of the atoms incorporated into the urea product.

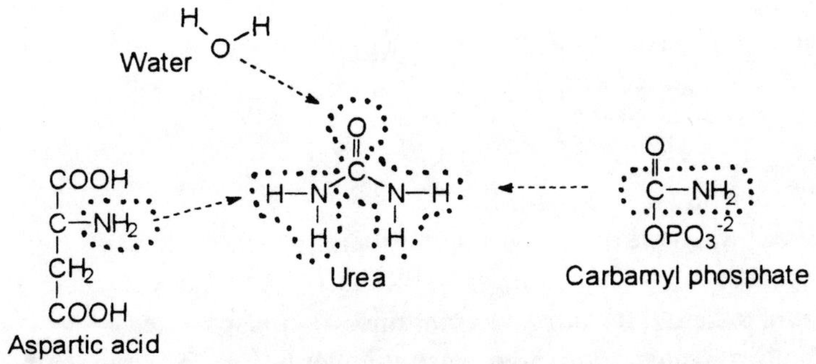

However, remember that the nitrogen supplied by aspartic acid may come from any of the amino acids by way of transamination reactions. Any amino acid can supply the nitrogen transferred from carbamyl phosphate, too, through a combination of transamination and oxidative deamination. As usual, much of the descriptive material in the chapter is reviewed in the problems at the end of the chapter.

## SELF-TEST

1. The principal digestive component of gastric juice is
    a. amylase   b. pepsinogen   c. carboxypeptidase   d. phosphofructokinase
2. The principal organ responsible for the degradation and synthesis of amino acids is the:
    a. brain   b. liver   c. stomach   d. pancreas   e. intestines
3. The biosynthesis of muscle protein from amino acids is classified as:
    a. anabolism   b. catabolism   c. digestion   d. transamination
4. The essential amino acids:
    a. can be synthesized in the body if nonessential amino acids are supplied in the diet
    b. are not present in the amino acid pool
    c. can be the limiting factor in determining the extent of protein biosynthesis
    d. All of the above are correct.
5. Amino acids are stored:
    a. with glycogen in liver and muscle
    b. in depots analogous to the fat storage areas
    c. in the nuclei of cells
    d. There are no storage facilities in the body for amino acids.
6. Which of the following amino acids is not an essential amino acid?
    a. Lys   b. Trp   c. Val   d. Glu   e. His
7. The amino acid pool can be supplied by:
    a. dietary amino acids
    b. the breakdown of tissue protein
    c. the biosynthesis of nonessential amino acids
    d. All of the above processes contribute to the population of the amino acid pool.
8. Protein biosynthesis:
    a. occurs only after prolonged starvation has severely depleted body protein
    b. is dictated by the genetic code
    c. occurs primarily in the small intestine
    d. All of the above are correct.
9. Growing children are in a state of:
    a. nitrogen balance   b. positive nitrogen balance   c. negative nitrogen balance

# Chapter 27 - Protein Metabolism

10. Proteins in the diet can be:
    a. used for the production of energy
    b. used in the replacement of tissue protein
    c. converted to carbohydrates
    d. converted to fats
    e All of the above are true.
11. A common protein deficiency disease is called:
    a. diarrhea    b. kwashiorkor    c. beriberi    d. scurvy    e. rickets
12. Which amino acid is most often metabolized via oxidative deamination?
    a. glutamate    b. α-ketoglutarate    c. ornithine    d. aspartate
13. In amino acid metabolism, which compound most commonly serves as the amine group acceptor during transamination reactions?
    a. glutamate    b. α-ketoglutarate    c. ornithine    d. aspartate
14. Which amino acid is converted to oxaloacetic acid through transamination?
    a. glutamate    b. α-ketoglutarate    c. ornithine    d. aspartate
15. Transaminases use a cofactor related to the vitamin:
    a. $B_1$    b. $B_2$    c. $B_6$    d. $B_{12}$    e. niacin
16. Glutamate oxaloacetate transaminase (GOT) is the enzyme that catalyzes:
    a. the conversion of phenylalanine to tyrosine
    b. the conversion of phenylalanine to phenylpyruvic acid
    c. the transfer of phosphate from creatine phosphate to ADP
    d. the transfer of an amine group from glutamic acid to oxaloacetic acid
17. The term Krebs cycle is not applied to:
    a. the citric acid cycle    b. the urea cycle    c. the fatty acid cycle
18. Glutamic acid is converted to α-ketoglutarate and _____ by oxidative deamination.
    a. asparate    b. ammonia    c. citrate    d. oxaloacetate    e. alanine
19. The urea formed in the urea cycle does not incorporate atoms contributed by:
    a. aspartic acid    b. carbamyl phosphate    c. ornithine    d. water
20. In addition to urea, which compound is formed by the cleavage of arginine in the urea cycle?
    a. arginine    b. aspartic acid    c. citrulline    d. ornithine
21. Which product is used by birds to excrete excess nitrogen?
    a. urea    b. ammonia    c. uric acid    d. glutamate
22. Which compounds can be converted to uric acid?
    a. adenine and guanine    b. adenine and thymine
    c. guanine and cytosine    d. uracil and thymine
23. Glycine is not an essential amino acid, because it can be produced from:
    a. phenylalanine    b. β-alanine    c. serine    d. glutamate
24. T / F  The amino acid pool is depleted to obtain material for the synthesis of nitrogen-containing compounds such as heme.
25. T / F  After proper chemical modification, all amino acids can contribute compounds to the tricarboxylic acid cycle.
26. T / F  It is possible for the composition of the amino acid pool to be adjusted to fit current needs of the body.
27. T / F  Infants suffering from PKU lack the normal pigmentation in skin, hair, and eyes.
28. T / F  In fish, nitrogen is excreted primarily as ammonia.
29. T / F  Nucleic acids are not required in the diet of human beings.
30. T / F  The precipitation of salts of uracil is characteristic of gout.

## ANSWERS

| | | |
|---|---|---|
| 1. b | 11. b | 21. c |
| 2. b | 12. a | 22. a |
| 3. a | 13. b | 23. c |
| 4. c | 14. d | 24. T |
| 5. d | 15. c | 25. T |
| 6. d | 16. d | 26. T |
| 7. d | 17. c | 27. F |
| 8. b | 18. b | 28. T |
| 9. b | 19. c | 29. T |
| 10. e | 20. d | 30. F |

# 28  Body Fluids

**KEY WORDS**

| | | | | |
|---|---|---|---|---|
| acidosis | alkalosis | blood pressure | osmotic pressure | systolic |
| diastolic | heparin | stroke | fibrin | sweat |
| lymphocytes | platelets | thrombocytes | osmosis | anemia |
| blood | fibrinogen | gamma globulins | kidneys | erythrocytes |
| thrombin | antigen | urine | leukocytes | immune |
| antibody | perspiration | sickle cell | albumins | hemoglobin |
| tears | plasma | capillaries | bilirubin | milk |
| serum | edema | lymph | hematocrit | shock |
| globulins | lysozyme | formed elements | autoimmune | Bohr effect |
| heat stroke | hypertension | hyperventilation | hypoventilation | leukemia |
| thrombosis | vaccine | interstitial fluid | | |

**SUMMARY**

28.1  Blood: Functions and Composition
  A. *Blood* is the principal transport system in the human body. It moves through an ~100,000 km long network of *capillaries* and blood vessels. It carries:
   1. Oxygen from the lungs to tissues
   2. Carbon dioxide from the tissues to the lungs
   3. Nutrients from the intestines to tissues
   4. Metabolic wastes to excretory organs
   5. Hormones from endocrine glands to target tissue
   6. Blood cells
  B. A 150 lb human has about 5 L of blood, of which about 40% is formed elements and 60% *plasma.*
   1. The formed elements include red blood cells, or erythrocytes (~4,000,000/mm$^3$); white blood cells, or leukocytes (~7000/mm$^3$); and thrombocytes, or platelets (~250,000/mm$^3$).
   2. The fluid portion of blood is called *plasma*.
   3. Blood *serum* is the fluid remaining after blood clotting has occurred.
  C. Plasma proteins
   1. Protein level is about 7–8 g/100 mL of blood plasma; most of these are synthesized in the liver.
   2. There are three major classes of plasma proteins.
      a. *Albumins* – (~55%) transport and osmotic pressure
      b. *Globulins* – (~40%)
         i) α- and β-globulins (~ 2 g/100 mL) are involved in transport.
         ii) γ-globulin ( ~ 1 g/100 mL) plays a role in antibody immune response.
      c. *Fibrinogen* (~5%) and prothrombin are blood-clotting proteins. Blood serum lacks fibrinogen and thus is unable to clot.
  D. The formed elements
   1. *Erythrocytes* (red blood cells) are formed in red bone marrow.
      a. A drop of blood may contain as many as 500 million erythrocytes. A *hematocrit* value of 45 implies that 45% of the blood cell volume is red cells.
      b. *Anemia* refers to any condition that lowers the percentage of red blood cells or hemoglobin in blood.

c. Erythrocytes do not have any aerobic metabolism and do not contain either mitochondria or a nucleus. They cannot reproduce, have no aerobic metabolism, and cannot synthesize fats or carbohydrates. They meet their energy needs from glucose metabolism via the glycolytic pathway and the pentose phosphate pathway.
d. The major function of erythrocytes is to carry hemoglobin, which transports oxygen from the lungs to the tissues.
e. Human red cells have a life span of about 4 months. It has been estimated that about 3 million red blood cells are destroyed every second and have to be replaced.
f. Blood type (A, B, AB, or O) is determined by which short-chain polysaccharide type (A or B, both or neither) is bound to the glycoproteins of the membranes of erythrocytes.

2. *Leukocytes* (white blood cells) are more like normal tissue cells. Leukocytes constitute the body's primary defense against foreign elements.
   a. *Lymphocytes* – synthesize and store antibodies
   b. Phagocytes (macrophages) – engulf and digest the invading organism
   c. Normal levels: ~ 7000 leukocytes/mm$^3$ blood. *Leukemia* is a cancer characterized by the uncontrolled production of immature white blood cells that are unable to destroy invading pathogens.

3. Thrombocytes *(platelets)* are instrumental in blood clotting. There are normally ~ 250,000 platelets/mm$^3$ blood.

E. Osmotic Pressure
   1. *Osmosis* refers to diffusion of solvent from a dilute solution into a more concentrated solution. This requires *osmotic pressure* to prevent the reverse process (diffusion) from dominating.
      a. The pressure of blood plasma proteins produces an osmotic pressure differential of about 25 mmHg.
      b. A pumping heart creates a *"blood pressure"* of about 32 mmHg at the arterial end of the capillaries; thus nutrients are forced from the blood into the *interstitial fluid*.
   2. Capillary diffusion
      a. Blood moves from the high-pressure arterial end of the capillaries to the low-pressure venous end.
      b. Interstitial fluid and blood plasma have similar electrolyte concentrations, but blood plasma also has proteins that cannot move through the capillary walls.
      c. At the arterial end of the capillaries, the hydrostatic pressure forces fluids, which carry nutrients to the cells, from the capillaries into the interstitial space.
      d. At the venous end of the capillaries, the hydrostatic pressure is low, so the high osmotic pressure of the blood causes fluid to diffuse back from the interstitial space into the blood.
   3. If the plasma protein concentration drops, the osmotic pressure also drops, causing swelling (*edema*) as fluids flow into the interstitial space.
   4. *Shock* is a physiological condition that results in a rapid decrease in blood volume and a dramatic drop in blood pressure.
   5. Electrolytes in plasma and erythrocytes
      a. Cations: Sodium ions are the main cations found in plasma. Potassium ions are found mainly in the *erythrocytes*. Calcium and magnesium ions are also found in blood plasma. ($Na^+$, $K^+$, $Ca^{++}$, $Mg^{++}$)
      b. Anions: The major anion electrolyte in blood plasma is chloride. In addition, sulfate ions and the buffer ions of bicarbonate and hydrogen phosphate are present ($Cl^-$, $SO_4^=$, $HCO_3^-$, $HPO_4^=$).
      c. The levels of many electrolytes are influenced by a number of disease states.

# Chapter 28 - Body Fluids

28.2 Blood Gases
   A. Oxygen and carbon dioxide transport: The *hemoglobin* (Hb) molecule:
      1. Four protein chains (2 α and 2 β subunits), each with a heme prosthetic group comprise Hb.
      2. Each of the four chains has several helical coils, which fold to produce a roughly globular structure with a hydrophobic crevice that binds the heme group.
      3. The heme group is a planar protophorphyrin ring containing an iron ($Fe^{2+}$) iron bound to 4 N's. A histidine side chain donates another N to fill the 5th binding site on iron.
      4. Oxygen binds to the 6th binding position on the heme iron. The binding of oxygen to hemoglobin is cooperative in that the binding of the 1st oxygen enhances the binding of the others. The release of oxygen at the tissues is also cooperative. The 15 g of Hb in 100 mL of blood combine with ~20 mL of oxygen (compared with only 0.3 mL of gaseous oxygen that could dissolve in 100 mL of saline solution).
      5. Oxygen pressure in the lungs is about 90–100 mmHg which is conducive to forming oxyHb. At respiring tissue, oxygen is about 25–40 mmHg. Also, the lower the pH, the greater the amount of oxygen released (*Bohr Effect*). Low oxygen tension and lower pH in respiring tissue results in more oxygen released and the formation of deoxyHb.
      6. $CO_2$ concentrations are high in respiring tissue. $CO_2$ is removed as bicarbonate ($HCO_3^-$; 70%) or bound to Hb (~20%).
   B. CO binds very tightly to Hb, interfering with $O_2$ transport. Nitrites can oxidize the heme iron from $Fe^{2+}$ to $Fe^{3+}$ to form methemoglobin.
   C. Red blood cells last approximately 120 days. The globin protein is broken down and its amino acids recycled. The heme portion breaks down into *bilirubin* products with yellow-brown pigments (jaundice).
   D. Many abnormal hemoglobins have been studied. Sickle cell Hb has a valine instead of a glutamic acid in the 6th amino acid position of the β chain. This small difference causes the red blood cells to sickle and be rapidly destroyed, which results in anemia. *Sickle cell anemia* is only one of many hereditary traits associated with metabolic or genetic diseases.

28.3 Blood Buffers
   A. Blood pH must be maintained within a very narrow range, around pH 7.4. The four major buffer systems in blood plasma are:
      1. Bicarbonate/carbonic acid system; $pK_a \sim 6.3$
      $$H^+ + HCO_3^- \leftrightarrow H_2CO_3 \leftrightarrow H_2O + CO_2$$
      2. Monohydrogen phosphate/dihydrogen phosphate system; $pK_a \sim 7.2$
      $$H^+ + HPO_4^{-2} \leftrightarrow H_2PO_4^-$$
      3. Plasma proteins
      4. Hemoglobin
   B. Hemoglobin has both acidic and basic forms. The release of oxygen and the picking up of hydrogen ions from respiring tissue helps convert the carbonic acids of respiration to the bicarbonate ion. The addition of oxygen and the release of hydrogen ions at the lungs promote the formation of carbonic acid, which is rapidly converted to carbon dioxide and exhaled.
   C. Respiratory rate also affects blood pH.
      1. *Hypoventilation*   = too slow   → *acidosis*
      2. *Hyperventilation*  = too fast   → *alkalosis*

28.4 Blood Clotting
   A. Blood clotting involves the formation of *fibrin* from fibrinogen by the action of a protease called *thrombin*.

1. The activation of thrombin from its proenzyme or zymogen form (prothrombin) involves $Ca^{++}$ ions and an elaborate cascade of activating proteins that are released when a tissue is cut or injured.
2. Blood serum is blood plasma without the *fibrinogen*.

B. Anticoagulants
 1. Plasmin is another protease that is activated later to dissolve the blood clot.
 2. *Heparin* is a sulfate-rich polysaccharide that appears to block the action of thrombin.
 3. Vitamin K is needed as a coenzyme to help an enzyme make γ-carboxyglutamate residues that in turn are needed by prothrombin to bind $Ca^{++}$ so that prothrombin can be converted to thrombin. Dicumarol is a drug antagonist of vitamin K that helps prevent blood clots from forming.
 4. Hemophilia is an inherited disorder characterized by inadequate production of clotting factors.

C. Blood clots can become lodged and cause cells in nearby tissue to become starved for oxygen and die.
 1. *Stroke*: tissue death occurs in the brain.
 2. Myocardial infarction (coronary *thrombosis*) – tissue death occurs in the heart.
 3. Aspirin may help prevent strokes and heart attacks by inhibiting thrombosis.

## 28.5 The Immune Response

A. The *gamma globulins* are produced by the body as part of the *immune* response. When a foreign body (*antigen*) invades the body, *antibody* proteins are produced to bind and incapacitate the antigen.
B. A *vaccine* primes this defense mechanism by presenting the body with a weakened or dead form of the antigen to practice on. The body can thus gain an immunity to that particular antigen.
C. Rejection of transplanted tissue is caused by the generation of antibodies, which attack the transplant.
D. A variety of diseases are associated with immune system defects.
 1. Multiple sclerosis is an *autoimmune* disease that destroys the myelin sheath of nerve cells.
 2. Some forms of arthritis are autoimmune diseases that destroy connective tissue.
 3. In AIDS (Acquired immune deficiency syndrome) the immune system is destroyed by the HIV virus, which leaves the victim without a defense system.

## 28.6 Blood Pressure

A. *Blood pressure* is the force exerted by the blood in the arteries.
 1. *Systolic* pressure is the maximum pressure achieved during contraction of the heart ventricles (~120 mmHg).
 2. *Diastolic* pressure is the lowest pressure that remains in the arteries before the next ventricular contraction (~80 mmHg).
B. High blood pressure (*hypertension*) can be influenced by larger blood volume, sodium ion levels, or blockage of arteries.

## 28.7 Urine and the Kidneys

A. The kidneys remove metabolic waste products from blood while helping to maintain a proper balance of water, electrolytes, and other components of body fluids.
 1. Blood flows from an artery into the capillary network of the *kidneys* and back out, into the vein. Waste products, such as urea, uric acid, and excess salts, are passed into collecting tubules and are eventually excreted as *urine*. A healthy adult passes 1.1–1.5 L of urine each day.
 2. About 170 L/day are filtered into the tubules. Many substances, such as glucose, are normally reabsorbed by the blood. However, most have a threshold level above which the excess is no longer absorbed. This happens with glucose and is often symptomatic of diabetes.

## Chapter 28 - Body Fluids

  B. Some materials have very low solubilities in the urine and precipitate out as "kidney stones." These usually consist of calcium phosphate, magnesium ammonium phosphate, and calcium carbonate or calcium oxalate.
  C. The "artificial kidney" (dialysis machine) was invented in the 1950s.
  In the early 1990s, ~60,000 people were on dialysis at an annual cost of approximately $1,800,000,000.

28.8 Sweat and Tears
  A. **Sweat**: The skin is an organ of excretion through which *sweat* passes. Sweat is 99% water but also contains electrolytes, urea, lipids, creatine, lactic acid, and pyruvic acid.
    1. Water loss via *perspiration*
       a. Insensible perspiration    – ~700 mL/day via skin and respiration
       b. Sensible perspiration      – 2.5 million sweat glands; output varies with activity
    2. Evaporation of sweat carries off ~540 cal heat/g water lost. **Heat stroke** occurs when our heat regulation system fails.
  B. *Tears* keep the eyes moist and contain *lysozyme*, an enzyme that ruptures bacterial cell walls to help prevent eye infections.
    1. Tears are multilayered: inner mucus layer, thin layer of lachrymal secretions, oily outer layer.
    2. Tear gases such as α-chloroacetophenone are specially designed eye irritants to induce copious flow of tears.

$$\text{Ph-C(=O)-CH}_2\text{-Cl} \quad \alpha-\text{Chloroacetophenone}$$

28.9 The Chemistry of Mother's Milk
  A. *Milk* is the secretion of the mammary glands. It contains fats, carbohydrates, proteins, minerals and vitamins.
    1. Casein – precipitated milk protein
    2. Lactose – milk sugar; a disaccharide of glucose and galactose
  B. Mother's milk probably is more nutritious for infants than is cow's milk, and it may also add to the infants immunological defenses against disease.
  C. Some people are unable to drink milk because their bodies cannot metabolize galactose. This condition is known as galactosemia and can cause mental retardation in infants.

(Optional Material ) *******************

28.10 Lymph: A Secondary Transport System
  A. The *lymph* system is composed of veins and capillaries, but no arteries. Interstitial fluid absorbed into the lymph capillaries moves into lymph veins, which empty into the veins of the blood circulatory system.
  B. The lymph system also absorbs fat from the intestine and produces some forms of white blood cells.

## DISCUSSION

Chapter 28 represents a change of pace. For some time now we have been examining various classes of biochemically important molecules. Our survey of these important compounds is now complete, and we are embarking on an examination of how these materials function in the body. In this chapter we've begun that examination by considering the properties of many of the fluids in which biochemical reactions take place. So in this chapter there's very little that's new as far as molecules are concerned. Hemoglobin

## Chapter 28 - Body Fluids

is the only compound structure considered in any detail, and we have encountered hemoglobin previously (Section 21.10).

Much of the material in the chapter, therefore, is descriptive. Many applications of chemical principles in living systems are provided. For the most part, this chapter does not require that you learn new scientific principles, but rather that you relate familiar chemistry to biological systems.

If you turn through the pages of Chapter 28, you'll notice that the only point at which we use chemical equations to any extent is in our discussion of the transportation of carbon dioxide by the blood. The transportation of $CO_2$ is thus more chemical than biological.

### In Capillaries Serving Metabolically Active Tissue

Carbon dioxide produced as a metabolic product in tissue cells migrates into the erythrocytes, where it combines with water.
$$CO_2 + H_2O \rightarrow H_2CO_3$$

The resulting carbonic acid protonates the conjugate base of the hemoglobin buffer in the erythrocyte.
$$H_2CO_3 + Hb^- \rightarrow HCO_3^- + HHb$$

The bicarbonate ion dissolves in the fluid within the erythrocyte and in the surrounding blood plasma. Any bicarbonate ion that migrates from the erythrocyte to the surrounding plasma is matched by the migration of a chloride ion from the surrounding plasma into the erythrocyte (chloride shift).

### In Capillaries Serving the Lungs

The conjugate acid of the hemoglobin buffer picks up oxygen to become the conjugate acid of the oxyhemoglobin buffer.
$$HHb + O_2 \rightarrow HHbO_2$$

The conjugate acid of the oxyhemoglobin buffer protonates the bicarbonate ion.
$$HHbO_2 + HCO_3^- \rightarrow HbO_2^- + H_2CO_3$$

The resulting carbonic acid dissociates into water and carbon dioxide.
$$H_2CO_3 \rightarrow H_2O + CO_2$$
The carbon dioxide gas is exhausted to the atmosphere.

The only numerical problems associated with Chapter 28 deal with concentrations of electrolytes. Try the problems that follow.

## Problems

1. Normal values for various electrolytes in urine collected over a 24-hour period are:
    - calcium ($Ca^{2+}$): 2.5 – 20 meq
    - chloride ($Cl^-$): 110 – 250 meq
    - magnesium ($Mg^{2+}$): 6.0 – 8.5 meq
    - potassium ($K^+$): 40 – 80 meq
    - sodium ($Na^+$): 80 – 180 meq

## Chapter 28 - Body Fluids

Analysis of a 24-hour urine sample detected the following amounts of electrolytes. Indicate for each ion whether the amount is normal or abnormal.

    calcium:         100 mg
    chloride:         3.55 g
    magnesium:    4.86 mg
    potassium:     1.955 g
    sodium:         6.9 g

2. Normal concentration ranges for various electrolytes in blood serum are:

    calcium:         4.5 – 5.3 meq/L
    chloride:         96 – 106 meq/L
    magnesium:    1.3 – 2.1 meq/L
    potassium:     3.5 – 5.0 meq/L
    sodium:         136 – 145 meq/L

The following concentrations were reported for a sample of serum. In each case indicate whether the value falls within the normal range.

    calcium:         4 mg%
    chloride:         355 mg%
    magnesium:    2.43 mg%
    potassium:     39.1 mg%
    sodium:         345 mg%

## SELF-TEST

1. Which material does blood serum not include?
    a. electrolytes    b. fibrinogen    c. proteins
2. Which of the plasma proteins is associated with the immune response of the body?
    a. albumin    b. fibrinogen    c. globulins    d. prothrombin
3. Which of the plasma proteins contributes most to the osmotic pressure of blood?
    a. albumin    b. fibrinogen    c. globulin    d. prothrombin
4. Which of the cations is not one of the principal electrolytes in blood plasma?
    a. $Na^+$    b. $K^+$    c. $Ca^{2+}$    d. $Mg^{2+}$    e. $Fe^{2+}$
5. Which of the anions is not one of the principal electrolytes in blood plasma?
    a. $Cl^-$    b. $HCO_3^-$    c. $CO_3^{2-}$    d. $HPO_4^{2-}$    e. $SO_4^{2-}$
6. Material moves in and out of the capillaries primarily through the process of:
    a. diffusion    b. evaporation    c. filtration
7. A normal hematocrit value (% volume of packed erythrocytes) would be about:
    a. 5    b. 10    c. 20    d. 45    e. 65
8. Which ion is necessary for blood clotting?
    a. $Cl^-$    b. $HCO_3^-$    c. $Ca^{2+}$    d. $Fe^{2+}$
9. Which ion is not part of the blood buffers?
    a. $HCO_3^-$    b. $HPO_4^{2-}$    c. $HSO_4^-$
10. A person of blood type _____ is a universal blood donor.
    a. A    b. B    c. AB    d. O    e. None is correct
11. Some doctors now recommend that older patients with histories of heart attack should take:
    a. vitamin C    c. vitamin E    c. vitamin $B_6$    d. aspirin    e. iron
12. A principal anticoagulating agent is:
    a. thrombin    b. zymogen    c. heparin    d. globulin    e. fibrin

## Chapter 28 - Body Fluids

13. Colloid osmotic pressure refers to:
    a. the pressure imparted to the blood by the pumping action of the heart
    b. the osmotic pressure of the blood resulting from dissolved electrolytes
    c. the osmotic pressure of the blood resulting from proteins present in colloidal dispersion
14. The hydrostatic pressure of the blood is:
    a. higher at the venous end of a capillary
    b. higher at the arterial end of a capillary
    c. approximately the same at both ends of a capillary
15. Which describes the conditions at the arterial end of the capillaries?
    a. colloid osmotic pressure exceeds hydrostatic pressure, and the net movement of fluid is into the interstitial space
    b. colloid osmotic pressure exceeds hydrostatic pressure, and the net movement of fluid is into the capillary
    c. hydrostatic pressure exceeds colloid osmotic pressure, and the net movement of fluid is into the interstitial space
    d. hydrostatic pressure exceeds colloid osmotic pressure, and the net movement of fluid is into the capillary
16. In edema, there is a net flow of fluid into the interstitial space because:
    a. the colloid osmotic pressure of the blood increases
    b. the colloid osmotic pressure of the blood decreases
    c. the hydrostatic pressure of the blood decreases
17. An antibody is:
    a. a foreign macromolecule that triggers the body's immune response
    b. a protein formed by the body to attack specific foreign particles
    c. a pathogenic microorganism
18. A vaccine contains:
    a. a weakened antigen   b. a weakened antibody   c. gamma globulin
19. Carbon dioxide produced in metabolic reactions is carried to the lungs chiefly as:
    a. free $CO_2$ gas    b. dissolved $CO_2$ gas    c. $HCO_3^-$    d. a prosthetic group of hemoglobin
20. Which reaction occurs in the capillaries of the lungs?
    a. $HHbO_2 \rightarrow HHb + O_2$
    b. $HHbO_2 + HCO_3^- \rightarrow HbO_2^- + H_2CO_3$
    c. $H_2CO_3 + Hb^- \rightarrow HHb + HCO_3^-$
21. To maintain electrical neutrality as the bicarbonate ions diffuse out of erythrocytes:
    a. potassium ions accompany the bicarbonate ions
    b. sodium ions accompany the bicarbonate ions
    c. chloride ions diffuse into the erythrocyte
22. Which is not true of a hemoglobin molecule?
    a. It is a conjugated protein.
    b. It incorporates iron in a +2 oxidation state.
    c. It contains four identical protein chains in a roughly tetrahedral arrangement.
    d. It incorporates four heme units, each of which can bind with an oxygen molecule.
23. Bilirubin is a product of:
    a. the breakdown of the prosthetic group of hemoglobin
    b. the breakdown of the alpha chain of hemoglobin
    c. the breakdown of the beta chain of hemoglobin

# Chapter 28 - Body Fluids

24. The abnormal hemoglobin that is characteristic of sickle cell anemia contains:
    a. iron in the +3 oxidation state
    b. a heme unit that has no iron ion
    c. peptide chains incorporating an incorrect amino acid
25. The constitution of the fluid within the lymphatic system is identical to:
    a. blood plasma      b. interstitial fluid      c. intracellular fluid
26. Lymph nodes are <u>not</u> involved in the manufacture of:
    a. antibodies      b. erythrocytes      c. leukocytes
27. Which type of nutrient is absorbed into the lymphatic system from the intestine?
    a. carbohydrate      b. fat      c. protein
28. Salts of which cation are <u>not</u> ordinarily found in kidney stones?
    a. $Na^+$      b. $Ca^{2+}$      c. $Mg^{2+}$
29. Which substance is <u>not</u> filtered out of the blood at the glomerulus?
    a. erythrocytes      b. glucose      c. urea      d. water
30. Insensible perspiration is the water lost:
    a. through the respiratory tract      b. from the sweat glands      c. from the lacrimal glands
31. Calcium oxalate is sometimes found in:
    a. kidney stones      b. tears      c. milk      d. sweat
32. T / F Heat is one of the major metabolic products carried off by perspiration.
33. T / F Lysozyme is a type of bacteria occasionally found in lacrimal fluid.
34. T / F Casein is the outer oily layer of tears.
35. T / F Mammals living in cold climates produce milk with high fat content.
36. T / F Anemia is a condition associated with abnormal concentrations of red blood cells.
37. T / F Plasma can be isolated if an anticoagulant is first added to freshly drawn blood.
38. T / F Sodium ions are found mainly in the plasma, and potassium ions are found mainly in the erythrocytes.
39. T / F Metabolic disorders more commonly produce the condition known as alkalosis rather than acidosis.
40. T / F The chloride shift refers to the loss of chloride ions to the urine when the chloride concentration threshold level is exceeded.

# ANSWERS

## Problems

1. calcium = 5 meq, normal
   chloride = 100 meq, slightly low
   magnesium = 0.4 meq, very low
   potassium = 50 meq, normal
   sodium = 300 meq, very high

2. calcium = 2 meq/L, low
   chloride = 100 meq/L, normal
   magnesium = 2 meq/L, normal
   potassium = 10 meq/L, high
   sodium = 150 meq/L, normal

## Self-Test

| | | | |
|---|---|---|---|
| 1. b | 11. d | 21. c | 31. a |
| 2. c | 12. c | 22. c | 32. T |
| 3. a | 13. c | 23. a | 33. F |
| 4. e | 14. b | 24. c | 34. F |
| 5. c | 15. c | 25. b | 35. T |
| 6. a | 16. b | 26. b | 36. T |
| 7. d | 17. b | 27. b | 37. T |
| 8. c | 18. a | 28. a | 38. T |
| 9. c | 19. c | 29. a | 39. F |
| 10. d | 20. b | 30. a | 40. F |

# PART II
Answers to Questions

# 1 Matter and Measurement

## ANSWERS TO REVIEW QUESTIONS

1.2  A distinguishing characteristic of science is the use of processes and methods. The scientific method makes use of data and observations gathered by experiment. The gathered data and observations are then explained through variable hypotheses. Further experiments are then proposed to test these proposals. Ultimately, by changing and restating hypotheses, scientific laws are formulated and theories are generated to explain experimental data and observations. Scientific theories are accepted or rejected on the basis of their ability to explain experimental data.

1.13  The SI base units for length, mass, and temperature are given in the following table.

| Unit | Name | Symbol |
|---|---|---|
| length | meter | m |
| mass | kilogram | kg |
| temperature | kelvin | K |

1.14  The basic unit of length in the SI system is the meter (m). The derived unit for area is the square meter ($m^2$), and the derived unit for volume is the cubic meter ($m^3$).

1.16  A young man who is 160 cm, or 5 ft 3 in. tall, and has a mass of 94 kg, or 210 lb, would be considered overweight.

$$160 \text{ cm} \times \frac{1 \text{ in}}{2.54 \text{ cm}} \times \frac{1 \text{ ft}}{12 \text{ in}} = 5.25 \text{ ft or 5 ft 3 in.}$$

$$94 \text{ kg} \times \frac{1 \text{ lb}}{0.454 \text{ kg}} = 207 \text{ lbs} \Rightarrow 210 \text{ lbs}$$

1.20  The Calorie (Cal) used by nutritionists is equivalent to 1000 cal or 1 kcal.

## ANSWERS TO PROBLEMS

### Some Fundamental Concepts

1.  The responses a., air, and c., the human body, are examples of matter. Both air and the human body have mass and occupy space.

3.  Since both samples are weighed under the same conditions, the gravitational force is the same. The mass of the samples would be proportional to their weight and the sample with twice the weight would have twice the mass. Sample B would be expected to have twice the mass of sample A.

5.  a. Physical change: The composition of the wool is not changed by shearing or spinning.
    b. Chemical change: Fertilizer and water become new substances as the grass grows.
    c. Chemical change: New compounds form as the milk sours.

## Chapter 1: Matter and Measurement

7.  a. Physical property: The composition of methanol does not change when its boiling point is observed.
    b. Chemical property: When phosphorus burns in air, new compounds are formed.
    c. Physical property: The composition of lead does not change when its color is observed.
    d. Physical property: The composition of sodium bicarbonate does not change when it is dissolved in water. Evaporation of the water leaves the solid sodium bicarbonate.

9.  Kinetic energy depends on mass and velocity according to the equation: $KE = \frac{1}{2}mv^2$. (a) Since the sprinter and the long-distance runner are assumed to have the same mass, the faster moving sprinter has greater kinetic energy. (b) The automobile has greater mass than the bicycle and is traveling faster. The automobile has greater kinetic energy.

11. The diver on the 10-m platform has greater potential energy.

### Elements, Compounds, and Mixtures

13. Hafnium (Hf) and chlorine (Cl) are elements. These are made of only one fundamental substance. The compound HF is made from the elements hydrogen and fluorine.

15. Helium is a substance. Maple syrup and smog are mixtures.

17. Gasoline is homogeneous, and iced tea is heterogeneous.

19. a. helium   b. nitrogen   c. fluorine   d. potassium   e. iron   f. copper

21. a. H   b. C   c. O   d. Zn   e. I   f. Hg

### Metric Measurement

23. a. 4.54 mg   b. 3.76 cm   c. 6.34 µg

25. a. cm   b. kg   c. dL   d. lb

### Significant Figures

27. a. 4   b. 3   c. 5   d. 4   e. 4   f. 2

29. a. 100.5 m   b. 153 g   c. 54.4 cm   d. 436 g   e. 111 mL   f. 2.4 cm

### Unit Conversions

31.  a. $50.0 \text{ km} \times \dfrac{1000 \text{ m}}{1 \text{ km}} = 5.00 \times 10^4 \text{ m}$     b. $5.46 \text{ mm} \times \dfrac{1 \text{ m}}{1000 \text{ mm}} = 0.546 \text{ m}$

     c. $97.5 \text{ kg} \times \dfrac{1000 \text{ g}}{1 \text{ kg}} = 9.75 \times 10^4 \text{ g}$     d. $47.9 \text{ mL} \times \dfrac{1 \text{ L}}{1000 \text{ mL}} = 4.79 \times 10^{-2} \text{ L}$

     e. $577 \mu\text{g} \times \dfrac{1 \text{ mg}}{1000 \mu\text{g}} = 0.577 \text{ mg}$     f. $237 \text{ mm} \times \dfrac{1 \text{ cm}}{10 \text{ mm}} = 23.7 \text{ cm}$

33.  a. $413 \text{ in.} \times \dfrac{1 \text{ yd}}{36 \text{ in.}} = 11.5 \text{ yd}$     b. $86.2 \text{ oz} \times \dfrac{1 \text{ lb}}{16 \text{ oz}} = 5.39 \text{ lb}$

     c. $64.0 \text{ fl oz} \times \dfrac{1 \text{ qt}}{32 \text{ fl oz}} = 2.00 \text{ qt}$     d. $12.6 \dfrac{\text{ft}}{\text{s}} \times \dfrac{1 \text{ mi}}{5280 \text{ ft}} \times \dfrac{3600 \text{ s}}{1 \text{ h}} = 8.59 \dfrac{\text{mi}}{\text{h}}$

35. a. $16.4 \text{ in} \times \dfrac{2.54 \text{ cm}}{1 \text{ in}} = 41.7 \text{ cm}$     b. $4.17 \text{ qt} \times \dfrac{1 \text{ L}}{1.057 \text{ qt}} = 3.95 \text{ L}$

    c. $1.61 \text{ kg} \times \dfrac{2.205 \text{ lb}}{1 \text{ kg}} = 3.55 \text{ lb}$     d. $9.34 \text{ g} \times \dfrac{1 \text{ oz}}{28.35 \text{ g}} = 0.329 \text{ oz}$

37. $90.0 \dfrac{\text{km}}{\text{h}} \times \dfrac{1000 \text{ m}}{1 \text{ km}} \times \dfrac{1 \text{ h}}{3600 \text{ s}} = 25.0 \dfrac{\text{m}}{\text{s}}$

**Density and Specific Gravity**

39. Density $(d) = \dfrac{m}{V}$     a. Density $(d) = \dfrac{87.5 \text{ g}}{75.0 \text{ mL}} = 1.17 \text{ g/mL}$

    b. $2.75 \text{ L} \times \dfrac{1000 \text{ mL}}{1 \text{ L}} = 2750 \text{ mL}$     Density $(d) = \dfrac{3460 \text{ g}}{2750 \text{ mL}} = 1.26 \text{ g/mL}$

41. Mass $(m) = V \times d$     a. Mass $(m) = 125 \text{ mL} \times \dfrac{0.962 \text{ g}}{1.00 \text{ mL}} = 1.20 \times 10^2 \text{ g}$

    b. Mass $(m) = 33.0 \text{ mL} \times \dfrac{0.660 \text{ g}}{1.00 \text{ mL}} = 21.8 \text{ g}$

43. Volume $(V) = \dfrac{m}{d}$     a. Volume $(V) = 475 \text{ g} \times \dfrac{1.00 \text{ cm}^3}{8.94 \text{ g}} = 53.1 \text{ cm}^3$

    b. Volume $(V) = 253 \text{ g} \times \dfrac{1.0000 \text{ mL}}{13.534 \text{ g}} = 18.7 \text{ mL}$

45. Total mass – mass of paper = mass of metal
$18.43 \text{ g} - 1.21 \text{ g} = 17.22 \text{ g}$

    Density $(d) = \dfrac{m}{V}$     Density $(d) = \dfrac{17.22 \text{ g}}{3.29 \text{ cm}^3} = 5.23 \text{ g/cm}^3$

47. Specific gravity $= \dfrac{\text{density of liquid}}{\text{density of water}} = \dfrac{1.02 \text{ g/mL}}{1.00 \text{ g/mL}} = 1.02$

**Energy: Temperature and Heat**

49. a. °C is larger than °F.     b. The unit Cal is larger than the unit cal

51. a. °F = 1.8 (°C) + 32     °F = 1.8 (37.0°C) + 32 = 98.6 °F

    b. °C = $\dfrac{(°F - 32)}{1.8}$     °C = $\dfrac{(5.5°F - 32)}{1.8} = -15 \text{ °C}$

    c. °F = 1.8 (°C) + 32     °F = 1.8 (273°C) + 32 = 523 °F

53. a. $2.75 \text{ kcal} \times \dfrac{1000 \text{ cal}}{1 \text{ kcal}} = 2750 \text{ cal}$

    b. $0.741 \text{ cal} \times \dfrac{1000 \text{ cal}}{1 \text{ Cal}} \times \dfrac{4.184 \text{ J}}{1 \text{ cal}} = 3.10 \times 10^3 \text{ J}$

    c. $8.63 \text{ kJ} \times \dfrac{1000 \text{ J}}{1 \text{ kJ}} \times \dfrac{1 \text{ cal}}{4.184 \text{ J}} = 2.06 \times 10^3 \text{ cal}$

55. Heat absorbed = mass × specific heat × $\Delta T$
$\Delta T$ = 50.0 °C − 20.0 °C = 30.0 °C
Heat absorbed = 50.0 g × 1.000 cal/(g·°C) × 30.0°C = 1500 cal

57. Heat released = mass × specific heat × $\Delta T$
$\Delta T$ = 20.0 °C − 90.0 °C = −70.0 °C     mass = 2.00 kg × 1000 g/kg = 2000 g
Heat released = 2000 g × 4.182 J/(g·°C) × (−70.0°C) = −585000 J
or 585 kJ <u>heat released</u>

59. $\Delta T = \dfrac{\text{heat absorbed}}{\text{mass} \times \text{specific heat}} = \dfrac{17.6 \text{ kcal} \times 1000 \text{ cal/kcal}}{454 \text{ g} \times 0.106 \text{ cal/(g} \cdot \text{°C)}} = 365.72 \text{ °C}$

$\Delta T = T_{\text{final}} - T_{\text{initial}} = 365.72$°C (The change in temperature has three significant figures since the mass and the heat absorbed are given to three significant figures.)
$T_{\text{final}} = T_{\text{initial}} + \Delta T = 22.5$ °C + 365.72 °C = 388 °C

# 2 Atoms

## ANSWERS TO REVIEW QUESTIONS

2.2    a. atomic    b. atomic    c. continuous    d. continuous
       e. atomic    f. continuous

2.7    a. A neutral atom that contains 11 protons contains 11 electrons.
       b. An atom that has a mass number of 23 and contains 11 protons must contain 12 neutrons.

2.10    The maximum number of electrons in a shell is given by the formula $2n^2$, where $n$ is the shell being considered. The maximum number of electrons in the third shell is then: $2(3)^2 = 18$ $e^-$. In an atom that contains 2 $e^-$ in the third shell, the total number of electrons in the atom would be:
$$n = 1;\ 2\ e^- \qquad n = 2;\ 8\ e^- \qquad n = 3;\ 2\ e^-$$
Total number of electrons = $[(2\ e^-) + (8\ e^-) + (2\ e^-)] = 12\ e^-$

2.11    a. silicon    $n = 1$    2 $e^-$      b. nitrogen    $n = 1$    2 $e^-$
         14 $e^-$      $n = 2$    8 $e^-$        7 $e^-$        $n = 2$    5 $e^-$
               $n = 3$    4 $e^-$

       c. sulfur     $n = 1$    2 $e^-$
         16 $e^-$      $n = 2$    8 $e^-$
               $n = 3$    6 $e^-$

2.12    a. The notation $2p^6$ represents six electrons.
        b. The $p$ orbital has a dumbbell shape.
        c. There are three $p$ orbitals.

## ANSWERS TO PROBLEMS

**Dalton's Atomic Theory**

1. The discovery of radioactivity contradicts the assumption in Dalton's atomic theory that atoms are indivisible.

3. These findings do not contradict Dalton's theory. Calcium and vanadium are different elements and have different masses.

5. These findings do contradict Dalton's atomic theory. Since these are both atoms of calcium, their masses should be the same according to Dalton's theory.

**The Nuclear Atom**

7. An electron has a unit negative charge, and a proton has a unit positive charge. The two unlike charges should attract each other.

290   Chapter 2: Atoms

9.  a. calcium    20 protons, 20 electrons    b. sodium    11 protons, 11 electrons
    c. fluorine    9 protons, 9 electrons    d. argon    18 protons, 18 electrons

11. a. zinc    30 protons, 32 neutrons    b. plutonium    94 protons, 147 neutrons
    c. technetium    43 protons, 56 neutrons    d. molybdenum    42 protons, 57 neutrons

13. Only the pair given in b. represents an isotope pair.

**Atomic Masses**

15. The isotope lithium-7 must be the predominant isotope, since the atomic mass is taken as a weighted average of the two isotopes.

17. $^{109}$Ag; Atomic mass of silver = 0.5(107 u) + 0.5($x$) = 107.8682 u
    0.5($x$) = 107.8682 u − [0.5(107 u)] = 54.3682 u    $x$ = 109 u

**Electronic Configurations**

19. a. silicon   $1s^2 2s^2 2p^6 3s^2 3p^2$      b. nitrogen   $1s^2 2s^2 2p^3$
    (14 electrons)      (7 electrons)

    c. sulfur   $1s^2 2s^2 2p^6 3s^2 3p^4$
    (16 electrons)

21. a. beryllium      b. nitrogen      c. aluminum

23. The ground state electron configuration for phosphorus is $1s^2 2s^2 2p^6 3s^2 3p^3$. The ground state electron configuration for the phosphide ion (P$^{3-}$) is $1s^2 2s^2 2p^6 3s^2 3p^6$ which is the same electron configuration as that of argon.

25. The valence shell of a Group 7A atom contains seven electrons.

27. An atom in its ground state is in its lowest energy form. In a ground state atom, the electrons fill the lowest energy levels available.

29. Light is emitted from an atom in an excited state when an exited state electron emits energy and returns to a lower energy shell.

31. The electron configuration of oxygen ($1s^2 2s^2 2p^4$) differs from the electron configuration of fluorine ($1s^2 2s^2 2p^5$) in that oxygen has four electrons in the 2p orbital and fluorine has five electrons in the 2p orbital.

**The Periodic Table**

33. Metals have a characteristic luster, are good conductors, and are generally malleable and ductile. Except for mercury, the metals are solids at room temperature and normal pressure. The metals (except for hydrogen) are located to the left of the diagonal stepped line that separates the elements B, Si, As, Te, and At from the elements Al, Ge, Sb, and Po. (See Figure 2.15 on page 58.)

35.   a. Group 4A      b. Group 2A      c. Group 2B      d. Group 7A
         period 2         period 4         period 5         period 3
         nonmetal         metal            metal            nonmetal

      e. Group 3A      f. Group 2A      g. Group 5A      h. Group 7A
         period 2         period 6         period 6         period 4
         nonmetal         metal            metal            nonmetal

37.   a. Ga            b. Cu            c. I

39.   b. At, and e. F

41.   b. Ne, d. He, and e. Xe

**Periodic Properties**

43.   a. Sulfur has the larger atomic radius. Atomic radii of the A-group elements decrease going from left to right across a period, and chlorine is to the right of sulfur in the third period of the periodic table.
      b. Magnesium has the larger atomic radius. Atomic radii of the A-group elements decrease going from left to right across a period, and aluminum is to the right of magnesium in the third period of the periodic table.

45.   a.  Al < Mg < Na                              Atomic radii of the A-group elements
          In order of increasing atomic radius      decrease going from left to right across
                                                    a period.

      b.  Mg < Ca < Sr                              Atomic radii of the A-group elements
          In order of increasing atomic radius      increase going from top to bottom
                                                    down a group of the periodic table.

47.   a.  Ba < Ca < Mg                              All three elements are in Group 2A.
          In order of increasing first ionization   First ionization energies decrease going
          energy                                    from top to bottom within a group in
                                                    the periodic table.

      b.  Al < P < Cl                               All three elements are in the third
          In order of increasing first ionization   period. In general, first ionization
          energy                                    energies increase going from left to
                                                    right within a group.

49.   The elements in Group 7A would be expected to have the highest electron affinity. In general, electron affinity increases going from left to right across the periodic table, except for Group 8A elements, which have a complete valence shell.

# 3 Chemical Bonds

## ANSWERS TO REVIEW QUESTIONS

3.1 The elements in Group 8A, the noble gases, are characterized by a particularly stable electron configuration. The noble gases have a stable octet ($ns^2np^6$) in their highest main energy level.

3.2 Ions are charged units in which the number of protons does not equal the number of electrons. A monatomic ion forms when an atom gains one or more electrons to form a negatively charged species or loses one or more electrons to form a positively charged species. A polyatomic ion is a group of atoms with an electrical charge that act as a unit.

3.3 The sodium atom contains 11 electrons, and the nucleus of the sodium atom contains 11 protons. The atom is neutral. The sodium ion nucleus also contains 11 protons. The sodium ion differs from the sodium atom, because the ion contains 10 electrons and carries a single positive charge. The neon atom, like the sodium ion, contains 10 electrons. The neon atom has 10 protons in the nucleus and is a neutral species making it different from the sodium ion. The number of neutrons in the sodium ion nucleus also differs from the number of neutrons in the neon atom nucleus.

3.5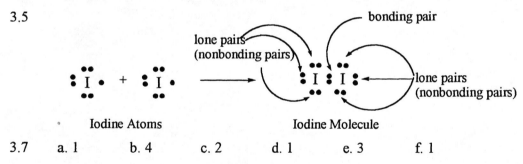

3.7   a. 1    b. 4    c. 2    d. 1    e. 3    f. 1

## ANSWERS TO PROBLEMS

**Lewis Structures: Elements**

1.   a. Na•

   b. ••
      •O•
      ••

   c. ••
      :F•
      ••

   d. •Ȧl•

Chapter 3: Chemical Bonds   293

**Monatomic Ions**

3.  a. Ba: $\longrightarrow$ Ba$^{2+}$ + 2e$^-$

    b. :Br• + 1e$^-$ $\longrightarrow$ :Br:$^-$

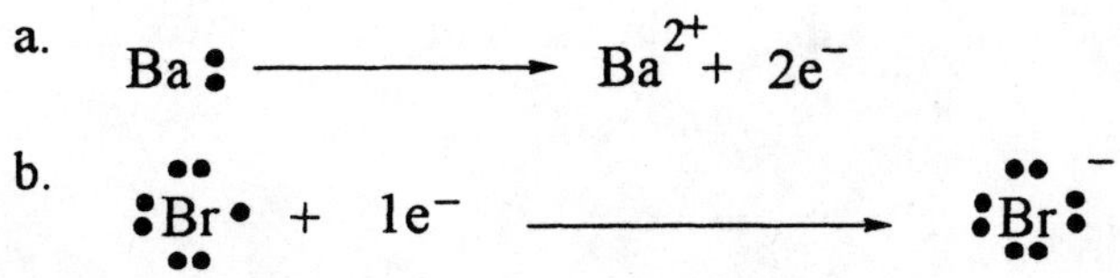

5.  a. 2+     b. 1−

**Lewis Structures: Ionic Compounds**

7.  a. Ca: + 2 :Br• $\longrightarrow$ Ca$^{2+}$ + 2 :Br:$^-$

    b. Mg: + •S• $\longrightarrow$ Mg$^{2+}$ + :S:$^{2-}$

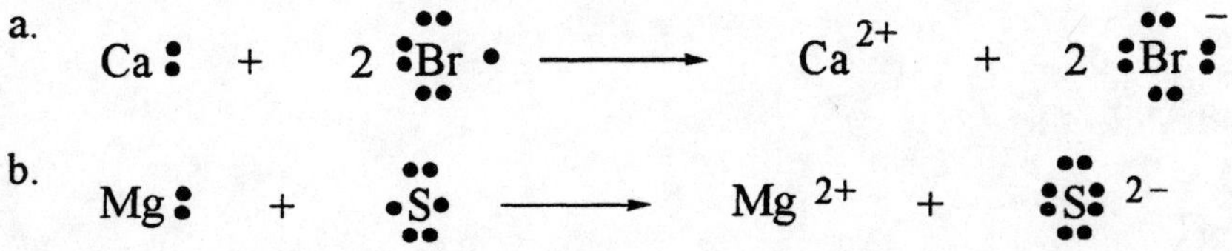

9.  a. Ca: and Ca$^{2+}$       b. •S• and :S:$^{2-}$

    c. Rb• and Rb$^+$         d. •P• and :P:$^{3-}$

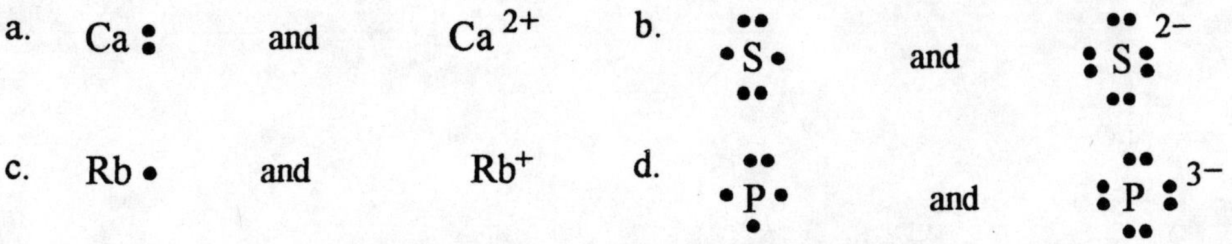

11. a. Na$^+$ :F:$^-$          b. K$^+$ :Cl:$^-$

    c. Na$^+$ :O:$^{2-}$ Na$^+$   d. :Cl:$^-$ Ca$^{2+}$ :Cl:$^-$

    e. :Br:$^-$ Mg$^{2+}$ :Br:$^-$

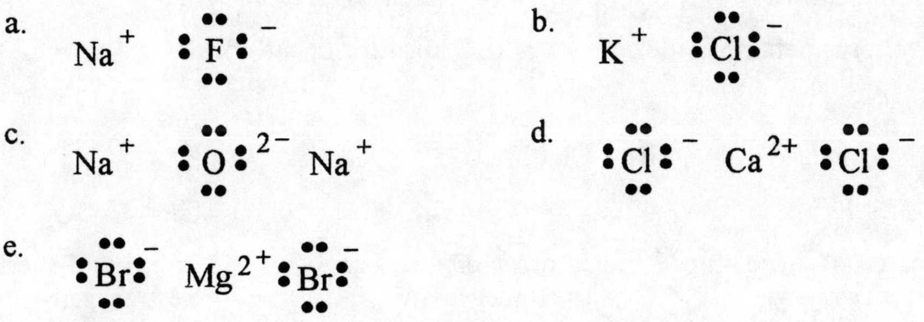

**Naming Ions and Ionic Compounds**

13. a. sodium ion              b. magnesium ion
    c. aluminum ion            d. chloride ion
    e. oxide ion               f. nitride ion

15. a. iron(III) ion           b. copper(II) ion
       (ferric ion)               (cupric ion)
    c. silver ion

17. a. Br$^-$                  b. Ca$^{2+}$
    c. K$^+$                   d. Fe$^{2+}$

19. a. sodium bromide          b. calcium chloride
    c. iron(II) chloride       d. lithium iodide
       (ferrous chloride)
    e. potassium sulfide       f. copper(I) bromide
                                  (cuprous bromide)

## 294  Chapter 3: Chemical Bonds

21. a. carbonate ion  b. hydrogen phosphate ion
    c. permanganate ion  d. hydroxide ion
23. a. $NH_4^+$  b. $HSO_4^-$
    c. $CN^-$  d. $NO_2^-$
25. a. $MgSO_4$  b. $NaHCO_3$
    c. $KNO_3$  d. $CaHPO_4$
27. a. $Fe_3(PO_4)_2$  b. $K_2Cr_2O_7$
    c. $CuI$  d. $NH_4NO_2$

**Covalent Bonds and Molecules**

29. a. H–P–H with lone pair on P, H below
    b. $CF_4$ Lewis structure with F atoms around C

31. a. $N_2O$  b. $P_4S_3$
    c. $PCl_5$  d. $SF_6$
33. a. carbon disulfide  b. dinitrogen tetrasulfide
    c. phosphorus pentafluoride  d. disulfur decafluoride

**Electronegativity**

35. a. N  b. Cl  c. F
37. a. B < N < F  b. Ca < As < Br  c. Ga < C < O
    In order of increasing electronegativity  In order of increasing electronegativity  In order of increasing electronegativity
39. a. ionic  b. polar covalent
    c. ionic  d. nonpolar covalent

**Lewis Structures**

41. a. H–C(H)(H)–O–H  b. H–C(=O)–H (formaldehyde)  c. H–N(H)–O–H (with lone pairs)
    d. H–N(H)–N(H)–H (hydrazine)  e. F–C(=O)–F  f. Cl–P(Cl)–Cl

43. a. •N=O•  b. :I–Be–I:  c. 

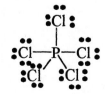

## Polyatomic Ions and Their Compounds

45.  a. [:Cl—O:]⁻    b. [:O–Cl–O:]⁻

 c. [O–P(O)(O)–O–H]²⁻    d. [:O–Br(O)–O:]⁻

47.  a. potassium nitrite    b. lithium cyanide
 c. ammonium iodide    d. sodium nitrate
 e. potassium permanganate    f. calcium sulfate

49.  a. sodium monohydrogen phosphate    b. ammonium phosphate
 c. aluminum nitrate    d. ammonium nitrate

## VSEPR Theory

51.  a. :Cl-Be-Cl:    two atoms bound to the central beryllium atom    linear
 no lone pairs on the central atom two electron sets

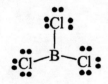

 b.     three atoms bound to the central boron atom    trigonal planar
 no lone pairs on the central atom three electron sets

 c.     two atoms bound to the central oxygen atom    bent
 two lone pairs on the central atom four electron sets

 d.     four atoms bound to the central silicon atom    tetrahedral
 no lone pairs on the central atom four electron sets

53.  a. :Cl-N-Cl: / :Cl:    three atoms bound to the central nitrogen atom    pyramidal
 one lone pair on the central atom four electron sets

 b. :Cl-S-Cl:    two atoms bound to the central sulfur atom    bent
 two lone pairs on the central atom four electron sets

## Polar and Nonpolar Molecules

55.    a.  polar                b.  polar             c.  nonpolar

57.    a.  $\overset{\delta+\ \ \delta-}{\text{H–O}}$        b.  $\overset{\delta+\ \ \delta-}{\text{N–Cl}}$        c.  nonpolar

59.    The $BeF_2$ molecule is nonpolar. The dipole electric fields in the bonds between the beryllium and fluorine atoms are directed in opposite directions and cancel each other out.

61.    a.  F–F < Cl–F < H–F        b.  H–H < H–Br < H–F
              In order of increasing polarity         In order of increasing polarity

63.    a.  $\overset{\ \ \ \ \ \ \ \ \ \ \delta+\ \delta-\ \ \delta+\ \delta-}{\text{F–F < Cl–F < H–F}}$        b.  $\overset{\ \ \ \ \ \ \ \ \ \ \delta+\ \delta-\ \ \delta+\ \delta-}{\text{H–H < H–Br < H–F}}$

# 4 Chemical Reactions

## ANSWERS TO REVIEW QUESTIONS

4.2 Two moles (or molecules) of hydrogen gas react with one mole (or molecule) of oxygen gas to produce two moles (or molecules) of water vapor.

4.3 a. balanced
b. not balanced
c. not balanced

4.6 The atomic mass of nitrogen refers to the mass of a single atom of nitrogen (N) given in amu. The atomic mass is determined as a weighted average of the isotopes of an atom, and these average atomic masses are given in the periodic table. The molecular mass of nitrogen refers to the mass of a molecule of nitrogen ($N_2$). The molecular mass of nitrogen is determined by multiplying the atomic mass of nitrogen by two, the number of atoms in a molecule of nitrogen.

4.8 The molecular mass of $CO_2$ is 44.010 u: the sum of the atomic mass of carbon and twice the atomic mass of oxygen. The molar mass of $CO_2$ is 44.010 g, which is the mass of one mole of $CO_2$.

4.9 $$1.00 \text{ mol } O_2 \times \frac{6.02 \times 10^{23} \text{ molecules } O_2}{1 \text{ mol } O_2} = 6.02 \times 10^{23} \text{ molecules } O_2$$

$$6.02 \times 10^{23} \text{ molecules } O_2 \times \frac{2 \text{ atoms } O}{1 \text{ molecule } O_2} = 1.20 \times 10^{24} \text{ atoms } O$$

4.10 $$1.00 \text{ mol } CaCl_2 \times \frac{6.02 \times 10^{23} \text{ Ca}^{2+} \text{ ions}}{1 \text{ mol } CaCl_2} = 6.02 \times 10^{23} \text{ Ca}^{2+} \text{ ions}$$

$$1.00 \text{ mol } CaCl_2 \times \frac{12.04 \times 10^{23} \text{ Cl}^- \text{ ions}}{1 \text{ mol } CaCl_2} = 1.20 \times 10^{24} \text{ Cl}^- \text{ ions}$$

4.11 At the molecular level: One molecule of methane gas will combine with two molecules of oxygen gas, forming a molecule of carbon dioxide gas and two molecules of water vapor. In terms of moles: A mole is by definition $6.02 \times 10^{23}$ molecules or atoms of a substance. The reaction can therefore also be interpreted as: One mole of methane gas reacts with two moles of oxygen gas to produce one mole of carbon dioxide gas and two moles of water vapor. In terms of mass: A mass of 16 g (one mole) of methane gas will react with 64 g (two moles) of oxygen gas, producing 44 g (one mole) of carbon dioxide gas and 36 g (two moles) of water vapor. Notice that a total of 80 g of reactants produces a total of 80 g of products as we expect according to the law of conservation of mass.

4.12 a. Two moles (or formula units) of solid potassium chlorate decompose to give two moles (or formula units) of solid potassium chloride and three moles (or molecules) of oxygen gas.

b. Two moles of solid aluminum react with six moles of hydrochloric acid in aqueous solution to give two moles of aluminum chloride in aqueous solution and three moles of hydrogen gas. (This equation could also be translated in terms of atoms, molecules, or formula units.)

4.14  a. Letter **C** represents the products.
b. Letter **B** represents the activation energy.
c. Letter **A** represents the products.
d. The reaction is exothermic since the energy of the reactants is greater than the energy of the products. Heat is given off in the reaction.

4.17  a. The rate of a reaction increases as the temperature increases.
b. Increasing reactant concentration increases the rate of a reaction.
c. A catalyst increases the rate of a reaction by providing a reaction pathway with lower activation energy.

# ANSWERS TO PROBLEMS

**Balancing Chemical Equations**

1. a. balanced    b. balanced    c. not balanced    d. not balanced

3.

| | | | | | | |
|---|---|---|---|---|---|---|
| a. | $Cl_2O_5$ | + | $H_2O$ | → | 2 $HClO_3$ | |
| b. | $V_2O_5$ | + | 2 $H_2$ | → | $V_2O_3$ | + 2 $H_2O$ |
| c. | 4 Al | + | 3 $O_2$ | → | 2 $Al_2O_3$ | |
| d. | Sn | + | 2 NaOH | → | $Na_2SnO_2$ | + $H_2$ |
| e. | $PCl_5$ | + | 4 $H_2O$ | → | $H_3PO_4$ | + 5 HCl |
| f. | $Na_3P$ | + | 3 $H_2O$ | → | 3 NaOH | + $PH_3$ |
| g. | $Cl_2O$ | + | $H_2O$ | → | 2 HClO | |
| h. | 2 $CH_3OH$ | + | 3 $O_2$ | → | 2 $CO_2$ | + 4 $H_2O$ |
| i. | 3 $Zn(OH)_2$ | + | 2 $H_3PO_4$ | → | $Zn_3(PO_4)_2$ | + 6 $H_2O$ |
| j. | $C_3H_8$ | + | 5 $O_2$ | → | 3 $CO_2$ | + 4 $H_2O$ |

**Volume Relationships in Chemical Equations**

5. $10.0 \text{ L } H_2 \text{ (g)} \times \dfrac{1 \text{ L } CH_4 \text{ (g) decomposed}}{2 \text{ L } H_2 \text{ (g) produced}} = 5.00 \text{ L } CH_4 \text{ (g) decomposed}$

7. $2 C_4H_{10} \text{(g)} + 13 O_2 \text{(g)} \rightarrow 8 CO_2 \text{(g)} + 10 H_2O \text{(g)}$

a. $0.529 \text{ L } C_4H_{10} \text{ (g)} \times \dfrac{10 \text{ L } H_2O \text{ (g) forms}}{2 \text{ L } C_4H_{10} \text{ (g) burned}} = 2.65 \text{ L } H_2O \text{ (g) forms}$

b. $16.1 \text{ L } C_4H_{10} \text{ (g)} \times \dfrac{13 \text{ L } O_2 \text{ (g) required}}{2 \text{ L } C_4H_{10} \text{ (g) burned}} = 105 \text{ L } O_2 \text{ (g) required}$

## Molecular Formulas and Formula Units

9.    a. 4      b. 4      c. 8      d. 6

11.   a. 12     b. 8

## Molecular Masses, Formula Masses, and Molar Masses

13.

| | | | |
|---|---|---|---|
| a. $C_6H_5Br$ | 6 × atomic mass of C = 6 × 12.011 u = | | 72.066 u |
| | 5 × atomic mass of H = 5 × 1.0079 u = | | 5.0395 u |
| | 1 × atomic mass of Br = 1 × 79.904 u = | | 79.904 u |
| | Molecular mass of $C_6H_5Br$ = | | 157.010 u |
| b. $H_3PO_4$ | 3 × atomic mass of H = 3 × 1.0079 u = | | 3.0237 u |
| | 1 × atomic mass of P = 1 × 30.974 u = | | 30.974 u |
| | 4 × atomic mass of O = 4 × 15.999 u = | | 63.996 u |
| | Molecular mass of $H_3PO_4$ = | | 97.994 u |
| c. $K_2Cr_2O_7$ | 2 × atomic mass of K = 2 × 39.098 u = | | 78.196 u |
| | 2 × atomic mass of Cr = 2 × 51.996 u = | | 103.992 u |
| | 7 × atomic mass of O = 7 × 15.999 u = | | 111.993 u |
| | Formula mass of $K_2Cr_2O_7$ = | | 294.181 u |

15.    a. Formula mass of $MnO_2$

       1 × atomic mass of Mn = 1 × 54.938 u =    54.938 u
       2 × atomic mass of O = 2 × 15.999 u =    31.998 u
       Formula mass of $MnO_2$ =   86.936 u
       Molar mass of $MnO_2$ = 86.936 g/mol

$$0.00500 \text{ mol } MnO_2 \times \frac{86.936 \text{ g}}{1 \text{ mol } MnO_2} = 0.435 \text{ g } MnO_2$$

   b. Formula mass of $CaH_2$

       1 × atomic mass of Ca = 1 × 40.078 u =    40.078 u
       2 × atomic mass of H = 2 × 1.00794 u =    2.01588 u
       Formula mass of $CaH_2$ =   42.094 u
       Molar mass of $CaH_2$ = 42.094 g/mol

$$1.12 \text{ mol } CaH_2 \times \frac{42.094 \text{ g}}{1 \text{ mol } CaH_2} = 47.1 \text{ g } CaH_2$$

   c. Molecular mass of $C_6H_{12}O_6$

       6 × atomic mass of C = 6 × 12.011 u =    72.066 u
       12 × atomic mass of H = 12 × 1.0079 u =    12.0948 u
       6 × atomic mass of O = 6 × 15.999 u =    95.994 u
       Molecular mass of $C_6H_{12}O_6$ = 180.155 u
       Molar mass of $C_6H_{12}O_6$ = 180.155 g/mol

$$0.250 \text{ mol } C_6H_{12}O_6 \times \frac{180.155 \text{ g}}{1 \text{ mol } C_6H_{12}O_6} = 45.0 \text{ g } C_6H_{12}O_6$$

17.  a. Molecular mass of $HNO_3$
   $1 \times$ atomic mass of H = $1 \times 1.0079$ u =     1.0079 u
   $1 \times$ atomic mass of N = $1 \times 14.0067$ u =   14.0067 u
   $3 \times$ atomic mass of O = $3 \times 15.999$ u =    47.997 u
   Molecular mass of $HNO_3$ = 63.012 u
   Molar mass of $HNO_3$ = 63.012 g/mol

$$98.6 \text{ g } HNO_3 \times \frac{1 \text{ mol } HNO_3}{63.012 \text{ g } HNO_3} = 1.56 \text{ mol } HNO_3$$

b. Molecular mass of $CBr_4$
   $1 \times$ atomic mass of C = $1 \times 12.011$ u =    12.011 u
   $4 \times$ atomic mass of Br = $4 \times 79.904$ u =   319.616 u
   Molecular mass of $CBr_4$ = 331.627 u
   Molar mass of $CBr_4$ = 331.627 g/mol

$$9.45 \text{ g } CBr_4 \times \frac{1 \text{ mol } CBr_4}{331.627 \text{ g } CBr_4} = 0.0285 \text{ mol } CBr_4$$

c. Formula mass of $FeSO_4$
   $1 \times$ atomic mass of Fe = $1 \times 55.845$ u =   55.845 u
   $1 \times$ atomic mass of S = $1 \times 32.066$ u =    32.066 u
   $4 \times$ atomic mass of O = $4 \times 15.999$ u =    63.996 u
   Formula mass of $FeSO_4$ = 151.907 u
   Molar mass of $FeSO_4$ = 151.907 g/mol

$$9.11 \text{ g } FeSO_4 \times \frac{1 \text{ mol } FeSO_4}{151.907 \text{ g } FeSO_4} = 0.0600 \text{ mol } FeSO_4$$

d. Formula mass of $Pb(NO_3)_2$
   $1 \times$ atomic mass of Pb = $1 \times 207.2$ u =    207.2 u
   $2 \times$ atomic mass of N = $2 \times 14.0067$ u =   28.0134 u
   $6 \times$ atomic mass of O = $6 \times 15.999$ u =    95.994 u
   Formula mass of $Pb(NO_3)_2$ = 331.2 u
   Molar mass of $Pb(NO_3)_2$ = 331.2 g/mol

$$11.8 \text{ g } Pb(NO_3)_2 \times \frac{1 \text{ mol } Pb(NO_3)_2}{331.2 \text{ g } Pb(NO_3)_2} = 0.0356 \text{ mol } Pb(NO_3)_2$$

**Mole and Mass Relationships in Chemical Equations**

19.  a. $2.09 \text{ mol } C_8H_{18} \times \dfrac{16 \text{ mol } CO_2 \text{ produced}}{2 \text{ mol } C_8H_{18} \text{ burned}} = 16.7 \text{ mol } CO_2 \text{ produced}$

   b. $4.47 \text{ mol } C_8H_{18} \times \dfrac{25 \text{ mol } O_2 \text{ required}}{2 \text{ mol } C_8H_{18} \text{ burned}} = 55.9 \text{ mol } O_2 \text{ required}$

# Chapter 4: Chemical Reactions

21. Write the balanced equation:
$$N_2 + 3H_2 \rightarrow 2NH_3$$
Calculate the moles of $H_2$ available:
$$440 \text{ g } H_2 \times \frac{1 \text{ mol } H_2}{2.0158 \text{ g } H_2} = 218.26 \text{ mol } H_2$$
Relate the moles of $H_2$ reacted to the moles of $NH_3$ produced:
$$218.26 \text{ mol } H_2 \text{ reacted} \times \frac{2 \text{ mol } NH_3 \text{ produced}}{3 \text{ mol } H_2 \text{ reacted}} = 145.50 \text{ mol } NH_3 \text{ produced}$$
Calculate the mass in grams of $NH_3$:
$$145 \text{ mol } NH_3 \times \frac{17.031 \text{ g}}{1 \text{ mol } NH_3} = 2480 \text{ g } NH_3$$

23. Write the balanced equation:
$$2H_2O_2 \rightarrow 2H_2O + O_2$$
Calculate the moles of $H_2O_2$ available:
$$24.0 \text{ g } H_2O_2 \times \frac{1 \text{ mol } H_2O_2}{34.0138 \text{ g } H_2O_2} = 0.706 \text{ mol } H_2O_2$$
Relate the moles of $H_2O_2$ reacted to the moles of $O_2$ produced:
$$0.706 \text{ mol } H_2O_2 \text{ reacted} \times \frac{1 \text{ mol } O_2 \text{ produced}}{2 \text{ mol } H_2O_2 \text{ reacted}} = 0.353 \text{ mol } O_2 \text{ produced}$$
Calculate the mass in grams of $O_2$:
$$0.353 \text{ mol } O_2 \times \frac{31.998 \text{ g}}{1 \text{ mol } O_2} = 11.3 \text{ g } O_2$$

25. Calculate the moles of $C_7H_8$ available:
$$829 \text{ g } C_7H_8 \times \frac{1 \text{ mol } C_7H_8}{92.1402 \text{ g } C_7H_8} = 9.00 \text{ mol } C_7H_8$$
Relate the moles of $C_7H_2$ reacted to the moles of $C_7H_5N_3O_6$ produced:
$$9.00 \text{ mol } C_7H_8 \text{ reacted} \times \frac{1 \text{ mol } C_7H_5N_3O_6 \text{ produced}}{1 \text{ mol } C_7H_8 \text{ reacted}} = 9.00 \text{ mol } C_7H_5N_3O_6 \text{ produced}$$
Calculate the mass in grams of $C_7H_5N_3O_6$:
$$9.00 \text{ mol } C_7H_5N_3O_6 \times \frac{227.131 \text{ g}}{1 \text{ mol } C_7H_5N_3O_6} = 2.04 \times 10^3 \text{ g } C_7H_5N_3O_6$$

27. Write the balanced equation:
$$NH_3 + 2\,O_2 \rightarrow HNO_3 + H_2O$$
Calculate the moles of $NH_3$ available:
$$971\text{ g }NH_3 \times \frac{1\text{ mol }NH_3}{17.0304\text{ g }NH_3} = 57.0\text{ mol }NH_3$$
Relate the moles of $NH_3$ reacted to the moles of $HNO_3$ produced:
$$57.0\text{ mol }NH_3\text{ reacted} \times \frac{1\text{ mol }HNO_3\text{ produced}}{1\text{ mol }NH_3\text{ reacted}} = 57.0\text{ mol }HNO_3\text{ produced}$$
Calculate the mass in grams of $HNO_3$:
$$57.0\text{ mol }HNO_3 \times \frac{63.0116\text{ g}}{1\text{ mol }HNO_3} = 3.59 \times 10^3\text{ g }HNO_3$$

## Structure, Stability, and Spontaneity

29. endothermic

31. A reaction mechanism is the step-by-step pathway by which a reaction proceeds.

## Le Châtelier's Principle

33. a. Equilibrium shifts to the left, producing more $H_2$ and $Cl_2$ and using up HCl.
    b. Equilibrium shifts to the right, producing more CO and $O_2$ and consuming $CO_2$.
    c. Equilibrium shifts to the right, producing more $O_3$ and consuming $O_2$.

# 5 Oxidation and Reduction

## ANSWERS TO REVIEW QUESTIONS

5.1   a. Oxygen occurs in the atmosphere mainly as $O_2$ molecules.
b. Oxygen forms compounds with nearly all other elements. Oxygen in the compound water makes up about 89% by mass of the water on Earth's surface. Combined with silicon and many other elements, oxygen makes up about 45.5% of Earth's solid crust.

5.2   Pure oxygen is prepared by fractional distillation of liquefied air.

5.3   a. Oxygen reacts with metals to form metal oxides. For example, magnesium reacts with oxygen to form magnesium oxide, a basic metal oxide.
$$2\ Mg\ (s)\ +\ O_2\ (g)\ \rightarrow\ 2\ MgO\ (s)$$
b. Nonmetals react with oxygen to form acidic oxides. For example, sulfur reacts with oxygen to form sulfur dioxide, an acidic oxide.
$$S\ (s)\ +\ O_2\ (g)\ \rightarrow\ SO_2\ (g)$$

5.4   a.  $C_3H_8\ (g)\ +\ 5\ O_2\ (g)\ \rightarrow\ 3\ CO_2\ (g)\ +\ 4\ H_2O\ (l)$

   Carbon dioxide and water are both nonmetal oxides.

   b.  $2\ H_2S\ (g)\ +\ 3\ O_2\ (g)\ \rightarrow\ 2\ SO_2\ (g)\ +\ 2\ H_2O\ (l)$

   Sulfur dioxide and water are both nonmetal oxides.

5.9   An atom of an element or combined in a compound that undergoes an increase in oxidation number is oxidized. An atom of an element or combined in a compound that undergoes a decrease in oxidation number is reduced. In the reaction of iron with copper(II) nitrate solution, the oxidation number of the iron increases from 0 to +2, and therefore the iron atom is oxidized. The copper undergoes a decrease in oxidation number from +2 to 0 and is reduced.
$$Fe\ (s)\ +\ Cu(NO_3)_2\ (aq)\ \rightarrow\ Fe(NO_3)_2\ (aq)\ +\ Cu\ (s)$$

## ANSWERS TO PROBLEMS

**Reactions of Oxygen**

1.

| | | | | | | |
|---|---|---|---|---|---|---|
| a. | C | + $O_2$ | $\rightarrow$ | $CO_2$ | | |
| b. | $2\ C_2H_6$ | + $7\ O_2$ | $\rightarrow$ | $4\ CO_2$ | + $6\ H_2O$ | |
| c. | $N_2$ | + $O_2$ | $\rightarrow$ | $2\ NO$ | | |
| d. | $C_3H_8$ | + $5\ O_2$ | $\rightarrow$ | $3\ CO_2$ | + $4\ H_2O$ | |

## Oxidation Numbers

3. a. 0  b. +4  c. -2  d. +6

5. Hydrogen is assigned an oxidation number of +1 when it is combined in a compound with a more electronegative element (nonmetals) such as oxygen or nitrogen. The oxidation number of hydrogen is –1 when it is combined in a compound with less electronegative elements (metals) such as calcium or lithium.

   Oxygen is normally assigned an oxidation number of –2; however, there are some exceptions. The oxygen atoms in peroxides such as hydrogen peroxide, $H_2O_2$, are given oxidation numbers of –1. Oxygen, when combined with the more electronegative fluorine atom, will also not have an oxidation number of –2. In the compound $OF_2$, for example, the oxidation number of oxygen is +2.

## Recognizing Redox Reactions

7. a. oxidation (Cl: +4 to +5)  b. oxidation (Mn: +2 to +4)
   c. reduction (Br: +1 to 0)  d. reduction (Sb: +3 to 0)

9. Some of the $SO_4^{2-}$ is reduced to give $SO_2$ (S: +6 to +4), and some of the $SO_4^{2-}$ does not undergo reduction or oxidation.

## Oxidizing Agents and Reducing Agents

11. a. Oxidizing agent: $O_2$  b. Oxidizing agent: $O_2$
    Reducing agent: Al  Reducing agent: $SO_2$

    c. Oxidizing agent: HCl  d. Oxidizing agent: $O_2$
    Reducing agent: Fe  Reducing agent: $CS_2$

13. a. S is oxidized; N is reduced.  b. I is oxidized; Cr is reduced.

15. $I^-$ is oxidized; $Cl_2$ is reduced.

17. reduced.

19. Acetylene is reduced; it is gaining hydrogens. ($H_2$ is being oxidized to the +1 state.)

21. $NO_2^-$ is reduced; ascorbic acid is a reducing agent.

# 6 Gases

## ANSWERS TO REVIEW QUESTIONS

6.1.   Troposphere; stratosphere.

6.2.   $N_2$: 78 %; $O_2$: 21 %; Ar: 1 %

6.6.   The mercury barometer measures the pressure of the system. It is constructed of a long glass tube (sealed at one end) inverted into a dish of mercury. The pressure of the atmosphere (or any system in which it is used) presses down on the mercury in the dish and drives mercury up into the tube. The greater the pressure, the greater the height of the mercury in the column. The average atmospheric pressure at sea level supplies force to support a mercury column 760 mm in height. Mercury is an ideal fluid to use for measuring atmospheric pressure because of its density. Compared to other, less dense fluids like water, a lower volume of mercury is required to "counterbalance" the pressure exerted by the atmosphere. This means that a column of an appropriate length (~76 cm) can be used for the apparatus. A barometer with water would be more than ten times longer.

6.9.   The high pressure compresses the gas into a very small volume. Thus, a greater amount of the gas can be stored in the given volume of a gas tank.

6.13.  Temperature: 0 °C (273 K); pressure: 1 atm. Often one would like to talk about the mass, moles, or volume of a gas without having to deal with the changes due to temperature and pressure. STP gives a constant set of parameters that approximate typical room conditions that can be used for these discussions.

6.16.  a. increase in pressure      b. increase in pressure      c. increase in pressure

6.19.  Not unless you like dead fish. The solubility of gases in water decreases with temperature. Boiling water will drive out nearly all of the dissolved oxygen needed by the fish to live.

6.22.  The air will be near body temperature (37°C), at which the vapor pressure of water is about 50 mmHg (estimate from Table 6.2).

## ANSWERS TO PROBLEMS

**Kinetic-Molecular Theory**

1.   a. The temperature is decreasing          b. The pressure decreases.

3.   a. Container A would have a greater density.
     b. The densities of A and B would be the same (but of different size).
     c. Container B would have a greater density.

**Chapter 6: Gases**

**Pressure**

5.   a. $\dfrac{0.985 \text{ atm}}{1} \times \dfrac{760 \text{ mmHg}}{1 \text{ atm}} = 749 \text{ mmHg}$

   b. $\dfrac{849 \text{ mm Hg}}{1} \times \dfrac{1 \text{ atm}}{760 \text{ mmHg}} = 1.12 \text{ atm}$

   c. $\dfrac{721 \text{ mm Hg}}{1} \times \dfrac{1 \text{ atm}}{760 \text{ mmHg}} = 0.949 \text{ atm}$

7.   213 mm (8.39 in.)

**Boyle's Law**

9.   a. $V_1 P_1 = V_2 P_2$    (1572 mmHg)(521 mL) = (752 mmHg)($V_2$)    $V_2 = 1090 \text{ mL}$

   b. $V_1 P_1 = V_2 P_2$    (1572 mmHg)(521 mL) = ($P_2$)(315 mL)    $P_2 = 2600 \text{ mmHg}$

11.   a. $\dfrac{750.0 \text{ mm Hg}}{1} \times \dfrac{1 \text{ atm}}{760 \text{ mmHg}} = 0.98684 \text{ atm}$

   $V_1 P_1 = V_2 P_2$.   (150 atm)(60.0 L) = (0.98684 atm)($V_2$)    $V_2 = 9.12 \times 10^3 \text{ L}$

   b. $\dfrac{9120 \text{ L}}{1} \times \dfrac{1 \text{ min}}{8 \text{ L}} = 1140 \text{ min} = 19 \text{ h}$

**Charles's Law**

13.   100 °C = 373 K     10 °C = 283 K

   $\dfrac{V_1}{T_1} = \dfrac{V_2}{T_2}$    $\dfrac{154 \text{ mL}}{373 \text{ K}} = \dfrac{V_2}{283 \text{ K}}$    $V_2 = 117 \text{ mL}$

15.   305 °C = 578 K

   $\dfrac{V_1}{T_1} = \dfrac{V_2}{T_2}$    $\dfrac{567 \text{ mL}}{578 \text{ K}} = \dfrac{425 \text{ mL}}{T_2}$    $T_2 = 433 \text{ K} = 160 \text{ °C}$

**Avogadro's Law and Molar Volume**

17.   5.0 g $H_2$ has the greatest number of molecules

   a. $\dfrac{5.0 \text{ g } H_2}{1} \times \dfrac{1 \text{ mol } H_2}{2 \text{ g}} = 2.5 \text{ mol } H_2$    b. $\dfrac{50 \text{ L } SF_6}{1} \times \dfrac{1 \text{ mol } SF_6}{22.4 \text{ L}} = 2.2 \text{ mol } SF_6$

   c. $\dfrac{1.0 \times 10^{24} \text{ molecules } CO_2}{1} \times \dfrac{1 \text{ mol } CO_2}{6.02 \times 10^{23} \text{ molecules}} = 1.7 \text{ mol } CO_2$

19.   $\dfrac{0.837 \text{ g Xe}}{1} \times \dfrac{\text{mol}}{131.3 \text{ g Xe}} \times \dfrac{22.4 \text{ L}}{\text{mol}} = 0.143 \text{ L} = 143 \text{ mL}$

## The Combined Gas Law

21. 25 °C = 298 K     755 °C = 1028 K

$$\frac{P_1V_1}{T_1} = \frac{P_2V_2}{T_2} \qquad \frac{721\,\text{mmHg}}{298\,\text{K}} = \frac{P_2}{1028\,\text{K}} \qquad P_2 = 2490\,\text{mmHg}$$

23. −15 °C = 258 K   25 °C = 298 K

$$\frac{P_1V_1}{T_1} = \frac{P_2V_2}{T_2} \qquad \frac{(191\,\text{mmHg})(2.53\,\text{m}^3)}{258\,\text{K}} = \frac{(1142\,\text{mmHg})(V_2)}{298\,\text{K}} \qquad V_2 = 0.489\,\text{m}^3$$

25. 15 °C = 288 K

$$\frac{P_1V_1}{T_1} = \frac{P_2V_2}{T_2} \qquad \frac{(760\,\text{mmHg})(4.65\,\text{mL})}{273\,\text{K}} = \frac{(756\,\text{mmHg})(V_2)}{288\,\text{K}} \qquad V_2 = 4.93\,\text{L}$$

## The Ideal Gas Law

27. $PV = nRT$     62 °C = 335 K

$(1.38\,\text{atm})(V) = (1.12\,\text{mol})(0.0821\,\text{L-atm-mol}^{-1}\text{-K}^{-1})(335\,\text{K})$     $V = 22.3\,\text{L}$

29. $PV = nRT$     29 °C = 302 K

$(P)(3.96\,\text{L}) = (4.64\,\text{mol})(0.0821\,\text{L·atm·mol}^{-1}\text{·K}^{-1})(302\,\text{K})$     $P = 29.1\,\text{atm}$

31. $PV = nRT$     45 °C = 318 K     968 mmHg = 0.918 atm

$(0.918\,\text{atm})(2.22\,\text{L}) = (n)(0.0821\,\text{L·atm·mol}^{-1}\text{·K}^{-1})(318\,\text{K})$     $n = 0.0781\,\text{mol Kr}$

## Gas Densities

33.  a. 1 mol CO = 28.01 g CO         $\dfrac{28.0\,\text{g CO}}{22.4\,\text{L}} = 1.25\,\text{g/L}$

   b. 1 mol AsH$_3$ = 77.95 g AsH$_3$     $\dfrac{77.95\,\text{g AsH}_3}{22.4\,\text{L}} = 3.48\,\text{g/L}$

   c. 1 mol Ar = 39.95 g Ar         $\dfrac{39.95\,\text{g Ar}}{22.4\,\text{L}} = 1.78\,\text{g/L}$

   d. 1 mol N$_2$ = 28.01 g N$_2$       $\dfrac{28.01\,\text{g N}_2}{22.4\,\text{L}} = 1.25\,\text{g/L}$

## Dalton's Law of Partial Pressures

35. From the table in the chapter, water has a partial pressure of 32 mmHg at 30°C.

$P_{\text{total}} = P_{O_2} + P_{H_2O}$     742 mmHg = $P_{O_2}$ + 32 mmHg     $P_{O_2}$ = 710 mmHg

37. 250 mmHg

39. $\dfrac{5.7\,\text{atm}}{6.0\,\text{atm}} \times 100 = 95\,\%$

# 7 Liquids and Solids

## ANSWERS TO REVIEW QUESTIONS

7.2. Both liquids and solids may have intermolecular forces; they are not easily compressed; and they have relatively high densities (compared to gases). Solids are different from liquids in that they have well-defined packing of atoms/molecules and that the forces that hold these atoms/molecules are sufficiently great to prevent movement of the atoms/molecules relative to one another.

7.4. A larger ionic charge and a smaller size cause the melting point to increase.

7.5. A polar liquid, that is, a liquid made up of polar molecules, contains molecules that can form relatively strong δ+ ··· δ− attachments between molecules. These attachments cause the molecules *not* to want to become isolated and go into the gaseous phase. Hence, it will take a higher temperature to vibrate the molecules sufficiently to separate them. Nonpolar liquids lack these attachments and the molecules are more free, at a lower temperature to escape to the gaseous phase.

7.6. $CH_4$ is a molecule with no dipole moment. There are few forces of attraction between the molecules to hold them together in a solid or liquid phase. They separate from one another and become a gas. $H_2O$ is a molecule with a strong dipole nature. There are great forces of attraction between the molecules, which hold them close to one another in a liquid phase.

7.7. Methanol. Methanol has stronger intermolecular forces because of its ability to hydrogen bond. The strong intermolecular forces lead to a greater tendency to stay in the condensed state.

7.8. a. Xe atoms are larger than Ne atoms and thus have greater dispersion forces.
b. Ethane. Ethane has a higher molecular weight and therefore stronger dispersion forces.

7.10. At low enough temperatures, the kinetic energy of $O_2$ molecules is sufficiently reduced so that the weakest of bonds, dispersion forces, can take place. These temporary bonds hold molecules near each other, creating a liquid state.

7.14. The heat of vaporization is the amount of heat required to vaporize a certain quantity of liquid. If there are great intermolecular forces holding the molecules in a liquid together, it will take more heat to disrupt these attractions in order to disperse the molecules into a gaseous state.

# Chapter 7: Liquids and Solids 309

# ANSWERS TO PROBLEMS

### Some General Considerations

1. Gases are easily compressed, while liquids and solids are not. Gases have great distances between individual molecules, and liquids and solids have tighter packing. Gases exhibit little intermolecular forces, while these forces in liquids and solids may be great.

3. Boiling an egg will take longer at higher altitudes because the temperature of the boiling water is lower (due to the decreased atmospheric pressure). Since the heat transfer (i.e., specific heat) will not change much with altitude, frying an egg will take about the same time.

5. a. melting            b. vaporization

### Intermolecular Forces

7. $CS_2$. Neither has a dipole moment, but $CCl_4$, a larger molecule, has much greater dispersion forces and hence will have a greater tendency to stay in the liquid state.

9. $C_2H_5OH$. $C_2H_5OH$ molecules have the ability to hydrogen bond to one another. This creates a network of molecules held by intermolecular forces that would tend to stay in the liquid phase at a higher temperature.

11. $H_2S$ (lowest), $H_2Se$, $H_2Te$ (highest). They are all relatively nonpolar molecules; the dispersion forces for the higher molecular weight compounds ($H_2Se$ and $H_2Te$) would be greater.

### Heat of Vaporization

13. $$\frac{45 \text{ cal}}{\text{g}} \times \frac{159.8 \text{ g}}{1 \text{ mol Br}_2} \times \frac{1 \text{ kcal}}{1000 \text{ cal}} = 7.2 \text{ kcal/mol}$$

15. $$\frac{5.81 \text{ kcal}}{\text{mol}} \times \frac{1 \text{ mol C}_2\text{H}_4\text{O}_2}{60.05 \text{ g}} \times \frac{1.0 \text{ g}}{1} = 9.68 \times 10^{-2} \text{ kcal}$$

17. $$\frac{1.00 \text{ cal}}{\text{g} \cdot {}^\circ\text{C}} \times \frac{1 \text{ kcal}}{1000 \text{ cal}} \times \frac{25.0 \text{ g H}_2\text{O}}{1} \times \frac{42.0 \, {}^\circ\text{C}}{1} = 1.05 \text{ kcal}$$

### Heat of Fusion

19. $$\frac{2.58 \text{ kcal}}{\text{mol}} \times \frac{1 \text{ mol C}_3\text{H}_6\text{O}}{58.08 \text{ g}} \times \frac{7.75 \text{ g}}{1} = 0.344 \text{ kcal}$$

21. $$\frac{355 \text{ g H}_2\text{O (s)}}{1} \times \frac{80 \text{ cal}}{\text{g}} \times \frac{1 \text{ kcal}}{1000 \text{ cal}} = 28.4 \text{ kcal}$$

# 8 Solutions

## ANSWERS TO REVIEW QUESTIONS

8.3. The ion-ion interactions will be overcome by ion-dipole interactions.

8.4. a. No. To dissolve NaCl, the individual sodium and chloride ions must be broken apart and stabilized by other forces. Since there are no polar attractions between the NaCl ions and benzene, the NaCl ions will remain ionically bonded together, and dissolution will not occur.
b. No. Motor oil cannot form bonds with water and thus cannot break into the strongly held water network.
c. Yes. Benzene molecules are loosely held together. Entropy (randomness) will be a sufficient force to allow the molecules of motor oil to disperse themselves throughout the benzene network.

8.5. Ethyl alcohol, because of its –OH groups, can hydrogen bond with water, while ethyl chloride cannot. Solubility in a solvent requires that the solute fit into the solvent network. In water this requires breaking some of the water-to-water hydrogen bonds, a process not favorable *unless* other bonds are made in exchange. Ethyl alcohol can make these bonds; ethyl chloride cannot.

8.8. Saturated. The amount of material that will precipitate out of solution will be exactly that amount by which the solution is over-saturated.

8.9. The partial pressure of carbon dioxide in soda (closed container) is relatively high because of the pressure over the solution in the headspace of the bottle. When the cap is removed, the pressure is lowered and the carbon dioxide solubility is lowered. Consequently, the carbon dioxide comes out of solution.

8.10. Fish need oxygen to live. Since the solubility of oxygen in cold water is greater than it is in warm water, cold water can support more aquatic life than can warm water. Hence, the major fisheries are located in cold waters.

## ANSWERS TO PROBLEMS

**Solubility of Ionic Compounds**

1. a. Soluble; both alkali metals and halides (outside "edges" of the periodic table) are generally soluble.
   b. Soluble; all nitrates are soluble.
   c. Insoluble; most carbonates are insoluble (except for alkali and ammonium salts).
   d. Insoluble; most phosphates are insoluble (except for alkali and ammonium salts).

**Dynamic Equilibria**

3. Dissolving of the solid and precipitation of the solute.

5. a. unsaturated                      b. 38°C

Chapter 8: Solutions 311

## Solubility of Covalent Compounds

7. Soluble: a, b; each has an oxygen atom, which can form hydrogen bonds to water.

## Molarity

9. a. $\dfrac{6.00 \text{ mol HCl}}{2.50 \text{ L}} = 2.40 \text{ M}$   b. $\dfrac{0.00700 \text{ mol Li}_2\text{CO}_3}{10.0 \text{ mL}} \times \dfrac{1000 \text{ mL}}{1 \text{ L}} = 0.700 \text{ M}$

11. a. $\dfrac{8.90 \text{ g H}_2\text{SO}_4}{100.0 \text{ mL}} \times \dfrac{1 \text{ mol H}_2\text{SO}_4}{98.08 \text{ g}} \times \dfrac{1000 \text{ mL}}{1 \text{ L}} = 0.907 \text{ M}$

    b. $\dfrac{439 \text{ g C}_6\text{H}_{12}\text{O}_6}{1.25 \text{ M}} \times \dfrac{1 \text{ mol C}_6\text{H}_{12}\text{O}_6}{180.2 \text{ g}} = 1.95 \text{ M}$

13. a. $\dfrac{1.00 \text{ mol NaOH}}{1 \text{ L}} \times \dfrac{40.00 \text{ g}}{1 \text{ mol NaOH}} \times \dfrac{2.00 \text{ L}}{1} = 80.0 \text{ g NaOH}$

    b. $\dfrac{4.25 \text{ mol C}_6\text{H}_{12}\text{O}_6}{1 \text{ L}} \times \dfrac{180.2 \text{ g}}{1 \text{ mol C}_6\text{H}_{12}\text{O}_6} \times \dfrac{1 \text{ L}}{1000 \text{ mL}} \times \dfrac{10.0 \text{ mL}}{1} = 7.65 \text{ g C}_6\text{H}_{12}\text{O}_6$

15. $\dfrac{1.25 \text{ mol NaOH}}{1 \text{ L}} \times \dfrac{1 \text{ L}}{6.00 \text{ mol NaOH}} = 0.208 \text{ L}$

17. $\dfrac{8.10 \text{ g KMnO}_4}{1} \times \dfrac{1 \text{ mol KMnO}_4}{158.0 \text{ g}} \times \dfrac{1 \text{ L}}{0.0250 \text{ mol KMnO}_4} = 2.05 \text{ L}$

## Percent Composition

19. a. $\dfrac{35.0 \text{ mL water}}{725 \text{ mL solution}} \times 100 = 4.83\%$   b. $\dfrac{78.9 \text{ mL acetone}}{1550 \text{ mL solution}} \times 100 = 5.09\%$

21. a. $\dfrac{4.12 \text{ g NaOH}}{104.12 \text{ g total}} \times 100 = 3.96\%$

    b. mass of ethanol = 5.00 mL × 0.789 g/mL = 3.945 g

    $\dfrac{3.945 \text{ g ethanol}}{53.95 \text{ g total}} \times 100 = 7.31\%$

23. 10.0% of 775 g = 77.5 g NaCl   You would add 77.5 g NaCl to 697.5 g of water.

25. 2.00% of 2.00 L = 0.0400 L acetic acid   You would add 0.0400 L (40.0 mL) of glacial acetic acid to 1.96 L of water.

27. 1.5% $MgSO_4 = \dfrac{1.5 \text{ g MgSO}_4}{100 \text{ mL solution}} \times \dfrac{250.0 \text{ mL}}{1} = 3.75$ g of $MgSO_4$ dissolved in 250 mL of solution.

29. 0.1% = 0.1 g/100 g = 100 mg/100 mL = 100 mg/dL

## Osmolarity and Colligative Properties

31.     Solute particles     a. two     b. one     c. three
        Osmol per mole     a. two     b. one     c. three

33.     a. 0.1 M $NaHCO_3$                 b. 1 M NaCl

35.     a. 0.5 mol                 b. 1.0 mol          c. 0.33 mol

37.     a. same                 b. 2 osmol/L glucose ($C_6H_{12}O_6$)

## Colloids

39. Colloidal dispersions are larger particles distributed evenly throughout a solution. Unlike true solutions, these solutions scatter light. Suspensions are temporarily nonhomogeneous mixtures of even larger sized particles. The particles will settle out with time or they can be filtered with filter paper.

# 9 Acids and Bases I

## ANSWERS TO REVIEW QUESTIONS

9.1. Acid: turns litmus to red, reacts with active metals like zinc and iron to produce $H_2$ gas, has a sour taste, produces a stinging feeling on the skin, and reacts with bases to form water and a salt. Base: Turns litmus to blue, feels slippery to the touch, tastes bitter, and reacts with acids to form water and a salt. None of these remains upon neutralization.

9.2 Hydronium ion ($H_3O^+$). Hydroxide ion ($OH^-$).

9.4 a. $H_2O$     b. $NH_3$     c. $H_3O^+$     d. $HCO_2^-$

9.5. a. acid (COOH group)     b. acid (ionizable H atom written first)
c. base (conjugate base of HCN)     d. acid (conjugate acid of an amine)
e. acid (COOH group)     f. base (an amine)

9.7. A polyprotic acid is a compound that can release more than one $H^+$ to the solution. $CH_4$ is not a polyprotic acid (nor is it an acid at all), because all the hydrogens are covalently bonded to the carbon, they never release into the solution.

9.9. An acidic anhydride is a substance that reacts with water to produce an acid. Nonmetal oxides are examples of this class. $SO_3$ is the acid anhydride of $H_2SO_4$. A basic anhydride is a substance that reacts with water to produce a base. Metal oxides are examples of this class. BaO is the base anhydride of $Ba(OH)_2$.

9.10. $H_2CO_3 \rightarrow H_2O + CO_2$

9.11. $H_2SO_4$

9.14. $NaHCO_3$, $CaCO_3$, $Al(OH)_3$, $Mg(OH)_2$, $MgCO_3$, $AlNa(OH)_2CO_3$

9.16. The poor absorption of $Mg^{2+}$ ions in the digestive tract causes a net movement of water molecules into the colon, which results in the laxative effect. Yes.

## ANSWERS TO PROBLEMS

**Brønsted-Lowry Acids and Bases**

1. a. $HClO_2 + H_2O \rightleftharpoons H_3O^+ + ClO_2^-$
   [Acid (1)]   [Base (2)]     [Acid (2)]   [Base (1)]

   b. $HSO_4^- + NH_3 \rightleftharpoons NH_4^+ + SO_4^{2-}$
   [Acid (1)]   [Base (2)]     [Acid (2)]   [Base (1)]

   c. $HCO_3^- + OH^- \rightleftharpoons H_2O + CO_3^{2-}$
   [Acid (1)]   [Base (2)]     [Acid (2)]   [Base (1)]

## Chapter 9: Acids and Bases I

**Strong and Weak Acids and Bases**

3.  a. neither          b. strong          c. weak          d. weak
4.  a. weak             b. neither         c. strong        d. strong
5.  a. salt             b. strong base     c. salt          d. weak acid
7.  a. strong base      b. strong acid
9.  an acid; weak

**Names of Acids and Bases**

11.  a. HCl     b. $H_2SO_4$     c. $H_2CO_3$     d. LiOH     e. $Mg(OH)_2$     f. KOH
12.  a. $HNO_3$     b. $H_2SO_3$     c. $H_3PO_4$     d. $H_2S$     e. $Ca(OH)_2$
15.  a. bromide ion and hydrobromic acid     b. nitrite ion and nitrous acid
17.  Hydroselenic acid

**Ionization of Acids and Bases**

19.  a.  $HI + H_2O \rightarrow H_3O^+ + I^-$

　　b.  $CH_3CH_2CO_2H + H_2O \rightleftharpoons CH_3CH_2CO_2^- + H_3O^+$

　　c.  $HNO_2 + H_2O \rightleftharpoons NO_2^- + H_3O^+$

　　d.  $H_2PO_4^- + H_2O \rightleftharpoons HPO_4^{2-} + H_3O^+$

21.  a.  $HNO_3 + H_2O \rightleftharpoons NO_3^- + H_3O^+$

　　b.  $KOH \rightarrow K^+ + OH^-$

　　c.  $HCO_2H + H_2O \rightleftharpoons HCO_2^- + H_3O^+$

　　d.  $CH_3NH_2 + H_2O \rightleftharpoons CH_3NH_3^+ + OH^-$

23.  a. base     b. acid     c. base

**Acidic and Basic Anhydrides**

25.  a. $H_2SO_4$; an acid          b. KOH; a base
27.  $SeO_3$

**Neutralization Reactions**

29.  $NaOH + HCl \rightarrow NaCl + H_2O$          Net: $OH^- + H^+ \rightarrow H_2O$

**Reactions of Acids with Carbonates and Bicarbonates**

33.  $HCO_3^- + H_3O^+ \rightarrow 2 H_2O + CO_2$

35.  $NaHCO_3 (s) + HCH_3CO_2 (aq) \rightarrow H_2O + CO_2 (g) + NaCH_3CO_2 (aq)$

　　$NaHCO_3 (s) + HCH_3CO_2 (aq) \rightarrow H_2O + CO_2 (g) + CH_3CO_2^- (aq) + Na^+ (aq)$

# 10 Acids and Bases II

## ANSWERS TO REVIEW QUESTIONS

10.1. Generally, a concentrated acid or base is the strongest concentration commercially available. The term concentrated indicates that the acid or base has not been diluted much with water and is at or near maximum strength.

10.2. Yes, the concentration is decreased by dilution, but the number of moles remains unchanged. They are just contained in a larger volume of solution.

10.5.  a. no (only 0.33 mol)   b. yes

10.6.  a. no (need 0.75 mol)   b. yes

10.8.  a. basic   b. acidic   c. neutral   d. acidic

10.9.  a. less   b. greater   c. less   d. greater

10.10. a $K_2SO_4$ is neutral; all of the others are basic.

10.11. HCl (lowest pH) < $CH_3CO_2H$ < $CH_3CO_2Na$ < KOH (highest pH)

10.14. decrease

10.15.  a. $HBO_2 + H_2O \leftrightarrow H_3O^+ + BO_2^-$
b. $HClO_2 + H_2O \leftrightarrow H_3O^+ + ClO_2^-$
c. $HC_9H_7O_4 + H_2O \leftrightarrow H_3O^+ + C_9H_7O_4^-$
d. $H_2Se + H_2O \leftrightarrow H_3O^+ + HSe^-$

10.18. $(HCO_2)_2Ca$. This will produce some formate ions ($HCO_2^-$), which will then combine with $H^+$ in solution to decrease the amount of ionization of $HCO_2H$.

10.20. $-COO^-$ groups react with acid and $-NH_3^+$ groups react with base.

10.22. too high

## ANSWERS TO PROBLEMS

**Dilution of Solutions**

1. a. $M_{conc} \times V_{conc} = M_{dil} \times V_{dil}$
(12.0 M) ($V_{conc}$) = (1.00 M) (2.00 L)
($V_{conc}$) = 0.167 L

   b. $M_{conc} \times V_{conc} = M_{dil} \times V_{dil}$
(1.04 M) ($V_{conc}$) = (1.00 M) (0.500 L)
($V_{conc}$) = 0.481 L

3. a. $M_{conc} \times V_{conc} = M_{dil} \times V_{dil}$
(18.0 M) ($V_{conc}$) = (6.00 M) (1.25 L)
($V_{conc}$) = 0.417 L

   b. $M_{conc} \times V_{conc} = M_{dil} \times V_{dil}$
(18.0 M) ($V_{conc}$) = (0.100 M) (0.575 L)
($V_{conc}$) = 0.00319 L

## Acid-Base Titrations

5. $\dfrac{33.2 \text{ mL}}{1} \times \dfrac{1 \text{ L}}{1000 \text{ mL}} \times \dfrac{0.150 \text{ mol NaOH}}{\text{L}} = 0.00498 \text{ mol NaOH} = 0.00498 \text{ mol HCl}$

$\dfrac{0.00498 \text{ mol HCl}}{0.0200 \text{ L}} = 0.249 \text{ M HCl}$

7. $\dfrac{28.2 \text{ mL}}{1} \times \dfrac{1 \text{ L}}{1000 \text{ mL}} \times \dfrac{0.0302 \text{ mol HCl}}{\text{L}} = 0.0008516 \text{ mol HCl}$

$0.0008516 \text{ mol HCl} \times \dfrac{1 \text{ mol Ca(OH)}_2}{2 \text{ mol HCl}} = 0.000426 \text{ M Ca(OH)}_2$

$\dfrac{0.000426 \text{ mol Ca(OH)}_2}{0.0185 \text{ L}} = 0.0230 \text{ M Ca(OH)}_2$

9. $\dfrac{0.10 \text{ mol NaOH}}{\text{L}} \times \dfrac{1 \text{ L}}{1000 \text{ mL}} \times \dfrac{25 \text{ mL}}{1} = 0.0025 \text{ mol NaOH} = 0.0025 \text{ mol HCl}$

$\dfrac{0.0025 \text{ mol HCl}}{0.020 \text{ L}} = 0.125 \text{ M HCl}$

11. $\dfrac{10.3 \text{ mL NaHCO}_3}{\text{L}} \times \dfrac{1 \text{ L}}{1000 \text{ mL}} \times \dfrac{0.404 \text{ mol NaHCO}_3}{\text{L}} = 0.00416 \text{ mol NaHCO}_3$

$0.00416 \text{ mol NaHCO}_3 \times \dfrac{1 \text{ mol H}_2\text{SO}_4}{2 \text{ mol NaHCO}_3} = 0.00208 \text{ mol H}_2\text{SO}_4$

$0.00208 \text{ mol H}_2\text{SO}_4 \times \dfrac{1 \text{ L}}{0.100 \text{ mol H}_2\text{SO}_4} = 0.0208 \text{ L H}_2\text{SO}_4 = 20.8 \text{ mL H}_2\text{SO}_4$

13. a. $\dfrac{25.00 \text{ mL KOH}}{1} \times \dfrac{1 \text{ L}}{1000 \text{ mL}} \times \dfrac{0.0365 \text{ mol KOH}}{\text{L}} \times \dfrac{1 \text{ mol HCl}}{1 \text{ mol KOH}} = 0.000913 \text{ mol HCl}$

$\dfrac{0.000913 \text{ M HCl}}{1} \times \dfrac{1 \text{ L}}{0.0195 \text{ mol HCl}} = 0.0468 \text{ L HCl} = 46.8 \text{ mL HCl}$

b.
$\dfrac{10.00 \text{ mL Ca(OH)}_2}{1} \times \dfrac{1 \text{ L}}{1000 \text{ mL}} \times \dfrac{0.0116 \text{ mol Ca(OH)}_2}{\text{L}} = 0.000116 \text{ mol Ca(OH)}_2$

$0.000116 \text{ mol Ca(OH)}_2 \times \dfrac{2 \text{ mol HCl}}{1 \text{ mol Ca(OH)}_2} = 0.000232 \text{ mol HCl}$

$\dfrac{0.000232 \text{ mol HCl}}{1} \times \dfrac{1 \text{ L}}{0.0195 \text{ mol HCl}} = 0.0119 \text{ L HCl} = 11.9 \text{ mL HCl}$

c.

$$\frac{20.00\,\text{mL NH}_3}{1} \times \frac{1\,\text{L}}{1000\,\text{mL}} \times \frac{0.0225\,\text{mol NH}_3}{\text{L}} \times \frac{1\,\text{mol HCl}}{1\,\text{mol NH}_3} = 0.000450\,\text{mol HCl}$$

$$\frac{0.000450\,\text{mol HCl}}{1} \times \frac{1\,\text{L}}{0.0195\,\text{mol HCl}} = 0.0231\,\text{L HCl} = 23.1\,\text{mL HCl}$$

**pH and pOH**

15. a. pH = –log [$H_3O^+$] = –log (1.0 × $10^{-2}$) = 2.00
    b. pH = –log [$H_3O^+$] = –log (1.0 × $10^{-4}$) = 4.00

17. a. pOH = –log [$OH^-$] = –log (1.0 × $10^{-2}$) = 2.00
    b. pOH = –log [$OH^-$] = –log (1.0 × $10^{-3}$) = 3.00

19. pH + pOH = 14.00       a. pOH = 12.00       b. pOH = 10.00

21. a. pH = –log [$H_3O^+$] = –log (3.3 × $10^{-3}$) = 2.48
    b. pH = –log [$H_3O^+$] = –log (5.7 × $10^{-5}$) = 4.24
    c. pH = –log [$H_3O^+$] = –log (8.1 × $10^{-4}$) = 3.09

23. pH = –log [$H_3O^+$] = –log (4.6 × $10^{-8}$) = 7.34

25. pH = –log [$H_3O^+$] = –log (2.0 × $10^{-12}$) = 11.7

27. pH = –log [$H_3O^+$] = 5.10       [$H_3O^+$] = 7.9 × $10^{-6}$ M

**Salt Solutions: Acidic, Basic, or Neutral?**

29. $CH_3CO_2^- + H_2O \rightleftharpoons CH_3CO_2H + OH^-$

31. a. neutral       b. basic       c. unable to say

33. Basic. The product is the salt of the weak acid and a strong base.

**Equilibria in Solutions of Weak Acids and Weak Bases**

35. a. $K_a = \dfrac{[H^+][OCl^-]}{[HOCl]}$   b. $K_a = \dfrac{[H^+][C_6H_7O_6^-]}{[HC_6H_7O_6]}$   c. $K_a = \dfrac{[H^+][HCO_2^-]}{[HCO_2H]}$

37. a. $K_a = \dfrac{[H^+][CH_3CO_2^-]}{[CH_3CO_2H]} = 1.8 \times 10^{-5}$       $\dfrac{(X)(X)}{(0.010 - X)} = 1.8 \times 10^{-5}$

   Assume X << 0.010, then  $\dfrac{X^2}{0.010} = 1.8 \times 10^{-5}$       $X^2 = 1.8 \times 10^{-7}$

   X = 4.2 × $10^{-4}$ M = [$H^+$]

   1 × $10^{-14}$ = [$H^+$][$OH^-$]       [$OH^-$] = 2.4 × $10^{-11}$ M

b. $K_a = \dfrac{[H^+][C_6H_5CO_2^-]}{[C_6H_5CO_2H]} = 6.3 \times 10^{-5}$ $\quad\quad \dfrac{(X)(X)}{(0.20-X)} = 6.3 \times 10^{-5}$

Assume $X \ll 0.20$, then $\dfrac{X^2}{0.20} = 6.3 \times 10^{-5}$ $\quad\quad X^2 = 1.26 \times 10^{-5}$

$X = 3.5 \times 10^{-3}$ M = $[H^+]$

$1 \times 10^{-14} = [H^+][OH^-]$ $\quad\quad$ $[OH^-] = 2.9 \times 10^{-12}$ M

c. $K_a = \dfrac{[H^+][CN^-]}{[HCN]} = 6.2 \times 10^{-10}$ $\quad\quad \dfrac{(X)(X)}{(0.50-X)} = 6.2 \times 10^{-10}$

Assume $X \ll 0.50$, then $\dfrac{X^2}{0.50} = 6.2 \times 10^{-10}$ $\quad\quad X^2 = 3.10 \times 10^{-10}$

$X = 1.8 \times 10^{-5}$ M = $[H^+]$

$1 \times 10^{-14} = [H^+][OH^-]$ $\quad\quad$ $[OH^-] = 5.7 \times 10^{-10}$ M

39. a. $K_b = \dfrac{[OH^-][NH_4^+]}{[NH_3]} = 1.8 \times 10^{-5}$ $\quad\quad \dfrac{(X)(X)}{(0.025-X)} = 1.8 \times 10^{-5}$

Assume $X \ll 0.025$, then $\dfrac{X^2}{0.025} = 1.8 \times 10^{-5}$ $\quad X^2 = 4.5 \times 10^{-7}$

$X = 6.7 \times 10^{-4}$ M = $[OH^-]$

$1 \times 10^{-14} = [H^+][OH^-]$ $\quad\quad$ $[H^+] = 1.5 \times 10^{-11}$ M

b. $K_b = \dfrac{[OH^-][CH_3NH_3^+]}{[CH_3NH_2]} = 4.2 \times 10^{-4}$ $\quad\quad \dfrac{(X)(X)}{(0.10-X)} = 4.2 \times 10^{-4}$

Assume $X \ll 0.10$, then $\dfrac{X^2}{0.10} = 4.2 \times 10^{-4}$ $\quad\quad X^2 = 4.20 \times 10^{-5}$

$X = 6.5 \times 10^{-3}$ M = $[OH^-]$

$1 \times 10^{-14} = [H^+][OH^-]$ $\quad\quad$ $[H^+] = 1.5 \times 10^{-12}$ M

c. $K_b = \dfrac{[OH^-][C_6H_5NH_3^+]}{[C_6H_5NH_2]} = 4.2 \times 10^{-10}$ $\quad\quad \dfrac{(X)(X)}{(0.10-X)} = 4.2 \times 10^{-10}$

Assume $X \ll 0.10$, then $\dfrac{X^2}{0.10} = 4.2 \times 10^{-10}$ $\quad\quad X^2 = 4.2 \times 10^{-11}$

$X = 6.5 \times 10^{-6}$ M = $[OH^-]$

$1 \times 10^{-14} = [H^+][OH^-]$ $\quad\quad$ $[H^+] = 1.5 \times 10^{-9}$ M

## Buffer Solutions

41. a. $K_a = \dfrac{[H^+][CN^-]}{[HCN]} = 6.2 \times 10^{-10} = \dfrac{(X)(0.25)}{(0.25)}$

   $X = 6.2 \times 10^{-10}$ M $= [H^+]$

   b. $K_a = \dfrac{[H^+][F^-]}{[HF]} = 6.6 \times 10^{-4} = \dfrac{(X)(0.20)}{(0.20)}$

   $X = 6.6 \times 10^{-4}$ M $= [H^+]$

   c. $K_a = \dfrac{[H^+][C_6H_5CO_2^-]}{[C_6H_5CO_2H]} = 6.3 \times 10^{-5} = \dfrac{(X)(0.045)}{(0.033)}$

   $X = 4.62 \times 10^{-5}$ M $= [H^+]$

43. $K_b = \dfrac{[NH_4^+][OH^-]}{[NH_3]} = 1.8 \times 10^{-5} = \dfrac{(X)(0.040)}{(0.40)}$

   $X = 1.8 \times 10^{-4}$ M $= [OH^-]$     $[H^+] = 5.6 \times 10^{-11}$ M

Use the Henderson-Hasselbalch equation $\left[ \text{pH} = pK_a + \log \dfrac{[A^-]}{[HA]} \right]$ to solve problems 45-54.

45. $pK_a = -\log K_a$     $pK_a = -\log (6.2 \times 10^{-10})$     $pK_a = 9.21$

   $\text{pH} = 9.21 + \log \dfrac{(0.040 \text{ M})}{(0.040 \text{ M})}$     $\text{pH} = 9.21 + 0 = 9.21$

47. $pK_a = -\log K_a$     $pK_a = -\log (6.3 \times 10^{-5})$     $pK_a = 4.20$

   $\text{pH} = 4.20 + \log \dfrac{(0.15 \text{ M})}{(0.15 \text{ M})}$     $\text{pH} = 4.20 + 0 = 4.20$

49. $pK_a = -\log K_a$     $pK_a = -\log (1.8 \times 10^{-4})$     $pK_a = 3.74$

   $\text{pH} = 3.74 + \log \dfrac{(0.60 \text{ M})}{(0.15 \text{ M})}$     $\text{pH} = 3.74 + 0.60 = 4.34$

51. $pK_a = -\log K_a$     $pK_a = -\log (1.3 \times 10^{-5})$     $pK_a = 4.89$

   $\text{pH} = 4.89 + \log \dfrac{(0.0786 \text{ M})}{(0.350 \text{ M})}$     $\text{pH} = 4.89 + (-0.65) = 4.24$

53. $pK_a = -\log K_a$     $pK_a = -\log (1.0 \times 10^{-10})$     $pK_a = 10.0$

   $\text{pH} = 10.0 + \log \dfrac{(0.10 \text{ M})}{(0.10 \text{ M})}$     $\text{pH} = 10.0 + 0 = 10.0$

# 11 Electrolytes

## ANSWERS TO REVIEW QUESTIONS

11.3. In molten NaCl the ions can move toward the appropriate electrode; a process needed for electrical current. In crystalline, solid NaCl the ion positions are fixed, and thus electricity is not conducted.

11.4. covalent

11.5 Ionization is the formation of ions from a compound when it dissolves in water. Ion dissociation is the separation of the ions of an ionic compound when it dissolves in water.

11.7. HCl ionizes in water by the following reaction:

$HCl (g) + H_2O (l) \rightarrow H_3O^+ (aq) + Cl^- (aq)$

11.8. Cathode. The cathode is the negatively charged electrode. You want to attract the positively charged, plating metal ions to this electrode, so that these ions can then be reduced to form the metallic coating.

## ANSWERS TO PROBLEMS

**Strong Electrolytes, Weak Electrolytes, and Nonelectrolytes**

1. a. strong   b. strong   c. weak   d. strong   e. strong

3. Lead chromate is an insoluble compound which does not liberate ions into solution.

5. Each mole of NaCl produces two moles of ions. Each ion or undissociated compound reduces the freezing point by an equal amount. The depression for the salt solution is not quite twice that for the sugar solution because of the interaction between ions when in a concentrated solution.

**Activity Series**

7. a. $Ca (s) + 2 HCl (aq) \rightarrow H_2 + CaCl_2 (aq)$
   b. $Ni (s) + 2 HCl (aq) \rightarrow H_2 + NiCl_2 (aq)$
   c. $Mg (s) + 2 HNO_3 (aq) \rightarrow H_2 + Mg(NO_3)_2 (aq)$

9. a. $2 Na (s) + 2 H_2O \rightarrow H_2 (g) + 2 NaOH (aq)$
   b. $Ba (s) + 2 H_2O \rightarrow H_2 (g) + Ba(OH)_2 (aq)$

11. a. $Mg (s) + Cu^{2+} (aq) \rightarrow Mg^{2+} (aq) + Cu (s)$
    b. $Ag (s) + Pb^{2+} (aq) \rightarrow$ no reaction
    c. $Fe (s) + Zn^{2+} (aq) \rightarrow$ no reaction
    d. $2 Al (s) + 3 Ni^{2+} (aq) \rightarrow 2 Al^{3+} (aq) + 3 Ni (s)$

## Solubility Product and Precipitation Criteria

13. $[Mg^{2+}] = 0.001$ M, $[CO_3^{2-}] = 0.001$ M: $(0.001 \text{ M})(0.001 \text{ M}) = 1 \times 10^{-6}$

    Since $1 \times 10^{-6}$ is greater than $3.5 \times 10^{-8}$ (the solubility constant of $MgCO_3$), precipitation will occur.

15. $[Pb^{2+}] = 1 \times 10^{-6}$ M, $[CrO_4^{2-}] = 1 \times 10^{-5}$ M: $(1 \times 10^{-6} \text{ M})(1 \times 10^{-5} \text{ M}) = 1 \times 10^{-11}$

    Since $1 \times 10^{-11}$ is greater than $2.8 \times 10^{-13}$ (the solubility constant of $PbCrO_4$), precipitation will occur.

# A  Inorganic Chemistry

## ANSWERS TO REVIEW QUESTIONS

1. a. Inorganic chemistry: the chemistry of all elements excluding carbon.
   b. Noble gas: relatively unreactive gasses that occupy the column at the far right side of the periodic table.
   c. Halogen: an element in Group 7A of the periodic table.
   d. Alkali metal: an element in Group 1A of the periodic table.
   e. Alkaline earth metal: an element in Group 2A of the periodic table.
   f. Transition element: metallic elements situated in the center portion of the periodic table in the B groups.

3. chlorophyll

5. Water with relatively high levels of $Ca^{2+}$, $Mg^{2+}$, or $Fe^{2+}$. These ions bind soap molecules and precipitate them, forming what we know as soap scum. This prevents the soap molecules from performing their cleansing action.

7. Be; it does not react with water, and it exists in metal form without the rapid oxidation observed in the other alkaline earth metals.

9. It is considerably lighter and corrodes less rapidly than iron.

11. Carbon monoxide binds tightly to hemoglobin, preventing it from carrying $O_2$ to the tissues.

13. Allotropes are different forms of the same element in the same state. Carbon (graphite and diamond) and oxygen (dioxygen and ozone)

15. Ozone in the stratosphere absorbs ultraviolet light, preventing a large portion of it from reaching the surface of the earth. This protects us from the harmful effects of ultraviolet light, that is, a higher rate of mutation in cells including, in humans, a greater incidence of skin cancer. Ozone in the troposphere can cause many different types of problems; two important ones are respiratory problems in animals and growth retardation in plants.

# 12 The Atomic Nucleus

## ANSWERS TO REVIEW QUESTIONS

12.1. Both X-rays and visible light are forms of electromagnetic radiation. X-rays have a much higher energy than does visible light.

12.2. a. beta particles   b. alpha particles

12.3. positron emission and electron capture

12.4. Mass number: no change   Atomic number: no change

12.5. Use shielding, maintain a distance from the radioactive source, and minimize time of exposure

12.6. The heavy alpha particles.

12.7. gamma

12.12. neutrons

12.14. There is a tremendous repulsion of protons as nuclei approach each other in nuclear fusion before they combine in a single nucleus. High temperatures are required to generate molecular speeds required to overcome this.

12.16. $^{131}$I

12.19. X-rays are detected in CT scans, and gamma rays are detected in PET scans.

## ANSWERS TO PROBLEMS

**Nuclear Symbols**

1. a. helium-4   b. beta particle   c. neutron   d. deuterium

**Nuclear Equations**

3. a. $^{209}_{82}Pb \rightarrow \,^{0}_{-1}e + \,^{209}_{83}Bi$   b. $^{225}_{90}Th \rightarrow \,^{221}_{88}Ra + \,^{4}_{2}He$

5. $^{31}_{16}S \rightarrow \,^{31}_{15}P + \,^{0}_{+1}e$

7. $^{87}_{35}Br \rightarrow \,^{1}_{0}n + \,^{86}_{35}Br$

9. $^{24}_{12}Mg + \,^{1}_{0}n \rightarrow \,^{1}_{1}H + \,^{24}_{11}Na$

11. a. $^{10}_{4}Be$   b. $^{1}_{1}H$   c. $^{4}_{2}He$

13. a. $^{2}_{1}H + \,^{2}_{1}H \rightarrow \,^{3}_{2}He + \,^{1}_{0}n$   b. $^{241}_{95}Am + \,^{4}_{2}He \rightarrow \,^{243}_{97}Bk + 2\,^{1}_{0}n$

    c. $^{121}_{51}Sb + \,^{4}_{2}He \rightarrow \,^{124}_{53}I + \,^{1}_{0}n$   $^{124}_{53}I \rightarrow \,^{125}_{52}Te + \,^{0}_{+1}e$

## Chapter 12: The Atomic Nucleus

**Half-Life**

15. One-eighth of its initial value means three half-lives. The length of time will be 24.12 days.

17. Five disintegrations per minute is 1/32 of the original 160 disintegrations per minute. This represents five half-lives, or 335 hrs.

# 13 Hydrocarbons

## ANSWERS TO REVIEW QUESTIONS

13.2. ethane, methane

13.3. a. 2   b. 7   c. 4   d. 9

13.6. Natural gas is mostly methane; bottled gas is propane and/or butanes.

13.8. The lighter alkanes dissolve and wash away body oils, causing dermatitis. Heavier chain alkanes (like mineral oils) act as skin softeners.

13.12. Ethylene and cyclopropane are both potent, quick-acting anesthetics.

13.15. Monomers for addition polymerization are all alkenes (contain a C=C bond).

13.18. a. para   b. ortho   c. meta

13.19. a. aromatic   b. aliphatic   c. aliphatic   d. aromatic

13.20. Many polycyclic aromatic hydrocarbons are carcinogenic.

## ANSWERS TO PROBLEMS

**Organic Versus Inorganic**

1. Organic: a, c   inorganic: b

3. a. NaOH   b. KCl

**Alkanes: Structures and Names**

5. a. $CH_3CH_2CH_2CH_2CH_2CH_2CH_3$

   b. $CH_3CH_2CHCH_2CH_3$
      $\phantom{CH_3CH_2C}|$
      $\phantom{CH_3CH_2C}CH_3$

   c. $\phantom{CH_3C}CH_3$
      $\phantom{CH_3C}|$
      $CH_3CCH_2CH_2CHCH_3$
      $\phantom{CH_3C}|\phantom{CH_2CH_2C}|$
      $\phantom{CH_3C}CH_3\phantom{CH_2CH}CH_3$

   d. $\phantom{CH_3CH_2C}CH_3$
      $\phantom{CH_3CH_2C}|$
      $CH_3CH_2CHCHCH_2CH_2CH_3$
      $\phantom{CH_3CH_2CHC}|$
      $\phantom{CH_3CH_2CHC}CH_2CH_3$

7. a. 3-methylpentane   b. 2,3-dimethylbutane

9. a. $CH_3CH_2-$   b. $CH_3CH-$
      $\phantom{CH_3CH}|$
      $\phantom{CH_3CH}CH_3$

11. Butane $CH_3CH_2CH_2CH_3$   isobutane or methylpropane  $CH_3CHCH_3$
    $\phantom{isobutane or methylpropane CH_3CH}|$
    $\phantom{isobutane or methylpropane CH_3CH}CH_3$

325

## Alkanes: Physical Properties

13. a. pentane  b. hexane  c. cyclohexane  d. nonane

## Cyclic Hydrocarbons

15. a. methylcyclopropane  b. 1,2-diethyl-4-methylcyclopentane
    c. cyclobutene  d. 3-ethylcyclohexene

17.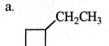

## Halogenated Hydrocarbons

19. a. $CH_3Cl$  b. $CHCl_3$

21.    $CH_3CHCH_3$        $BrCH_2CH_2CH_3$
           | 
          Br

    isopropyl bromide      propyl bromide
    (2-bromopropane)      (1-bromopropane)

## Saturated Versus Unsaturated

23. a. unsaturated  b. unsaturated  c. saturated

## Alkenes and Alkynes

25. a. $HC\equiv CH$  b.  c. $HC\equiv CCHCH_2CH_2CH_3$ with $CHCH_3$ / $CH_3$  d. $(H_3C)(H_3C)C=C(CH_3)(CH_3)$

27. a. $(H_3C)(H_3C)C=CHCH_2CH_3$  b. $H_2C=CHCH_2CH_2CHCH_3$ with $CH_3$

29. a. 2-methyl-1-pentene  b. 2-methyl-2-pentene  c. 2,5-dimethyl-2-hexene

## Isomers

31. a. same  b. same  c. isomers
33. a. isomers  b. same  c. isomers

## Chemical Reactions of Alkenes

35. a. 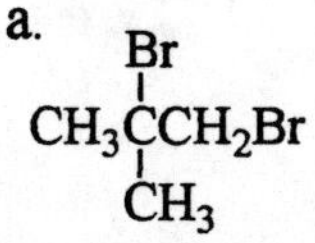 CH₃CBr(CH₃)CH₂Br

b. 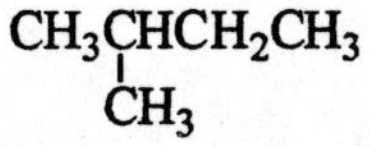 CH₃CH(CH₃)CH₂CH₃

c. 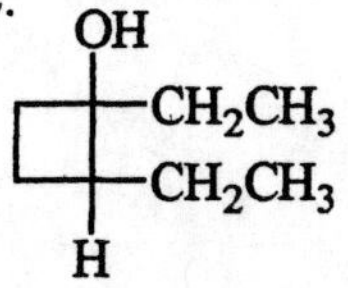 1-hydroxy-1,1-diethylcyclobutane (OH and H on cyclobutane with two CH₂CH₃ groups)

37. a. H₂, Ni      b. H₂O, H⁺

39. a. 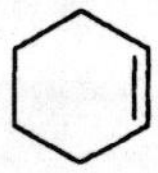 cyclohexene      b. 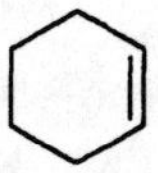 cyclohexene

## Aromatic Compounds

41. a. 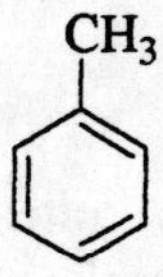 toluene (CH₃-benzene)

b. 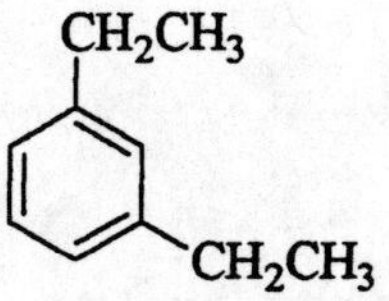 1,3-diethylbenzene

c. 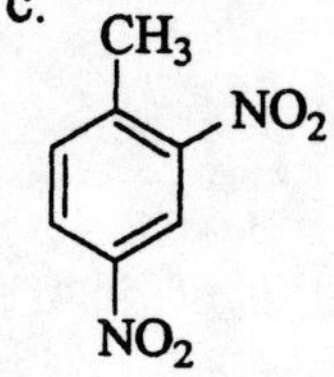 1-methyl-2,4-dinitrobenzene

43. a. ethylbenzene      b. isopropylbenzene

c. 2-nitrotoluene      d. 3,5-dichlorotoluene

# 14 Alcohols, Phenols, and Ethers

## ANSWERS TO REVIEW QUESTIONS

14.1. A functional group is a group of atoms in an organic molecule that confers characteristic properties to the molecule.
a. C=C, double bond   b. R-OH, hydroxyl group   c. R-O-R, alkoxy group

14.2. Alcohols and phenols both contain a hydroxyl group (-OH). In an alcohol the hydroxyl group is attached to an aliphatic group. In the phenol the hydroxyl group is attached directly to an aromatic ring.

14.3. Methanol has a smaller nonpolar group than does 1-hexanol.

14.4. Because of its compact shape, *tert*-butyl alcohol experiences weaker intermolecular attractions than straight chain alcohols. Thus, it has a lower boiling point than 1-hexanol and 1-butanol.

14.11. Ethanol. It ties up the liver enzymes that convert methanol to formaldehyde.

14.17. Antiseptics and disinfectants are substances that kill bacteria and other microorganisms. Antiseptics are intended for use on living tissue, while disinfectants are intended for nonliving objects.

## ANSWERS TO PROBLEMS

### Alcohols: Names and Structural Formulas

1. a. 1-hexanol              b. 2-hexanol

3. a. 4,4-dichloro-2-butanol   b. 3,3-dibromo-2-methyl-2-butanol

5. a.
$$CH_3CH_2\underset{}{\overset{OH}{CH}}CH_2CH_2CH_3$$
b.
$$CH_3\underset{}{\overset{OH}{CH}}-\underset{\underset{CH_3}{|}}{\overset{\overset{CH_3}{|}}{C}}-CH_3$$

7. a.
$$CH_3CH_2\underset{}{\overset{OH}{CH}}\underset{\underset{CH_3}{|}}{\overset{\overset{CH_3}{|}}{CH}}\overset{CH_3}{CH}CH_2CH_3$$
b.
$$\text{Ph}-\underset{}{\overset{CH_2CH_3}{CH}}CH_2CH_3 \text{ with HOCH-}$$

# Chapter 14: Alcohols, Phenols, And Ethers

## Physical Properties

9. methanol < ethanol < 1-propanol

11. 1-octanol < 1-butanol < methanol

## Preparation of Alcohols

13. a. CH₃CH=CH₂  b. (cyclopentene with -CH₃)  c. CH₃C(CH₃)=CH₂

## Chemical Reactions of Alcohols

15. a. oxidation    b. dehydration    c. hydration

17.
CH₃CH₂CH₂CH₂OH ⟶ CH₃CH₂CH₂COOH

CH₃CH₂CH(OH)CH₃ ⟶ CH₃CH₂C(=O)CH₃

CH₃CH(CH₃)CH₂OH ⟶ CH₃CH(CH₃)COOH

(CH₃)₃COH ⟶ no reaction

19. (tetrahydropyran structure)

21. a. CH₃CH(OH)CH₂CH₃    b. cyclohexanecarboxylic acid

23. a. H⁺, H₂O    b. K₂Cr₂O₇, H⁺    c. conc. H₂SO₄, 140°C, excess alcohol

## Polyhydric Alcohols

25. a. HOCH₂CH₂CH₂CH(OH)CH₃    b. HOCH₂CH(OH)CH₃

## Phenols

27. a. 2-nitrophenol (*o*-nitrophenol)  b. 4-bromophenol (*p*-bromophenol)

29. a. 3-iodophenol (*m*-iodophenol structure: OH on benzene with I at meta position)

    b. 4-methylphenol (benzene with OH and CH₃ para)

31. a. C₆H₅OH + NaOH ⟶ C₆H₅O⁻Na⁺ + H₂O

    b. no reaction

## Ethers: Names and Structural Formulas

33. a. dipropyl ether            b. diphenyl ether

35. a. $CH_3CH_2OCH_3$    b. C₆H₅—OCH₂—C₆H₅

# 15 Aldehydes and Ketones

## ANSWERS TO REVIEW QUESTIONS

15.1. Aldehydes and ketones both contain a carbonyl group (C=O). Ketones have two carbon chains attached to the carbonyl carbon. Aldehydes have a carbon chain and a hydrogen attached to the carbonyl carbon.

15.2. The boiling point of 1-butanal is lower than that of 1-butanol because aldehydes do not have the ability to form hydrogen bonds. The boiling point of 1-butanal is higher than that of diethyl ether because the carbonyl group is more polar than the ether linkage.

15.3. The aldehydes that are formed by the oxidation of primary alcohols are themselves easily oxidized further to carboxylic acids. Ketones formed from the oxidation of secondary alcohols cannot be further oxidized.

15.4. 
| Reaction | Aldehyde Product | Ketone Product |
|---|---|---|
| reduction | primary alcohol | secondary alcohol |
| hydration | hydrate | hydrate |
| addition of alcohol | hemiacetal | hemiketal |

15.7. Anhydrous conditions favor formation of the acetal, and aqueous conditions favor formation of the aldehyde.

15.8. Examples: vanillin, cinnamaldehyde, benzaldehyde

## ANSWERS TO PROBLEMS

### Names and Structural Formulas

1.   a. benzaldehyde      b. 3-hydroxypropanal
    c. 4,4-dimethylpentanal      d. 2-chlorobenzaldehyde

3.   a. 5-methyl-3-hexanone      b. cyclopentanone
    c. 2-pentanone      d. 4-bromo-2,2-dimethyl-3-pentanone

5.   a. $CH_3CH_2CH_2CHO$      b. $CH_3CH_2CH_2CH_2CH(CH_3)CH_2CHO$      c. $O_2N\text{-}C_6H_4\text{-}CHO$ (para)

7.   a. $CH_3COCH_2CH_2CH_2CH_3$      b. $CH_3COCH(Br)CH_2CH_2CH_3$      c. 4-methylcyclohexanone

### Physical Properties

9. 2-Propanol; alcohols can hydrogen bond.

11. Acetaldehyde; aldehydes are more polar than ethers.

## Chapter 15: Aldehydes and Ketones

**Preparation of Aldehydes and Ketones**

13. a. 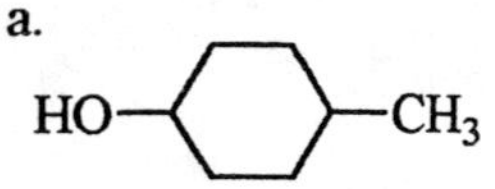 HO—⟨cyclohexane⟩—CH₃

    b. 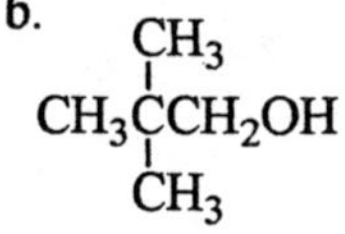 CH₃C(CH₃)(CH₃)CH₂OH

    c. 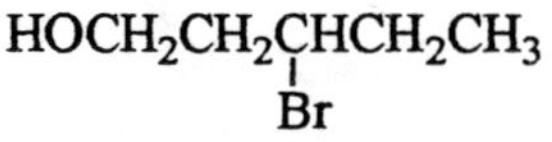 HOCH₂CH₂CH(Br)CH₂CH₃

**Chemical Reactions**

15. a. CH₃CHO + CH₃OH ⟶ CH₃C(OH)(H)(OCH₃)

    b. 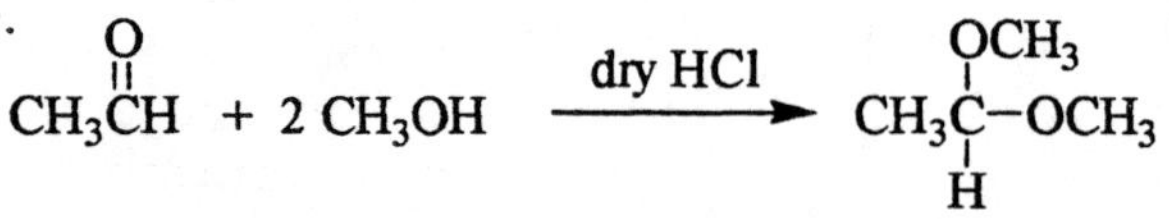 CH₃CHO + 2 CH₃OH —dry HCl→ CH₃C(OCH₃)(H)(OCH₃)

    c. 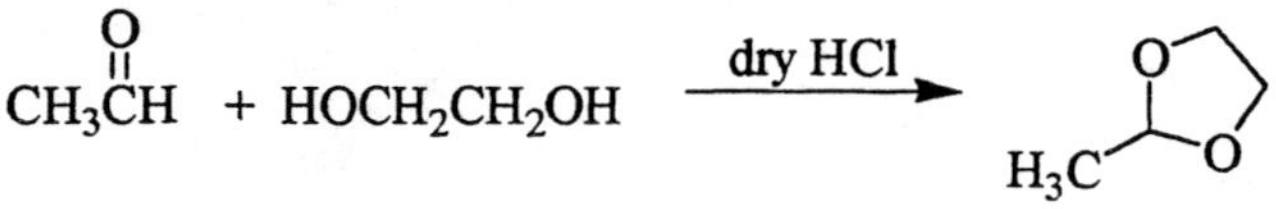 CH₃CHO + HOCH₂CH₂OH —dry HCl→ (1,3-dioxolane with H₃C substituent)

17. a.  CH₃COCH₃ + CH₃OH ⟶ CH₃C(OH)(CH₃)(OCH₃)

    b. 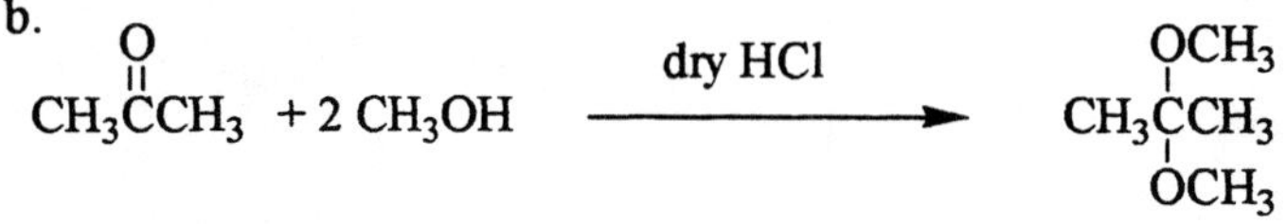 CH₃COCH₃ + 2 CH₃OH —dry HCl→ CH₃C(OCH₃)(CH₃)(OCH₃)

    c. 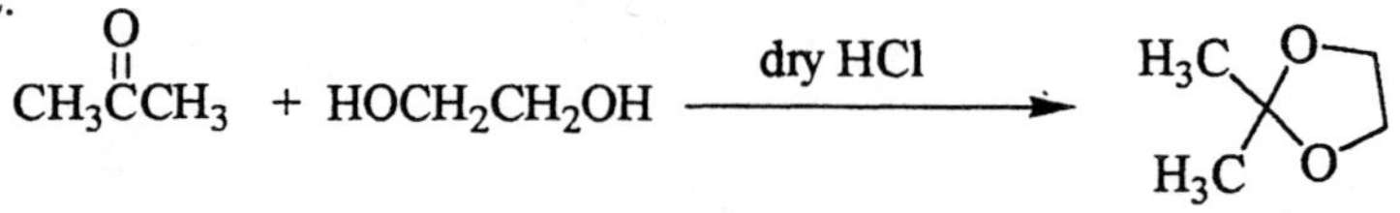 CH₃COCH₃ + HOCH₂CH₂OH —dry HCl→ (1,3-dioxolane with two H₃C substituents)

19. a. Yes, only pentanal would test positive.
    b. No, both would test negative.
    c. Yes, only pentanal would test positive.
    d. Yes, only pentanal would test positive.
    e. No, both would test negative.

21. $K_2Cr_2O_7$ and $H^+$ will oxidize 2-pentanol; a greenish precipitate of chromium(III) compounds would form.

23. a. $Ag^+$, $NH_3$    b. $CrO_3$, HCl, pyridine, $CH_2Cl_2$   c. 2 $CH_3OH$, dry HCl

**Hydrates, Hemiacetals, and Acetals**

25. Hemiacetals: a, d

27. Hydrates: c, e

# 16 Carboxylic Acids and Derivatives

## ANSWERS TO REVIEW QUESTIONS

16.2. a. formic acid (methanoic acid)     b. acetic acid (ethanoic acid)
c. propionic acid (propanoic acid)     d. butyric acid (butanoic acid)

16.3. a. 1-pentanol     b. pentanal

16.4. Butanoic acid is more soluble in water than 1-butanol. The carboxyl group readily hydrogen bonds with water, and acids having four carbon atoms or less are completely soluble in water.

16.7. Carboxylic acids have characteristically unpleasant odors. Esters have characteristically pleasant odors.

16.9. Esterification is the conversion of a carboxylic acid to ester. Neutralization forms salts – ionic compounds; esterification forms covalent compounds.

16.10. a. Acidic hydrolysis of an ester yields the corresponding carboxylic acid and alcohol. Basic hydrolysis yields the corresponding carboxylate salt and alcohol.

b. Basic hydrolysis gives complete conversion to the carboxylate salt. Under acidic conditions, the equilibrium is unfavorable for ester hydrolysis.

16.12. Alcohols and phosphoric acid.

16.13. Lower; *N,N*-dimethylacetamide cannot form intermolecular hydrogen bonds, whereas acetamide can.

16.14. 1,6-Hexanediamine and 1,10-decanedioic acid.

16.15. Acidic hydrolysis yields a carboxylic acid and an ammonium salt. Basic hydrolysis yields a carboxylate salt and ammonia or an amine

# ANSWERS TO PROBLEMS

**Carboxylic Acids: Names and Structural Formulas**

1. a.
$$CH_3CH_2CH_2CH_2CH_2CH_2COH$$ (with C=O)

   b. 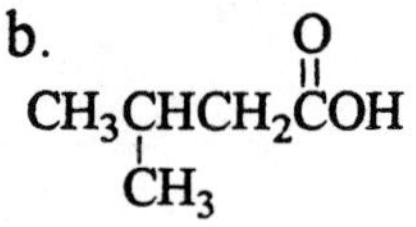
$$CH_3CHCH_2COH$$ (with C=O and CH_3 branch)

   c. (benzene ring with COOH and two Br substituents at adjacent positions)

   d. 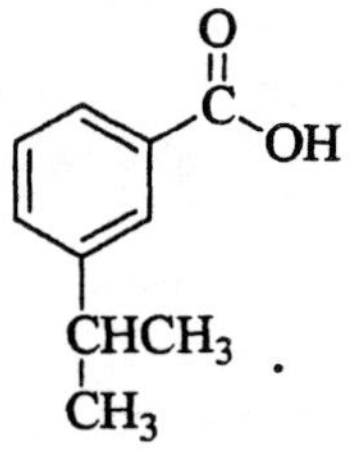 (benzene ring with COOH and CH(CH_3)_2 substituent at meta position)

3. a.
$$HO-\overset{O}{\underset{}{C}}-\overset{O}{\underset{}{C}}-OH$$

   b. 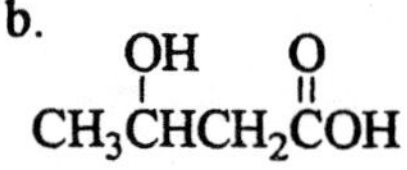
$$CH_3\underset{OH}{CH}CH_2COH$$ (with C=O)

5. a. 3-methylbutanoic acid  
   b. 3,4,4-trimethylpentanoic acid  
   c. 4-hydroxybutanoic acid  
   d. 2,4-dimethylpentanoic acid

**Salts: Names and Structural Formulas**

7. a.
$$CH_3CO^-K^+$$ (with C=O)

   b.
$$(CH_3CH_2CO^-)_2Ca^{2+}$$ (with C=O)

**Esters: Names and Structural Formulas**

9. a.
$$CH_3COCH_3$$ (with C=O)

   b. 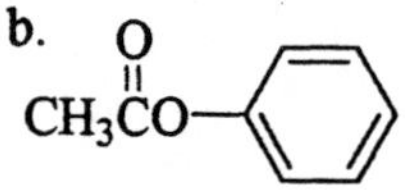
$$CH_3CO-\text{(phenyl)}$$ (with C=O)

11. a. 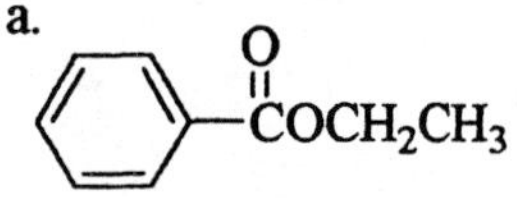 (phenyl)—COCH_2CH_3 (with C=O)

    b. 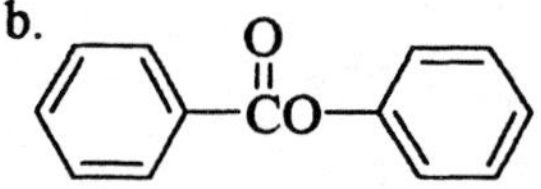 (phenyl)—CO—(phenyl) (with C=O)

13. a. methyl benzoate  
    b. methyl formate (or methyl methanoate)  
    c. ethyl propionate (or ethyl propanoate)

**Amides: Names and Structural Formulas**

15. a.
$$CH_3CH_2CH_2CNH_2$$ (with C=O)

    b.
$$CH_3CH_2CH_2CH_2CH_2CNH_2$$ (with C=O)

    c.
$$CH_3\underset{H}{CNCH_3}$$ (with C=O)

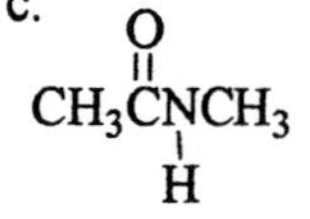

17. a. benzamide  
    b. 2-methylbutanamide  
    c. acetamide

**Physical Properties**

19. II; butanoic acid has the ability to form dimers, which are hydrogen bonded to one another. The ether cannot hydrogen bond.

21. I; the amide can hydrogen bond, while the ester cannot.

# Chapter 16: Carboxylic Acids and Derivatives 335

23. I; the acid can both hydrogen bond with water and partially ionize in water. This helps its solubility. The alkane is insoluble in water because of its lack of polarity.

25. I; the longer hydrocarbon chain of II will make it less able to mix with the polar water.

**Chemical Reactions**

27. a. $CH_3CH_2CH_2\overset{O}{\underset{\|}{C}}OH + NaOH \longrightarrow CH_3CH_2CH_2\overset{O}{\underset{\|}{C}}O^-Na^+ + H_2O$

    b. $CH_3CH_2CH_2\overset{O}{\underset{\|}{C}}OH + NaHCO_3 \longrightarrow CH_3CH_2CH_2\overset{O}{\underset{\|}{C}}O^-Na^+ + CO_2 + H_2O$

29. $CH_3\overset{O}{\underset{\|}{C}}OCH_2CH_3 + H_2O \xrightarrow{H^+} CH_3\overset{O}{\underset{\|}{C}}OH + CH_3CH_2OH$

31. Ph–C(O)NH$_2$ + H$_2$O $\xrightarrow{H^+}$ Ph–COOH + NH$_4^+$

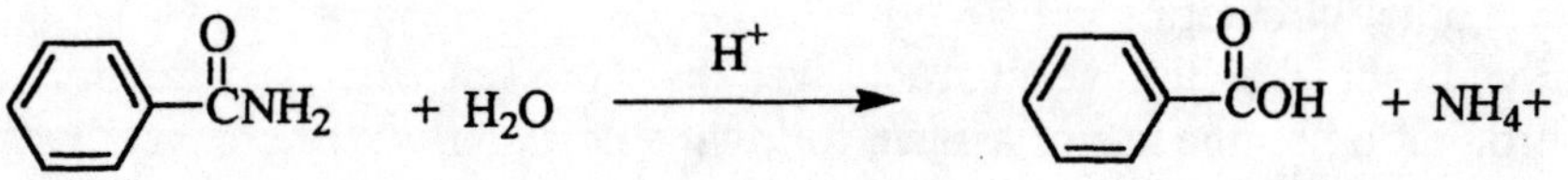

33. a. $CH_3CH_2\overset{O}{\underset{\|}{C}}O^-Na^+ + H_2O$

    b. phthalate disodium salt + H$_2$O (benzene ring with two –CO$^-$Na$^+$ groups ortho)

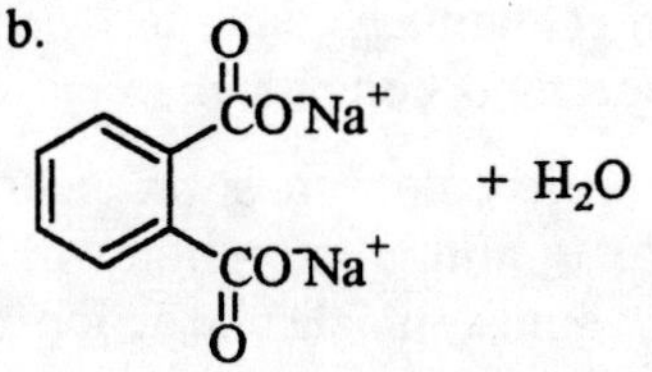

35. a. Ph–CO$^-$Na$^+$ + HOCH$_2$CH$_2$CH$_3$

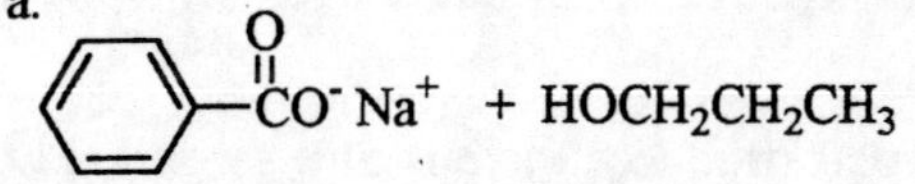

    b. cyclohexyl–OC(O)CH$_3$ + H$_2$O

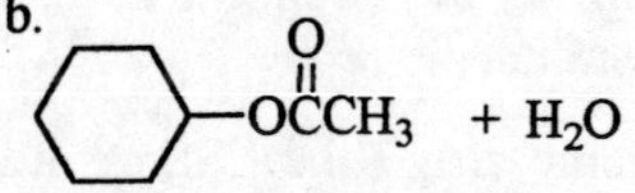

37. a. $CH_3\overset{O}{\underset{\|}{C}}OCH_2CH_2CH_3 + H_2O$

    b. $CH_3O\overset{O}{\underset{\|}{C}}CH_2\overset{O}{\underset{\|}{C}}OCH_3 + 2\ H_2O$

39. a. $CH_3\overset{O}{\underset{\|}{C}}OH + NH_4Cl$

    b. Ph–CO$^-$Na$^+$ + NH(CH$_3$)$_2$

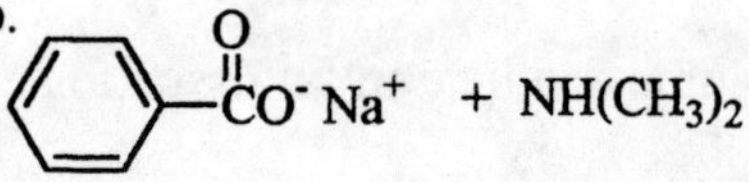

41. a. K$_2$Cr$_2$O$_7$, H$^+$   b. K$_2$Cr$_2$O$_7$, H$^+$   c. NaOH

43. a. $CH_3\overset{O}{\underset{\|}{C}}OH$, H$^+$   b. LiOH

**Phosphorus Compounds**

45. a. $CH_3CH_2O\underset{\underset{OH}{|}}{\overset{\overset{O}{\|}}{P}}OCH_2CH_3$

    b. $CH_3O\underset{\underset{OH}{|}}{\overset{\overset{O}{\|}}{P}}OH$

    c. $HO-\underset{\underset{OH}{|}}{\overset{\overset{O}{\|}}{P}}-O-\underset{\underset{OH}{|}}{\overset{\overset{O}{\|}}{P}}-O-\underset{\underset{OH}{|}}{\overset{\overset{O}{\|}}{P}}-OH$

## B Drugs: Some Carboxylic Acids, Esters, and Amides

## ANSWERS TO REVIEW QUESTIONS

1. a. Drug: any chemical substance that affects an individual in such a way as to bring about physiological, emotional, or behavioral change.
   b. Analgesic: a pain-relieving drug.
   c. Antipyretic: a fever-reducing drug.
   d. Pyrogen: fever-producing compounds that are produced by and released from leukocytes and other circulating cells.
   e. Reye's syndrome: a disease characterized by vomiting, lethargy, confusion, and irritability, brought on by the use of aspirin following certain flues or chickenpox.

3. An alkaloid is a nitrogen-containing, plant-derived compound that causes physiological effects. Examples are morphine, codeine, nicotine, and cocaine.

5. Morphine blocks specific receptor sites on neurons in the brain and spinal cord. Meperidine and methadone mimic morphine, and are therefore morphine agonists. Naloxone, a morphine antagonist, blocks the action of morphine.

7. Methadone blocks the euphoria normally caused by heroin but prevents the withdrawal symptoms from occurring.

9. A hallucinogenic drug (like LSD) is a drug that produces visions and sensations that are not part of reality.

11. a. It has been difficult to assess both long-term toxic effects of LSD and its potential for causing chromosomal damage.
    b. It is disputed whether marijuana causes brain damage and long-term psychoses.

13. It is relatively fat soluble and held in fat tissue for extended periods.

## ANSWERS TO PROBLEMS

15. Carboxylic acid

17. a. $CH_3OH$, $H^+$     b. NaOH

# 17 Amines and Derivatives

## ANSWERS TO REVIEW QUESTIONS

17.5.   Amines have a lone pair of electrons on a nitrogen atom; they can act as proton acceptors.

17.6.   Lemon juice donates H$^+$ ions to the amines to generate the ionized form. The ionized (protonated) forms of amines are much less volatile, thus reducing the fishy odors.

## ANSWERS TO PROBLEMS

**Classification of Compounds**

1. a. amide  b. neither  c. both

3. a. alcohol (1°)  b. amine (1°)  c. alcohol (2°)  d. amine (1°)  e. ether  f. phenol

**Amines: Structures and Names**

5. a. CH$_3$NHCH$_3$  b. CH$_3$CH$_2$NCH$_2$CH$_3$ with CH$_3$  c. cyclobutane with OH and NH$_2$  d. HOCH$_2$CH$_2$NH$_2$

7. a. phenyl-NH$_2$  b. Br-phenyl-NH$_2$  c. pyrimidine  d. phenyl-NHCH$_2$CH$_3$

9. a. propylamine  b. methylisopropylamine
   c. triethylamine  d. 2-aminopentane

**Names and Formulas of Amine Salts**

11. a. phenyl-NH$_3^+$ Br$^-$  b. (CH$_3$)$_4$N$^+$ Cl$^-$

13. a. diethylammonium bromide  b. tetraethylammonium iodide

337

## Chapter 17: Amines and Derivatives

**Physical Properties**

15. Butylamine. It can hydrogen bond (while pentane cannot), causing a more tightly held liquid phase.

17. Propylamine. Tertiary amines have no hydrogens bonded to the nitrogen; therefore, they do not hydrogen bond with one another.

19. $CH_3CH_2NH_2$. It has the ability to hydrogen bond with water.

21. 
$$\underset{CH_2CH_2CHCH_2CH_2}{NH_2 \quad NH_2 \quad NH_2}$$
The more hydrogen bonding groups placed on a compound, the more soluble it is in water.

**Chemical Reactions**

23. a. $CH_3NH_3^+ \, Br^-$  b. $[(CH_3)_3NH^+]_2SO_4^{2-}$

25. a. 
$$CH_3CH_2CH_2CH_2CH_2\underset{H}{\overset{\overset{O}{\|}}{C}N}CH_2CH_2CH_2CH_3$$

b. Ph–C(=O)–N(H)–Ph

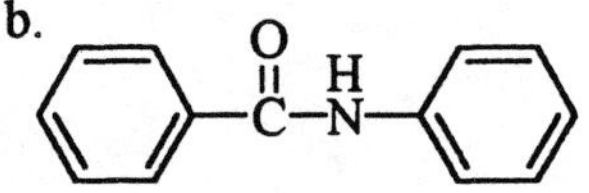

27. a. $CH_3CH_2\underset{CH_3}{NH}$ and $\overset{O}{\underset{}{HOCCH_2CH_3}}$  b. Ph–$NH_2$ and $\overset{O}{\underset{}{HOCCH_2CH_3}}$

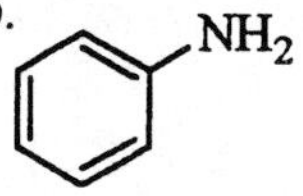

29. Ph–C(=O)–$NH_2$

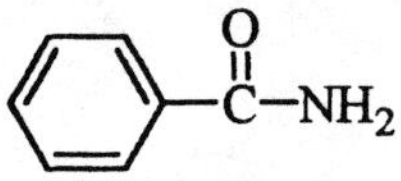

31. a. HCl  b. $HNO_3$

# C   Brain Amines and Related Drugs

## ANSWERS TO REVIEW QUESTIONS

1. a. Neurons: nerve cells.
   b. Synapse: a small gap between a neuron and a second cell (perhaps a second neuron).
   c. Neurotransmitter: chemical released by a neuron into a synapse to signal the next cell.

3. a. Tryptophan (precursor of serotonin) and tyrosine (precursor of norepinephrine and dopamine).
   b. It has been proposed that ingestion of these amino acids (as part of protein foods) will directly affect the level of neurotransmitter into which they may be transformed. Ingestion of glucose has also been tied to increased serotonin levels.

5. Cocaine prevents reuptake of dopamine by neurons, thus prolonging dopamine's stimulatory effects.

7. A general anesthetic is a depressant that acts on the brain to produce unconsciousness, as well as insensitivity to pain.

9. a. Amphetamines cause stimulation or bring an individual "up."
   b. Barbiturates cause central nervous system depression or bring an individual "down".

11. Mania is thought to be due to high levels of biological amines in certain areas of the brain. Blockage of norepinephrine release will result in less of this neurotransmitter in the synapse and, thus, less of the mania-type state.

## ANSWERS TO PROBLEMS

**Structural Formulas and Functional Groups**

13. Barbiturates all contain a six-membered barbituric acid ring structure.

    Groups are added to a carbon located between a pair of carbonyl groups to alter the drug's properties such as effectiveness and length of duration.

15. Amide, ketone, ether, alcohol, amine.

**Toxicities**

17. cocaine

# 18 Stereoisomerism

## ANSWERS TO REVIEW QUESTIONS

18.1. Polarized light is light that is vibrating in a single plane.

18.2. a. An optically active substance rotates a beam of plane-polarized light.
b. A dextrorotatory substance rotates plane-polarized light in a clockwise direction.
c. A levorotatory substance rotates plane-polarized light in a counterclockwise direction.

18.5. No, to be enantiomers, the molecule and its mirror image must be nonsuperimposable.

18.8. a. identical  b. identical  c. identical in magnitude, different in sign (+,−)
d. identical  e. identical  f. may react at different rates

18.9 (+)-Menthol melts at 43 °C, boils at 212 °C, has a density of 0.890 g/cm³ and a specific rotation of +50°.

18.10. a. yes  b. no  c. no

## ANSWERS TO PROBLEMS

**Chirality**

1. Yes. No.

3. a. the carbon of the –CH– group    b. none

5. a.    b.    c.

   d.    e.    f.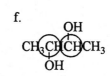

7. a. yes   b. no   c. yes   d. no

9. a.
```
    CH3              CH3
    |                |
H – C – OH      HO – C – H
    |                |
    CH2CH3           CH2CH3
```

b.
```
    CO2H             CO2H
    |                |
H – C – CH3     H3C – C – H
    |                |
    CH3              CH3
```

340

11.

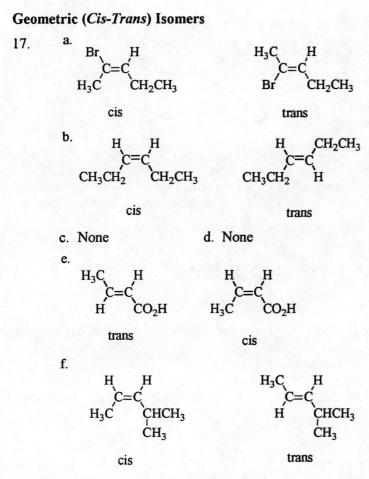

**Multiple Chiral Centers**

13.  a, d

**Van't Hoff's Rule**

15.  a. 4    b. 8

**Geometric (*Cis-Trans*) Isomers**

17.  a. cis / trans
  b. cis / trans
  c. None    d. None
  e. trans / cis
  f. cis / trans

# 19 Carbohydrates

## ANSWERS TO REVIEW QUESTIONS

19.4. D and L serve to signify the absolute configuration of the penultimate (next to last) carbon atom with respect to glyceraldehyde. In a Fischer projection, a D sugar has the H on the left and the OH on the right. An L sugar has the H on the right and the OH on the left.

19.8. If one measures the specific rotation of a solution with polarimetry, one will find that the values will change from +112° (pure α-D-glucose) to +52.7° (an equilibrium mixture of alpha and beta forms).

19.12.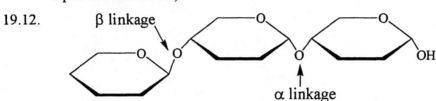

19.13. Starch serves as fuel for the plant, particularly for the seedling. Cellulose is the structural support of the plant. It is what makes up the bulk of plant cell walls in the stems, leaves, and trunk.

## ANSWERS TO PROBLEMS

**Monosaccharides: Terminology, Stereochemistry, and Cyclic Structures**

1. a. aldose/hexose    b. aldose/pentose    c. ketose/hexose    d. aldose/hexose

3. a. D    b. L

5. The D,L designations are not related to the rotation of plane-polarized light, but indicate the relationship between the stereochemistry at the chiral center farthest from the carbonyl group and that of D- and L-glyceraldehyde.

7.
```
      CHO                        CHO
   H—C—OH                     HO—C—H
  HO—C—H                      HO—C—H
   H—C—OH     D-Glucose        H—C—OH     D-Mannose
   H—C—OH                      H—C—OH
     CH₂OH                       CH₂OH
```

342

# Chapter 19: Carbohydrates

9. [structure: pyranose ring with CH₂OH, OH groups; arrow labeled "anomeric carbon"]

11. [structure: pyranose ring with CH₂OH and OH groups]

## Properties and Reactions of Monosaccharides

13. a, b, and c will give positive Benedict's tests.

15. 
```
   CO₂H              CHO              CO₂H
  H-C-OH            H-C-OH           H-C-OH
  HO-C-H            HO-C-H           HO-C-H
  H-C-OH            H-C-OH           H-C-OH
  H-C-OH            H-C-OH           H-C-OH
   CH₂OH             CO₂H             CO₂H

D-Gluconic Acid   D-Glucuronic Acid   D-Glucaric Acid
```

## Disaccharides

17. a. beta           b. alpha

19. a. for (a) it is alpha; there is no anomeric carbon in (b).
    b. (b) is *not* a reducing sugar

21. [structure: disaccharide with two pyranose rings linked by oxygen, showing CH₂OH and OH groups]

23. H⁺ or maltase

## Polysaccharides

25. Amylopectin contains an occasional α-1,6-glycosidic linkage, while amylose does not. They both have α-1,4-glycosidic linkages between glucose as the primary polymeric linkage.

27. Both contain α-1,4- and α-1,6-glycosidic linkages, but glycogen has more frequent α-1,6 linkages than amylopectin.

# 20 Lipids

## ANSWERS TO REVIEW QUESTIONS

20.5. Fats are the primary energy reserve of the body. In addition, they perform a number of vital functions in the body including maintenance of body temperature and insulation of organs and tissues against mechanical and electrical shocks and are integral components of biological membranes.

20.6.

$$\begin{array}{c}
H_2C-OC(CH_2)_7CH=CH(CH_2)_7CH_3 \\
| \\
HC-OC(CH_2)_7CH=CH(CH_2)_7CH_3 \\
| \\
H_2C-OC(CH_2)_7CH=CH(CH_2)_7CH_3
\end{array} \xrightarrow{H_2,\ Ni} \begin{array}{c}
H_2C-OC(CH_2)_{16}CH_3 \\
| \\
HC-OC(CH_2)_{16}CH_3 \\
| \\
H_2C-OC(CH_2)_{16}CH_3
\end{array}$$

triolein → tristearin

20.9.

$$\begin{array}{c}
H_2C-OC(CH_2)_{14}CH_3 \\
| \\
HC-OC(CH_2)_7CH=CH(CH_2)_7CH_3 \\
| \\
H_2C-OPOCH_2CH_2NH_3^+ \\
| \\
O^-
\end{array}$$

20.14. A saponifiable lipid can be hydrolyzed under alkaline conditions, whereas a nonsaponifiable lipid cannot undergo hydrolysis because there are no ester linkages in the molecule.

## ANSWERS TO PROBLEMS

### Fatty Acids

1. a. saturated, 6 carbons   b. unsaturated, 18 carbons   c. saturated, 18 carbons

3. a. $CH_3(CH_2)_{10}COH$   b. $CH_3(CH_2)_5CH=CH(CH_2)_7COH$   c. $CH_3(CH_2)_{14}CO^-\ K^+$

5. The *cis* configuration of the double bond in unsaturated fatty acids causes severe kinks or bends in the long hydrocarbon chain. This prevents molecules from packing tightly together, weakening the attractions between adjacent fatty acids.

Chapter 20: Lipids    345

**Fats, Oils, and Soaps**

7.  a.  
$$\begin{array}{l} H_2C-OC(CH_2)_{16}CH_3 \\ \phantom{H_2C-}\overset{\|}{O} \\ HC-OC(CH_2)_{16}CH_3 \\ \phantom{HC-}\overset{\|}{O} \\ H_2C-OC(CH_2)_{16}CH_3 \\ \phantom{H_2C-O}\overset{\|}{O} \end{array}$$

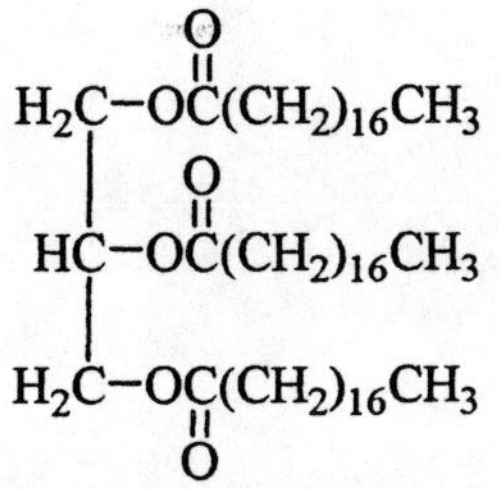

b.  
$$\begin{array}{l} H_2C-OC(CH_2)_{14}CH_3 \\ \phantom{H_2C-}\overset{\|}{O} \\ HC-OC(CH_2)_7CH=CH(CH_2)_7CH_3 \\ \phantom{HC-}\overset{\|}{O} \\ H_2C-OC(CH_2)_{16}CH_3 \\ \phantom{H_2C-O}\overset{\|}{O} \end{array}$$

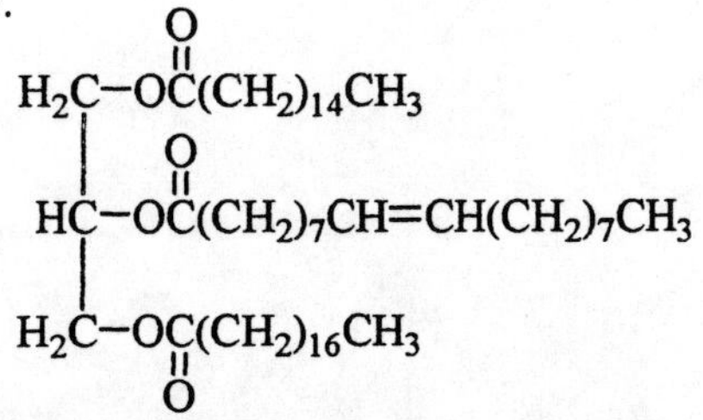

(other answers possible)

9.  
$$\begin{array}{l} H_2C-OC(CH_2)_{14}CH_3 \\ HC-OC(CH_2)_{14}CH_3 \\ H_2C-OC(CH_2)_{14}CH_3 \end{array} \xrightarrow{NaOH} \begin{array}{l} H_2C-OH \\ HC-OH \\ H_2C-OH \end{array} + 3\ CH_3(CH_2)_{14}CO^-\ Na^+$$

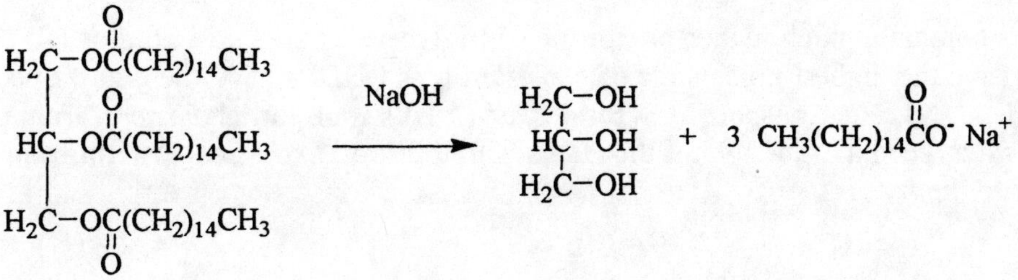

11. Corn oil has a higher degree of unsaturation, thus has a higher iodine number.

13.  
$$\begin{array}{l} H_2C-OC(CH_2)_{10}CH_3 \\ HC-OC(CH_2)_{10}CH_3 \\ H_2C-OC(CH_2)_{10}CH_3 \end{array} \xrightarrow{NaOH} \begin{array}{l} H_2C-OH \\ HC-OH \\ H_2C-OH \end{array} + 3\ CH_3(CH_2)_{10}CO^-\ Na^+$$

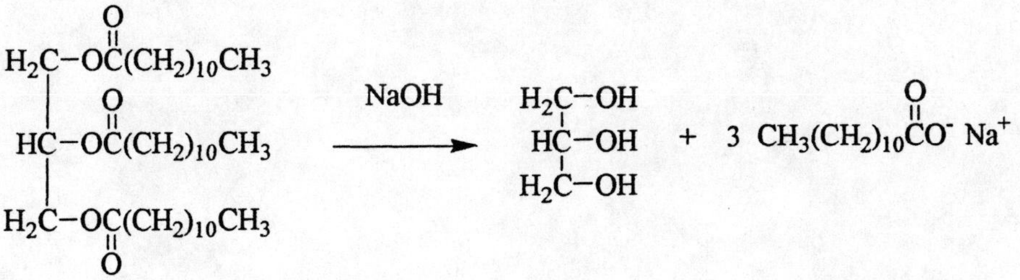

**Membrane Lipids and Cell Membranes**

15. phospholipid: a,c;           glycolipid: b;           sphingolipid: b,c

17.  
$$CH_3(CH_2)_{12}CH=CHCHOH$$
$$CH-NHC(CH_2)_{14}CH_3$$

with sugar ring: CH₂OH, OH, OH, OH attached to pyranose linked via O–CH₂

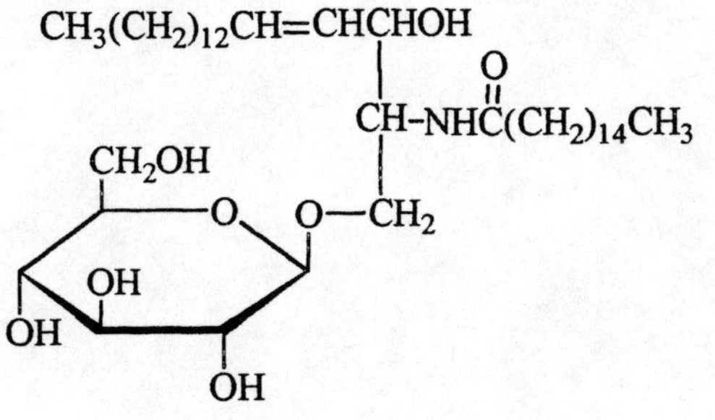

## Steroids

21.

```
      ___
     / C \___
    /___\ D /
   / B \___/
  / A \___/
 /___\
```

## Cholesterol and Cardiovascular Disease

23. VLDLs contain a much higher proportion of triglycerides than do LDLs or HDLs, while LDLs have the highest percentage of cholesterol, and HDLs have the highest percentage of protein. VLDLs transport triglycerides, and LDLs transport cholesterol from the liver to cells that need it. One role of the HDLs is to transport excess cholesterol from various tissues to the liver.

# E  Hormones

## ANSWERS TO REVIEW QUESTIONS

1. Paracrine factors are chemical messengers that move from one cell to another within a single tissue. Hormones are chemical messengers that are released in one tissue and are transported through the circulatory system to one or more other tissues.

3. hypothalamus

5. a. Estradiol, progesterone, testosterone, aldosterone.
   b. Epinephrine, norepinephrine.
   c. Prolactin, vasopressin, oxytocin, insulin.

7. a. Hormone: a chemical messenger that is secreted into the blood by an endocrine gland. Example: insulin.
   b. Androgen: a male sex hormone. Example: testosterone.
   c. Estrogen: a female sex hormone. Example: estradiol.
   d. Progestin: a synthetic progesterone agonist. Example: norethindrone.

9. The female sex hormone analogs used in oral contraceptives cause the body to shut down FSH and LH release at an inappropriate time for egg development. Thus the egg fails to develop correctly, and the fertilization and growth process is halted.

11. Arachidonic acid serves as the precursor for the synthesis of the prostaglandins.

# 21 Proteins

## ANSWERS TO REVIEW QUESTIONS

**21.2.** a. asparagine  b. glycine  c. glycine  d. arginine

**21.7.** a.
$$H_3N^+CHC(=O)-NHCHCO^-$$
with side chains CHCH$_3$/CH$_3$ and CHCH$_3$/OH

b.
$$H_3N^+CHC(=O)-NHCHCO^-$$
with side chains CH$_2$–C$_6$H$_4$–OH and CH$_2$CH$_2$C(=O)NH$_2$

**21.13.** The predominant force is hydrogen bonding between the carboxyl oxygen of one amino acid and the amide hydrogen of another amino acid.

**21.16.** 
$$NH_3^+CHCH_2S-SCH_2CHNH_3^+$$
with $-C(=O)O^-$ groups on each α-carbon

**21.22.** a. Secondary, tertiary, and quaternary.  b. Denaturation is rarely reversible.

## ANSWERS TO PROBLEMS

### Properties and Reactions of Amino Acids

**1.** a. –CH(OH)CH$_3$  b. –(CH$_2$)$_4$NH$_3^+$  c. –CH$_2$–C$_6$H$_4$–OH

**3.** a. $^+NH_3CH_2CO_2^-$  b. $^+NH_3CH(CH_3)CO_2^-$  c. $^+NH_3CH(CH(CH_3)_2)CO_2^-$

**5.** a. lysine   b. proline, histidine, tryptophan
c. phenylalanine   d. aspartic acid, glutamic acid

**7.** a. CH$_3$–CH(NH$_3^+$)–C(=O)–OH
b. CH$_3$–CH(NH$_3^+$)–C(=O)–O$^-$
c. CH$_3$–CH(NH$_2$)–C(=O)–O$^-$

## The Peptide Bond and the Sequence of Amino Acids

11.   NH₃⁺CHCNHCH₂CNHCHCO⁻ with O double bonds on each C, side chain CH₂-phenyl on first residue, CH₃ on third residue

$$NH_3^+\text{-}CH(CH_2C_6H_5)\text{-}CO\text{-}NH\text{-}CH_2\text{-}CO\text{-}NH\text{-}CH(CH_3)\text{-}CO^-$$

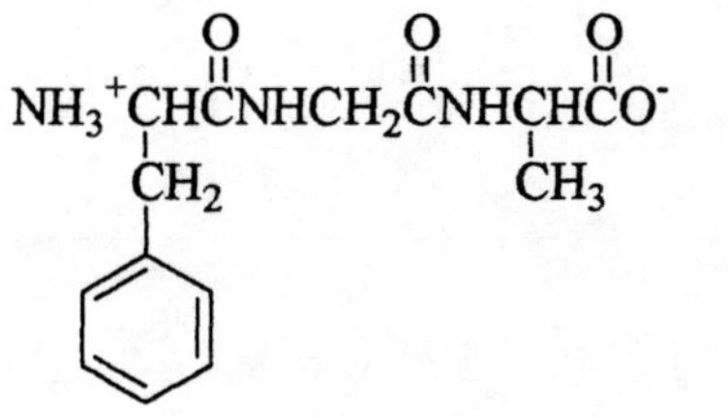

13. Ala-Ser-Cys-Phe

## Peptide Hormones

15.  a. Angiotension II acts as a vasoconstrictor to maintain blood pressure.
b. Overproduction of angiotension II may lead to some forms of hypertension. Captopril prevents the overproduction of angiotension II.

## Classification of Proteins and Protein Structure

17. Wool has an arrangement of polypeptide chains with the polypeptide backbone coiling itself in a helical manner. Hydrogen bonds form between adjacent groups vertically in the helix. Wool is elastic because these hydrogen bonds can be stretched (elongated) and compressed (shortened) without disrupting its structure. This arrangement is called an α-helix.

19.  a. salt linkages          b. dispersion forces
     c. hydrogen bonding       d. disulfide linkages

21.  a. globular               b. fibrous              c. globular

## Electrochemical Properties of Proteins

23.  a. low pH                 b. high pH              c. isoelectric pH

## Denaturation of Proteins

25.  globular

# 22  Enzymes

## ANSWERS TO REVIEW QUESTIONS

22.1.   a. lactose    b. cellulose    c. peptides (and proteins)    d. lipids

22.2.   Hydolases require water to carry out the hydrolysis reaction. Lyases remove a group without having water participating as a reactant.

22.8.   Urease is more specific because it catalyzes the hydrolysis of one compound (urea), while carboxypeptidases will catalyze the removal of the C-terminal amino acid from a wide variety of peptides and proteins.

22.14.  a. Penicillin binds irreversibly to an enzyme called transpeptidase, an important enzyme responsible for a step in bacterial cell wall synthesis. Inhibition of this enzyme by penicillin prohibits formation of the cell wall and, thus, bacterial death.
b. Resistant strains of bacteria produce penicilllinase, an enzyme that breaks down penicillin before it can cause harm to the transpeptidase.

## ANSWERS TO PROBLEMS

### Classification and Characteristics of Enzymes

1.  a. lyase        b. hydrolase        c. transferase

3.  a. lactase        b. hydrolase

5.  a. Ethanol is the substrate, zinc is the cofactor, and alcohol dehydrogenase without the zinc is the apoenzyme
    b. No, coenzymes are organic molecules, and $Zn^{2+}$ is inorganic.

7.  It is synthesized as trypsinogen so that trypsin will not be active in the pancreas (where it is synthesized) and, thus, degrade important proteins in that tissue.

### Mode of Enzyme Action and Specificity

9.  Enzymes function by first combining with a substrate molecule to form an intermediate compound referred to as an enzyme-substrate complex. (This intermediate is more reactive than the substrate alone.) In a subsequent step this intermediate reacts further (usually with another reactant) to form products and to regenerate the enzyme. Thus, the enzyme can run the reaction repeatedly.

11. a. Asp or Glu        b. Asp or Glu        c. Lys        d. Lys
    e. Ser, Thr, or Tyr    f. Cys                g. Val        h. none        i. Phe

13. The turnover number tells you how rapidly an enzyme catalyzes a particular reaction.

## Factors That Influence Enzyme Activity

15. Above the optimum temperature, excessive heat causes the denaturation of the enzyme. Heat, as a denaturing agent, disrupts those interactions responsible for maintaining the unique conformation of the enzyme. Below the optimum temperature, chemical reactions occur much more slowly because of a decrease in kinetic energy of the molecules. Slow-moving molecules collide less frequently, and therefore the probability of a favorable reaction (product formation) decreases.

17. The enzyme will become less active due to the change in ionization of a key group in the active site.

## Enzyme Inhibition and Chemotherapy

19. A competitive inhibitor is a compound that structurally resembles the substrate and competes with the substrate for the active site of the enzyme. The inhibitory effect is reversible by increasing substrate concentration. A noncompetitive inhibitor is a compound that forms strong bonds with either the free enzyme or the enzyme-substrate complex. It bonds to the enzyme at a site remote from the active site, but in doing so, alters the conformation of the active site.

21. It is possible to synthesize penicillin-like compounds that resist cleavage by penicillinase; a penicillinase inhibitor (like clauvulinic acid) can be combined with the penicillin to prevent its degradation.

# F  Vitamins

## ANSWERS TO REVIEW QUESTIONS

1. a. Minerals are inorganic while vitamins are organic molecules.
   b. Both minerals and vitamins are essential to our diet.
   c. Neither minerals nor vitamins are needed in large amounts in our bodies.

3. An excess of fat-soluble vitamins is more dangerous. We generally excrete water-soluble vitamins readily in the urine. Fat-soluble vitamins build up in fatty tissues (including the brain). Large excess can cause a variety of toxic effects.

5. Water-soluble vitamins have polar groups attached, which make them soluble, while fat-soluble vitamins are comprised primarily of nonpolar structures made up of carbons and hydrogens.

7. The water-soluble vitamins (vitamin C excluded) act as or are transformed into coenzymes that we need for normal metabolism.

9. ascorbic acid-vitamin C;  ergocalciferol-vitamin $D_2$;  thiamine-vitamin $B_1$;  retinol-vitamin A;  tocopherol-vitamin E.

11. a. scurvy     b. pellagra     c. pernicious anemia

13. a. niacin (vitamin $B_3$) and riboflavin (vitamin $B_2$)     b. biotin
    c. thiamine (vitamin $B_1$)     d. methylcobalamin     e. folic acid

15. Despite its high molar mass, vitamin $B_{12}$ has a high proportion of electronegative nitrogens and oxygens (primarily amide groups) that can hydrogen bond with water, making the vitamin water-soluble.

17. Yes. Vitamin C, ascorbic acid, is a single chemical whether it is made naturally or synthetically. Vitamin C tablets can contain all sorts of other ingredients along with the ascorbic acid. Perhaps some of these help prevent the deterioration of vitamin C and help the absorption or action of the vitamin C once in the body.

# 23  Nucleic Acids and Protein Synthesis

## ANSWERS TO REVIEW QUESTIONS

23.9. Each is paired with a daughter strand.

23.10. DNA and RNA

23.12. mRNA is the "information" molecule carrying the instructions for the sequence of amino acids in a protein to the ribosome. tRNA molecules are involved in transporting amino acids to the ribosome in an order dictated by mRNA. These amino acids then become incorporated into a growing protein chain.

23.13. mRNA contains the codon; tRNA contains the anticodon.

23.15. a. 5-Bromouracil is incorporated into a new DNA strand in place of thymine, but it base pairs with guanine instead of adenine.
b. chemical mutagen.

## ANSWERS TO PROBLEMS

**Nucleotides**

1. The sugar unit in RNA is ribose. The sugar unit in DNA is 2-deoxyribose.

3. a. neither           b. nucleoside        c. nucleotide

5. a. ribose            b. ribose            c. deoxyribose

7. a. $N^6$-Methyladenine (The $N^6$ nomenclature is used to show that the placement of the methyl group is on the amino group on carbon number 6 of the purine ring system.)

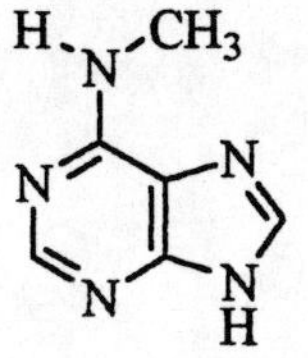

b. 5-Methylcytosine

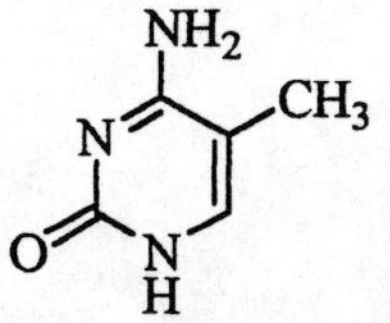

c. 7-Methylguanine

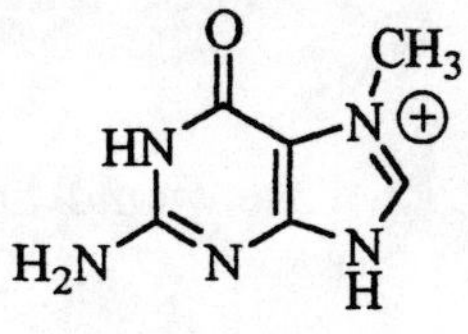

354  Chapter 23: Nucleic Acids and Protein Synthesis

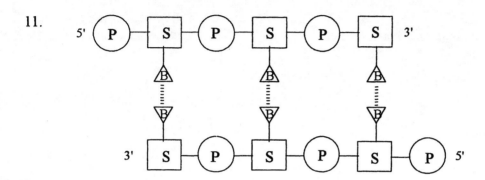

d. 4-thiouracil (These structures are tautomers. They differ only by the position of the hydrogens and the double bonds.)

**Primary and Secondary Structure of Nucleic Acids**

9. Phosphoric acid is a major structural component of the backbone of nucleic acid molecules. At physiological pH, the hydrogens of the acid are dissociated, and the phosphate group is negatively charged.

11.

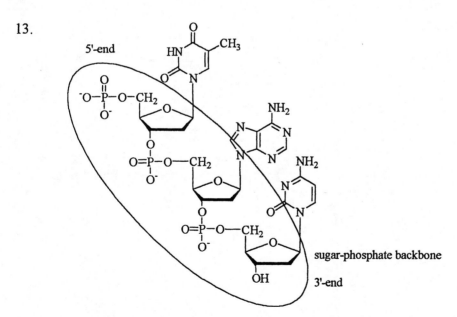

13.

15. a. guanine    b. uracil    c. cytosine    d. adenine

17. 12 [ 3 A-T (3 × 2) + 2 G-C (2 × 3)]

19. No, since RNA is a single strand, the molar quantity of each base would have been different from the others.

## Replication and Transcription

21. a. Both replication and transcription utilize one DNA strand as a template and synthesize the new strand in a 5' to 3' direction, using specific base-pairing to determine the sequence of nucleotides.

    b. In replication the entire DNA template strand is copied and the two strands remain bound together, while in transcription only a segment of the strand is transcribed and the RNA strand dissociates from the template. In replication deoxynucleotides are used, while ribonucleotides are used in transcription.

23. a. 3' d-TACCGTTAGGAGTTTGCGACA 5'

    b. 5' AUGGCAAUCCUCAAACGCUGU 3'

## Protein Synthesis and the Genetic Code

25. a. 3' AAC 5'      b. 3' CUU 5'      c. 3' AGG 5'      d. 3' GUG 5'

27. 27 nucleotides; 9 amino acids x 3 nucleotides/amino acid.

29. a. Phe           b. His            c. Ser            d. Pro

31. Met-Ser-Asp-Phe-Ala-Gly-Leu

## Applications of Molecular Biology

33. a. substitution
    b. The fourth amino acid would be Leu, rather than Phe..

35. No, the mutation must occur in the DNA of a sperm or egg cell for it to be passed on to any children.

# G  Viruses

## ANSWERS TO REVIEW QUESTIONS

1. a. Nucleic acid (DNA or RNA) surrounded by a protein coat.
   b. A DNA viruses contain DNA as their genetic material, while RNA viruses contain RNA.

3. Viral origin: AIDS, chicken pox, hepatitis, influenza, polio. Bacterial origin: cholera, diphtheria, syphilis, tetanus, tuberculosis (see Table G.1 for others).

5. HIV is known as a retrovirus because the RNA strands of the virus synthesize DNA in the host cell, a process that is the opposite of the normal process.

7. Benign tumors do not invade neighboring tissues, while malignant tumors can spread and infect other tissues.

9. Safrole (from sassafras) and aflatoxins (produced by molds on foods such as mushrooms, peppers, and peanuts), to name a few.

11. Oncogenes are genes we (and other animals) carry in our genomes that, through mutations, become converted to genes that cause cell proliferation. Oncogenes are not dependent on specific signals and, in the absence of suppressor genes, promote tumor growth.

13. Cancer is caused by mutations in specific genes. If a DNA repair enzyme cannot function properly, this increases the number of mutations in the DNA and increases the likelihood of a mutation occurring in key genes that lead to the development of cancer.

15. Cisplatin binds to DNA and prohibits replication.

# 24 Metabolism and Energy

## ANSWERS TO REVIEW QUESTIONS

24.3. The energy of hydrolysis of ATP is used by many enzymes and other biomolecules to drive unfavorable reactions.

24.12. Two carbons are removed from the Krebs cycle as molecules of carbon dioxide ($CO_2$)

24.15. a. 2.5 ATP  b. 1.5 ATP

24.16. Actomysin is the structural protein of muscle. It is able to hydrolyze ATP and couple this energy to drive the contraction process.

## ANSWERS TO PROBLEMS

### ATP: Universal Energy Currency

1. AMP contains a single phosphate group attached to the ribose sugar, ADP has two phosphate groups hooked together onto the ribose sugar, and ATP contains three phosphate groups.

3. a, c, d

### Digestion and Absorption of Major Nutrients

5. Mucin is a glycoprotein that attaches to food particles. This lubricates them for easier swallowing.

7. Bile salts act like soap molecules by breaking down large, water-insoluble lipids into smaller micelles. The lipases can act on these smaller micelles more efficiently than the larger lipid particles.

9. a. pepsin: Gly-Ala-Phe and Tyr    chymotrypsin: Gly-Ala-Phe and Tyr
      trypsin: Gly-Ala-Phe-Tyr
   b. pepsin: Ala-Ile-Tyr and Ser  chymotrypsin: Ala-Ile-Tyr and Ser      trypsin: Ala-Ile-Tyr-Ser
   c. pepsin: Val-Phe And Arg-Leu    chymotrypsin: Val-Phe and Arg-Leu
      trypsin: Val-Phe-Arg And Leu
   d. pepsin: Leu And Thr-Glu-Lys    chymotrypsin: Leu-Thr-Glu-Lys
      trypsin: Leu-Thr-Glu-Lys

### Krebs Cycle

11. a. citrate        b. oxaloacetate + acetyl-CoA        c. L-malate
    d. isocitrate dehydrogenase

13. a. 11a        b. 11c        c. 11a

## Chapter 24: Metabolism and Energy

15. Oxidation half-reactions:

Step 3

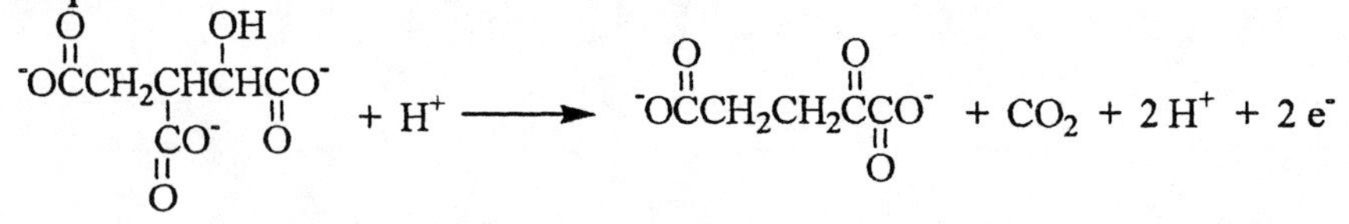

$$^-OCCH_2CHCHCO^- + H^+ \longrightarrow {}^-OCCH_2CH_2CCO^- + CO_2 + 2\,H^+ + 2\,e^-$$

(with side $CO^-$ group as oxaloacetate-type intermediate; $\alpha$-ketoglutarate → succinyl precursor)

Step 4

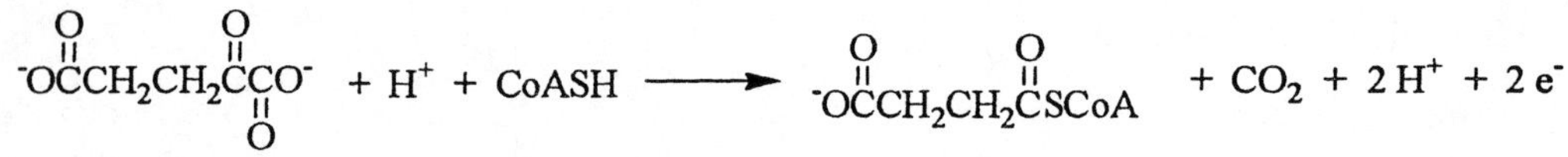

$$^-OCCH_2CH_2CCO^- + H^+ + CoASH \longrightarrow {}^-OCCH_2CH_2CSCoA + CO_2 + 2\,H^+ + 2\,e^-$$

Step 6

$$^-OCCH_2CH_2CO^- \longrightarrow {}^-OCCH=CHCO^- + 2\,H^+ + 2\,e^-$$

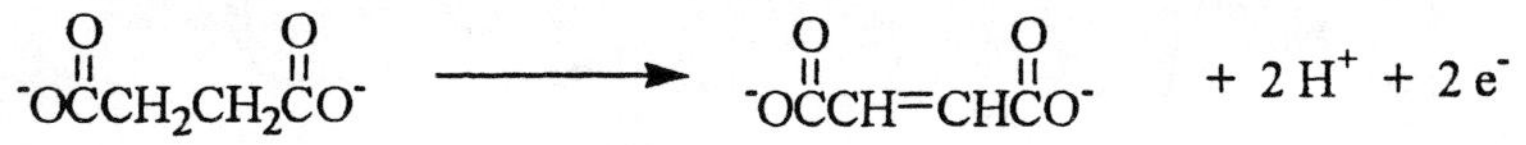

Reduction half-reactions:

Steps 3 & 4: $NAD^+ + 2\,H^+ + 2\,e^- \rightarrow NADH + H^+$

Step 6: $FAD + 2\,H^+ + 2\,e^- \rightarrow FADH_2$

## Cellular Respiration

17. $O_2$; $H_2O$

19. a. The reduced coenzymes NADH and FADH$_2$ are oxidized by the electron transport chain only if ADP is simultaneously phosphorylated to ATP.

b. The transport of H$^+$ out of the mitochondrial matrix forms a difference in H$^+$ concentration across the inner mitochondrial matrix. This proton gradient provides a pathway for H$^+$ to interact with ATP synthase leading to a structural change of the enzyme and the synthesis and release of ATP.

## Muscle Power

21. a. Type I: slow twitch; Type IIB: fast twitch.
    b. Type I: aerobic oxidation; Type IIB: anaerobic metabolism.

23. The high catalytic activity of actomysin in Type IIB fibers suggests that the tissue can hydrolyze ATP at high rates and, thus, is important for bursts of vigorous physical activity.

# 25    Carbohydrate Metabolism

## ANSWERS TO REVIEW QUESTIONS

25.1.    a. Glyceraldehyde-3-phosphate is oxidized to 1,3-bisphosphoglycerate
         b. $NAD^+$

25.4.    During the reduction of pyruvate to lactate, NADH is oxidized to $NAD^+$, thus reoxidizing NADH in the absence of oxygen. The $NAD^+$ allows glycolysis to continue.

25.5.    Two molecules of ATP are produced from one molecule of glucose in anaerobic glycolysis.

25.11.    Insulin can bind to liver, muscle, and adipose cells and activate them to take up glucose from the blood, lowering blood-sugar levels.

## ANSWERS TO PROBLEMS

**Glycolysis and Metabolism of Pyruvate**

1.    a. glyceraldehyde-3-phosphate and dihydroxyacetone phosphate
     b. phosphoenolpyruvate             c. triose phosphate isomerase
     d. glucose-6-phosphate

3.    a. 1b          b. 2c          c. 1c and 2b          d. 2c

5.    $NAD^+$

7.    Lactic acid, pyruvic acid, acetaldehyde, and ethanol are the only nonphosphorylated metabolites of glycolysis and fermentation. All the other metabolites are phosphorylated.

9.    a. 4          b. 2          c. 2

11.    a. NADH is reoxidized in muscle cells in the reaction in which pyruvate is reduced to lactate.
      b. In yeast cells, NADH is reoxidized in the reaction in which acetaldehyde (formed from the decarboxylation of pyruvate) is reduced to ethanol.

**ATP Yield from Glycolysis/Gluconeogenesis**

13.    a. 32          b. (i) 2; (ii) 2; (iii) 28          c. 35%

15.    Pyruvate, lactate, oxaloacetate, and some amino acids.

## Blood Glucose and Glycogen Metabolism

17. The normal blood sugar level ranges from 70 - 100 mg/dL of blood. The renal threshold value, ranging from 150 - 170 mg/dL, is the point when the kidneys begin secreting glucose.

19. UDP-glucose

21. Only 2 ATP/glucose are obtained when glucose is the starting material; however, 3 ATP/glucose are obtained from the degradation of glycogen because each glucose obtained from glycogen degradation is already phosphorylated (glucose-1-phosphate which is readily converted to glucose-6-phosphate), thus eliminating the need to hydrolyze an ATP to form glucose-6-phosphate from glucose.

## Regulation of Carbohydrate Metabolism

23. Glucagon binds to a specific receptor, which leads to the formation of cAMP. Release of cAMP leads to the activation of phosphorylase (and thus, glycogen degradation) and inhibition of glycogen synthase (and thus, glycogen synthesis), which leads to an increase in the amount of glucose.

25. The binding of the hormone to its receptor activates adenylate cyclase, which is bound to the inner cell membrane. This enzyme catalyzes the conversion of ATP to cAMP within the cell.

27. a. glycolysis  b. glycogenesis  c. gluconeogenesis;
    d. glycogenolysis  e. glycolysis

# 26 Lipid Metabolism

## ANSWERS TO REVIEW QUESTIONS

26.1. The essential fatty acids are those that contain more than one double bond.

26.3. The carbon atom beta to the carboxyl group of the fatty acid undergoes successive oxidations.

26.4.   a. 8    b. 7    c. 7

26.5.   a. 120    b. 376.5 - 378.5

26.8.   acetyl-CoA

## ANSWERS TO PROBLEMS

**Storage and Mobilization of Fats**

1. The oxidation of 1.0 g of carbohydrate liberates about 4.2 kcal. The oxidation of 1.0 g of lipid liberates about 9.5 kcal.

3. a. glycogen in the liver    b. fat reserves in adipose tissue

5. 
$$H_2C-OC(CH_2)_{14}CH_3 \atop HC-OC(CH_2)_{14}CH_3 \atop H_2C-OC(CH_2)_{14}CH_3 \xrightarrow{lipase} H_2C-OH \atop HC-OC(CH_2)_{14}CH_3 \atop H_2C-OC(CH_2)_{14}CH_3 + CH_3(CH_2)_{14}COH \xrightarrow{lipase} H_2C-OH \atop HC-OC(CH_2)_{14}CH_3 \atop H_2C-OH + CH_3(CH_2)_{14}COH$$

$$\xrightarrow{lipase} H_2C-OH \atop HC-OH \atop H_2C-OH + CH_3(CH_2)_{14}COH$$

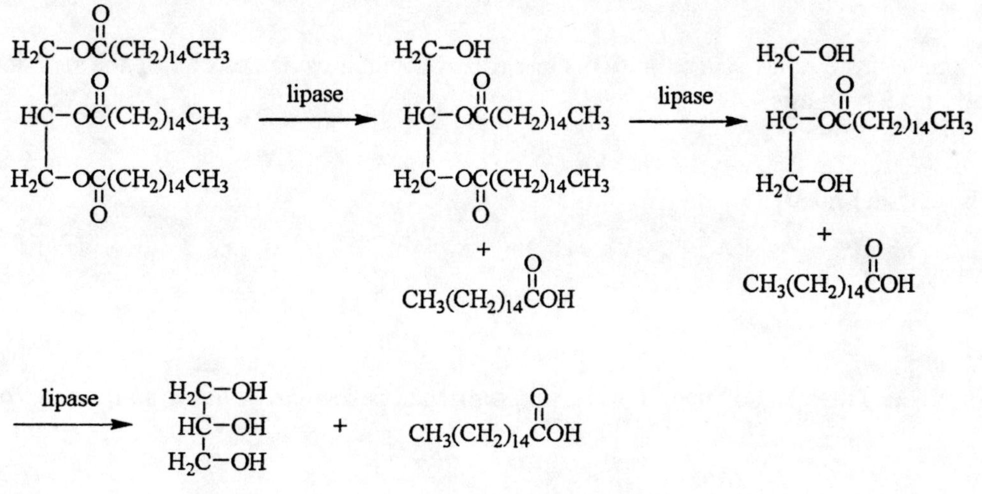

7. 
glycerol $\xrightarrow[\text{ATP} \quad \text{ADP}]{\text{glycerol kinase}}$ L-glycerol 3-phosphate $\xrightarrow[\text{NAD}^+ \quad \text{NADH + H}^+]{\text{glycerol phosphate dehydrogenase}}$ dihydroxyacetone phosphate

glycerol: $CH_2OH$ — $HO-C-H$ — $CH_2OH$

L-glycerol 3-phosphate: $CH_2OH$ — $HO-C-H$ — $CH_2OPO_3^{2-}$

dihydroxyacetone phosphate: $CH_2OH$ — $O=C$ — $CH_2OPO_3^{2-}$

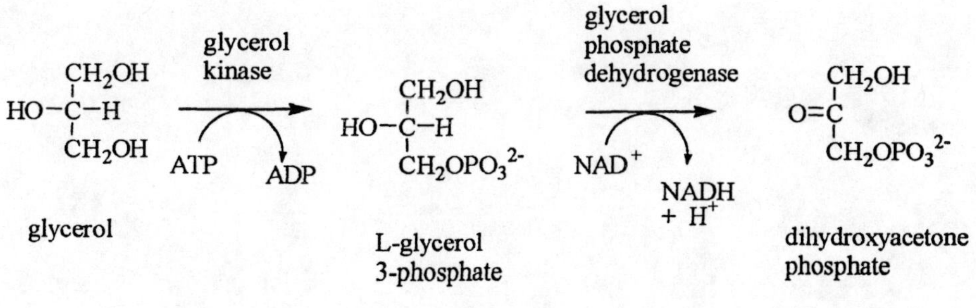

## Fatty Acid Oxidation

9. Oxidation-reduction, hydration, and lysis (cleavage).

11. a. 6 turns      b. 12 turns

13. a. 9      b. 8      c. 8

15. a. 92      b. 294.5      c. 324.5

## Ketosis

17. In starvation, cells rely on fatty acid oxidation to supply their energy needs. This results in higher amounts of acetyl-CoA that cannot be oxidized in the Krebs cycle due to a decrease in the concentration of oxaloacetate, needed for glucose synthesis. The acetyl-CoA is then used to synthesize ketone bodies, two of which are weak acids.

19. a. kidney and heart      b. brain

## Fatty Acid Synthesis

21. a. Fatty acid oxidation occurs in the mitochondria, while fatty acid synthesis occurs in the cytosol of the cytoplasm.
    b. Fatty acid oxidation utilizes $NAD^+$ and FAD, while fatty acid synthesis utilizes NADPH.
    c. In fatty acid oxidation, carbons are removed in units of two, while in fatty acid synthesis they are added in units of two.

23. 14

25. A single protein has several active sites, each of which catalyzes one of the reactions in fatty acid synthesis.

## Obesity, Exercise, and Diets

27. a. 22, normal      b. 18, underweight      c. 28, overweight

29. 42%

31. 2 h

33. These diets may be deficient in necessary nutrients, particularly in B vitamins and iron.

# 27 Protein Metabolism

## ANSWERS TO REVIEW QUESTIONS

27.3.

$$\text{Alanine} + \text{Oxaloacetate} \xrightarrow{\text{transaminase}} \text{Pyruvate} + \text{Aspartate}$$

27.6. uric acid

27.9. Starvation is a condition in which the body is totally deprived of food, while kwashiorkor is a disease that is the result of insufficient protein in the diet.

## ANSWERS TO PROBLEMS

### Amino Acid Metabolism

1. a. The essential amino acids contain carbon chains, or aromatic rings, that are not present as intermediates of carbohydrate or lipid metabolism. The inability to synthesize these amino acids results from the animal's inability to manufacture the correct carbon skeleton.
   b. Since all amino acids required for the construction of a particular protein must be present at the time of its synthesis, a deficiency in one or more essential amino acids will prevent protein synthesis from proceeding.

3. a. There is an inadequate intake of protein, while the body is degrading protein to provide precursors for glucose synthesis.
   b. Large amounts of protein are being synthesized during pregnancy, so very little is being excreted.

### Catabolism of Amino Acids

5. 
[Transamination reaction: phenylalanine + pyruvate → phenylpyruvate + alanine]

7. a. Transamination between glutamate and oxaloacetate to form α-ketoglutarate and aspartate.
   b. Transamination between glutamate and pyruvate to form α-ketoglutarate and alanine.

9. a. both          b. ketogenic          c. glucogenic

## Storage and Excretion of Nitrogen

11. a. glutamine

    b.

    glutamate + $NH_4^+$ →(ATP → ADP, synthetase)→ glutamine + $H_2O$

13. The nitrogens originate from ammonium ion and aspartate, and the carbon comes from bicarbonate ion.

## Synthesis of Nonessential Amino Acids and Amino Acid Derivatives

15. a. methionine          b. phenylalanine

17. a. Histidine →(histidine decarboxylase)→ Histamine + $CO_2$

    b. Tyrosine →(tyrosine decarboxylase)→ Tyramine + $CO_2$

    c. 3,4-Dihydroxyphenylalanine →(dopa decarboxylase)→ Dopamine + $CO_2$

## Relationships Among the Metabolic Pathways

19. a. insulin          b. glucagon          c. insulin

21. The brain requires less glucose because it is using ketone bodies for some of its energy needs.

# 28 Body Fluids

## ANSWERS TO REVIEW QUESTIONS

28.1.  (1) To help maintain osmotic relationships between the tissues.
(2) To transport oxygen and nutritive materials to the cells and waste products to the excretory organs.
(3) To regulate the body temperature.
(4) To control the pH of the body.
(5) To protect the organism against infection.

28.4.  Arterial blood contains dissolved nutrients, oxygen, hormones, and vitamins. Venous blood contains metabolic waste products, has a lower oxygen content, and higher carbon dioxide content.

28.7.  
a. bicarbonate buffer  
b. phosphate buffer  
c. hemoglobin  
d. bicarbonate buffer

28.12. formed elements

## ANSWERS TO PROBLEMS

### Blood: Functions and Composition

1. $Na^+$, $K^+$, $Ca^{2+}$, $Mg^{2+}$, $HCO_3^-$, $Cl^-$, $HPO_4^{2-}$, $SO_4^{2-}$

3. Most are synthesized in the liver.

5. a. leukocytes    b. erythrocytes    c. platelets

### Blood Gases

7. Hemoglobin is a protein composed of four polypeptide chains, each bound to a molecule of heme.

9. a. +2    b. +2    c. +3    d. +2

### Blood Buffers

11. Bicarbonate buffer, phosphate buffer, plasma proteins, and hemoglobin.

13.

## Blood Clotting

15. Blood clotting factors circulate as proenzymes so they are present when needed but cannot cause unnecessary clotting.

17. Several blood-clotting factors require Vitamin K for the reaction in which glutamate side chains are carboxylated to form γ–carboxyglutamate.

## The Immune Response

19. An antibody is composed of four polypeptide chains: two heavy chains and two light chains that are linked by disulfide bonds.

21. An antigen is a substance that triggers the synthesis of antibodies.

## Blood Pressure

23. The systolic pressure is the maximum pressure achieved during contraction of the heart ventricles. When the ventricles relax, the blood pressure drops, and the lowest pressure that remains in the arteries before the next ventricular contraction is called the diastolic pressure.

25. Diuretics reduce the blood volume and thus, the pressure exerted against the walls of the arteries.

## Urine and the Kidneys

27. In the kidneys the glomerulus filters most components (except proteins and formed elements) into the kidney tubules. As this fluid passes through the tubules, there is a selective reuptake into the blood of important constituents that the body will save. Waste products are not reabsorbed and are excreted as urine.

29. Liquid intake, amount of perspiration, presence of fever or diarrhea.

## Sweat and Tears

31. Inorganic: $Na^+$, $Cl^-$, $Ca^{2+}$; Organic: urea, amino acids, lipids.

33. Lysozymes perform an antibacterial action; they cleave cell walls of bacteria.

## The Chemistry of Mother's Milk

35. a. Reindeer milk contains much higher amounts of fat and protein. This is important because of the colder climate in which the reindeer live.

    b. Porpoise milk contains much higher amounts of fat and protein. These provide more energy and insulation from colder temperatures.